AF352969

Quality Evaluation of Bee Products

Quality Evaluation of Bee Products

Editors

Liming Wu
Qiangqiang Li

Basel • Beijing • Wuhan • Barcelona • Belgrade • Novi Sad • Cluj • Manchester

Editors
Liming Wu
Chinese Academy of
Agricultural Sciences
Beijing, China

Qiangqiang Li
Chinese Academy of
Agricultural Sciences
Beijing, China

Editorial Office
MDPI
St. Alban-Anlage 66
4052 Basel, Switzerland

This is a reprint of articles from the Special Issue published online in the open access journal *Foods* (ISSN 2304-8158) (available at: https://www.mdpi.com/journal/foods/special_issues/beeproduct_quality).

For citation purposes, cite each article independently as indicated on the article page online and as indicated below:

Lastname, A.A.; Lastname, B.B. Article Title. *Journal Name* **Year**, *Volume Number*, Page Range.

ISBN 978-3-0365-9618-1 (Hbk)
ISBN 978-3-0365-9619-8 (PDF)
doi.org/10.3390/books978-3-0365-9619-8

Contents

About the Editors

Liming Wu

Liming Wu, Ph.D., professor and Deputy Director of the Institute of Apicultural Research, Chinese Academy of Agricultural Sciences, Beijing, China.

He is a leading talent in scientific and technological innovation for the Organization Department of the CPC Central Committee, as well as a young and middle-aged scientific and technological innovation leading talent for the Ministry of Science and Technology. Additionally, he serves as a post-scientist for the National Bee Industry Technology System, as the Chief Scientist of the Scientific and Technological Innovation Project "Bee Product Quality and Risk Assessment" Team at the Chinese Academy of Agricultural Sciences, and as a group leader of the Apiculture Standardization Working Group within the Chinese Apiculture Society. He has achieved significant breakthroughs in the areas of bee product production process control, quality, and functional evaluation, resulting in numerous original accomplishments. As a result of his contributions, he was awarded second prize for a national technological invention (as the first winner) and has twice been awarded first prize at the provincial and ministerial levels.

Qiangqiang Li

Qiangqiang Li, Ph.D., associate professor at the Institute of Apicultural Research, Chinese Academy of Agricultural Sciences, Beijing, China.

She is mainly engaged in the quality control and functional evaluation of bee products. She is currently leading one National Natural Science Foundation project and two basic scientific research projects of central public welfare research institutes and contributing to eight important national projects, including a National Key Research and Development project and a National Modern Bee Industrial Technology System project, as well as several National Natural Science Foundation projects. As the first/corresponding author, she has published more than 30 papers in *Food Chemistry*, *Environment International*, the *Journal of Agricultural and Food Chemistry*, etc. Additionally, she holds nine valid national patents for inventions. She also serves as an editor and reviewer of several SCI journals.

Article

A Novel, Rapid Screening Technique for Sugar Syrup Adulteration in Honey Using Fluorescence Spectroscopy

Sha Yan [1,2], Minghui Sun [2], Xuan Wang [2], Jihao Shan [3] and Xiaofeng Xue [2,*]

[1] College of Food Science and Engineering, Shanxi Agricultural University, Taigu 030801, China; yanshawell@163.com
[2] Institute of Apiculture Research, Chinese Academy of Agricultural Sciences, Beijing 100093, China; smh4601112@126.com (M.S.); wangxuancas@163.com (X.W.)
[3] Institute of Agro-Products Processing Science and Technology, Chinese Academy of Agricultural Sciences, Beijing 100081, China; shanjihao2007@163.com
* Correspondence: xue_xiaofeng@126.com

Abstract: The adulteration of honey with different sugar syrups is common and difficult to detect. To ensure fair trade and protect the interests of apiarists, a rapid, simple and cost-effective detection method for adulterants in honey is needed. In this work, fluorescence emission spectra were obtained for honey and sugar syrups between 385 and 800 nm with excitation at 370 nm. We found substantial differences in the emission spectra between five types of honey and five sugar syrups and also found differences in their frequency doubled peak (FDP) intensity at 740 nm. The intensity of the FDP significantly declined ($p < 0.01$) when spiking honey with $\geq$10% sugar syrup. To validate this method, we tested 20 adulterant-positive honey samples and successfully identified 15 that were above the limit of detection. We propose that fluorescence spectroscopy could be broadly adopted as a cost-effective, rapid screening tool for sugar syrup adulteration of honey through characterization of emission spectra and the intensity of the FDP.

Keywords: honey; sugar syrup; adulteration; fluorescence spectroscopy

Citation: Yan, S.; Sun, M.; Wang, X.; Shan, J.; Xue, X. A Novel, Rapid Screening Technique for Sugar Syrup Adulteration in Honey Using Fluorescence Spectroscopy. *Foods* **2022**, *11*, 2316. https://doi.org/10.3390/foods11152316

Academic Editor: M. Carmen Seijo

Received: 1 July 2022
Accepted: 27 July 2022
Published: 3 August 2022

Publisher's Note: MDPI stays neutral with regard to jurisdictional claims in published maps and institutional affiliations.

1. Introduction

Honey is an economically important food product worldwide. It is prized by many cultures for its pleasing aroma and flavor, and high nutritional value and potential healing properties. Given its many valuable nutritional and culinary properties, the price of natural honey is much higher than that of syrup, such as refined corn syrup (CS), rice syrup (RS), beet syrup (BS) and maltose syrup (MS). In the honey market, it is common for some merchants to attempt to take economic advantage of the differences in price by diluting honey with these less expensive sugar syrups [1,2]. Honey adulteration not only affects the quality of the honey but also harms the bee-keeping industry by negatively impacting sales and product reputation [3]. Therefore, there is an urgent need to develop effective methods for the detection of sugar syrup-adulterated honey in order to reassure consumers and ensure fair competition.

Several review papers have summarized recent studies dealing with honey adulteration, and a number of different analytical techniques are currently employed for the detection of syrup in adulterated honey [3–6]. These methodologies include common analytical techniques such as thin-layer chromatography (TLC) [7], stable carbon isotopic ratio analysis (SCIRA) [8], gas chromatography (GC) [9], high-performance anion exchange chromatography (HPAEC) [10], high-performance liquid chromatography (HPLC) [11,12] and high-performance liquid chromatography/quadrupole time of flight mass spectrometry (HPLC-Q-TOF) [3], among other emerging technologies [13–15]. Although these techniques offer detailed and accurate results, the processes are time-consuming and expensive and can only be performed in well-equipped laboratories by highly trained analysts.

Some advanced spectroscopic techniques have also been used for detecting honey adulteration, such as infrared spectroscopy (IR) [16,17], Raman spectroscopy (RP) [18], inductively coupled plasma optical emission spectroscopy (ICP-OES) [19] and nuclear magnetic resonance (NMR) [4,20,21], which are rapid and relatively easy but must be used in conjunction with multivariate analysis. However, the shortcoming of these methods is that they require a high level of expertise to set the chemometric models necessary to distinguish pure honey from its adulterated counterparts. Although adulteration is a global issue for the honey market, the levels and types of adulteration vary between regions. In developed nations and affluent cities, the quality of honey is generally good, and adulteration is less pervasive, since honey quality can be rigorously monitored by well-equipped testing agencies and inspectors with analytical expertise. However, in less developed regions, the regulatory infrastructure often lacks advanced analytical equipment and experts, resulting in widespread honey adulteration and often large quantities of glucose–fructose syrup [3]. There is therefore a critical need for the honey market to develop a cost-effective, simple, rapid and easy-to-use screening method for monitoring the adulteration of honey.

Fluorescence spectroscopy, a common analytical technique with relatively low-cost instrumentation that is ubiquitous in biological, food science and toxicology labs, offers several advantages for the characterization of chemical constituents, such as high sensitivity and specificity [22]. Several early studies successfully used fluorescence spectroscopy to distinguish the geographical and botanical origins of honey samples through front-face and synchronous fluorescence spectroscopy [23–26]. Fluorescence spectroscopy, in combination with chemometrics, has also been used to discriminate between true and artificial honey samples produced by feeding bee colonies with sucrose [22,27]. However, few studies have been reported on the detection of multiple adulterant sugar syrups in honey using fluorescence spectroscopy.

In the process of finding the fluorescence emission spectra of honey and different sugar syrups, we found a significant difference in the intensity of the frequency doubled peak (FDP) between honey and sugar syrup samples. The second-order diffraction of the monochromator produces the FDP. When the monochromator is set to transmit 740 nm, a small fraction of 370 nm excitation light will also be transmitted through the emission monochromator. Due to the chemical differences between syrup and honey [28], the FDP intensity at 740 nm can be influenced by different honey and sugar samples and may serve as a potential indicator to distinguish honey from sugar syrups.

We envisioned that, based on the differences in fluorescence spectra between honey and syrup, a rapid, low-cost and easy-to-use syrup-adulterated honey screening approach could be developed. In this study, we report the first application, to our knowledge, of fluorescence spectroscopy in combination with FDP intensity data for rapid screening of sugar syrup adulterants in commercial and spiked honey samples. By using the apex wavelength of characterized emission spectra in conjunction with FDP intensities as profiles, we show that natural honey from different botanical origins can be readily distinguished from honey adulterated with a range of different syrups without the need for multivariate analysis. Our findings represent a critical and clear step toward the application of common fluorescence spectroscopy in the screening for honey adulteration.

2. Materials and Methods

2.1. Sample Collection

To examine the fluorescence emission spectra and FDP of pure honey from different botanical origins, we sampled five common types of commercial honey produced across China and derived from *Robinia pseudoacacia* L. (acacia), *Brassica napus* L. (rape), *Tilia tuan Szyszy* L. (medlar), *Lycium barbarum* L. (linden) and *Vitex negundo* L. (chaste). A total of 112 honey samples, including acacia honey (32, labeled A1–A32), rape honey (32, labeled R1–R32), medlar honey (18, labeled J1–J18), linden honey (15, labeled L1–L15) and chaste honey (15, labeled V1–V15), were obtained from twenty-five collaborating beekeepers located in Zhejiang Province, Sichuan Province, Shandong Province, Shaanxi Province, Jilin Province, Xinjiang Province and Beijing. The entirety of the collection process, for all honey samples, was monitored and recorded to ensure sample authenticity. All honey samples were collected from capped combs, ensuring all samples were ripe and mature. The moisture content of the samples was ≤19.0%.

Thirty-one sugar syrup samples, produced by different companies, were purchased from different markets located in Sichuan, Henan, Shandong and Jiangsu Provinces. These samples included ten high-fructose corn syrups (HFCS1−HFCS10), six rice syrups (RS1−RS6), five beet syrups (BS1−BS5), five cassava syrups (CS1−CS5) and five maltose syrups (MS1−MS5).

Twenty adulterant-positive acacia honey samples (sam1–sam20) were graciously provided by an independent testing lab of the Shandong Bee Industry and Bee Products Quality Monitoring. The glucose, fructose, sucrose and maltose contents were determined by the high-performance liquid chromatography-refractive index detection (HPLC-RID) method, as described in [3,5]. According to [29], the stable carbon isotopic ratio analysis was performed using an isotope ratio mass spectrometer (SerCon EA, Wistaston, UK) to determine $^{13}C/^{12}C$ ratios. The TLC test was performed following the procedures in [7,10].

2.2. Sample Preparation

HFCS, RS, BS, CS and MS were mixed and added to pure honey samples in a range of concentrations (10%, 20%, 30%, 90%, w/w) for analysis.

For sample pretreatment, 10 g of honey sample and 10 mL of deionized water were added to a 50 mL centrifuge tube and mixed by vortex for 3 min. The resulting sample solutions were directly used for subsequent fluorescence spectroscopy analysis.

2.3. Fluorescence and Frequency Doubled Peak (FDP) Measurements

The fluorescence measurements were conducted with three brands of fluorescence spectrophotometers including a Hitachi F-4500 (Tokyo, Japan), a Shimadzu RF-5301PC (Shimadzu, Japan) and a Shanghai Lingguang F97 (Shanghai, China), which were all equipped with a plotter unit and a 1 cm quartz cell. The fluorescence of each sample solution was directly measured with an excitation wavelength of 370 nm and an emission range from 385 to 800 nm. The scanning speed was 12,000 nm/min, and spectrometer slits were set for a 5 nm band-pass. Fluorescence measurements of each sample were carried out in triplicate. The pattern of fluorescence spectra and the FDP intensity of all samples were recorded at room temperature.

2.4. Statistical Analyses

The results were expressed using the means ± standard deviations. SPSS 19.0 (IBM, Stamford, CT, USA) was used to analyze significant differences, and ANOVA (LSD test) was performed to determine significant differences.

3. Results and Discussion

3.1. Fluorescence Spectra Characteristics of Honey and Syrup

To first determine if the FDP is a common phenomenon among fluorescence spectrophotometers from different manufacturers, and to test for variation between instruments, the fluorescence spectra for each honey type and sugar syrup were measured by the three brands of fluorescence spectrophotometers (Figure 1). By removing the specific optical filter during the acquisition process, the results show that the FDP is a common phenomenon among the different manufacturers. Furthermore, the application of fluorescence spectroscopy in combination with measurements of the FDP intensity can be used as a suitable indicator for the adulteration of honeys with a variety of sugar syrups.

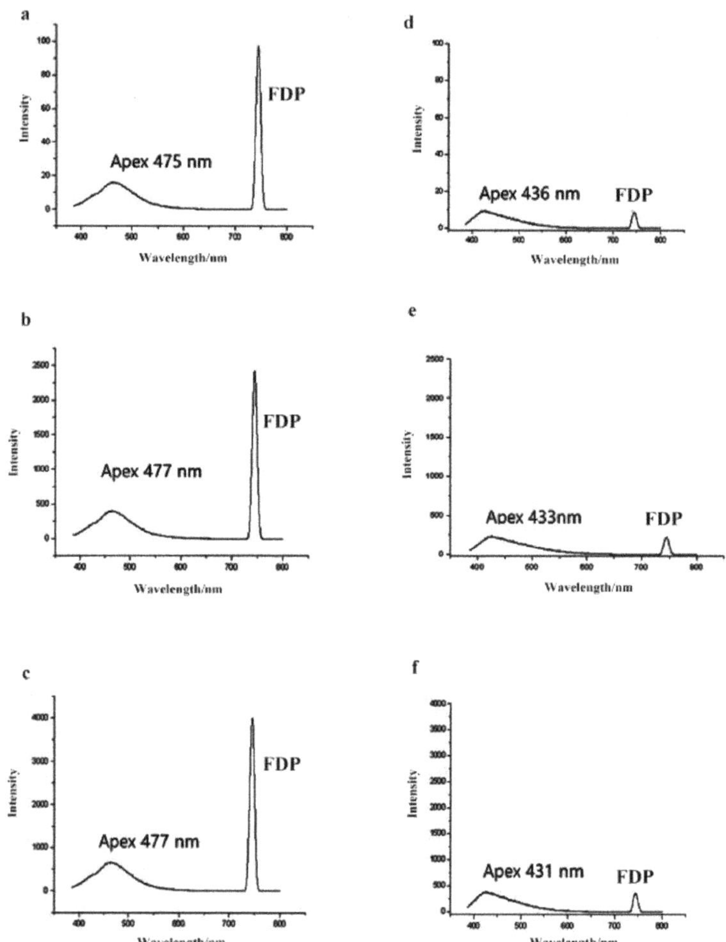

Figure 1. Comparison of honey and syrup fluorescence spectra generated by fluorescence spectrophotometers from three different manufacturers ((**a**) pure honey analyzed by a Hitachi F-4500; (**b**) pure honey analyzed by a Lingguang F97; (**c**) pure honey analyzed by a Shimadzu RF-5301PC; (**d**) syrup analyzed by a Hitachi F-4500; (**e**) syrup analyzed by a Lingguang F97; (**f**) syrup analyzed by a Shimadzu RF-5301PC).

Although we found differences in the absolute, or quantitative, measurements between instruments from the different manufacturers, the apex and the area ratio of fluorescence emission spectra and the area ratio of the FDP between honey and syrup were uniform across all three instruments (Table S2). Data in our study were collected using a Hitachi F-4500 (Tokyo, Japan).

3.2. The Fluorescence Spectra Profiles of Different Honeys and Syrups

To aid in the characterization of honey and syrup profile spectra, we acquired published information on fluorescence spectra from previous research on honey and syrup samples [22–26]. Compositionally, honey consists of sugars and water with small amounts of proteins, free amino acids, peptides, minerals, vitamins, organic acids, flavonoids and other phenolic compounds and aroma compounds. These constituents, and especially many of the minority components, exhibit fluorescent properties [30]. For example, a peak with an excitation of 280 nm and emission of 340 nm can potentially indicate fluorescence from aromatic amino acids present in the honey [22]. A fluorescence peak with an emission of 450 nm and excitation at 250 nm can be attributed to non-flavonoid phenolic compounds in the honey sample [31]. The Maillard reaction products, such as hydroxymethylfurfural and furosine, can have characteristic fluorescence peaks with emission wavelengths ranging from 420 to 470 nm and excitation wavelengths between 340 and 380 nm [22]. We compared the fluorescence emission spectra excitation at 250~380 nm. The results show that all honeys had a fixed apex wavelength at 468 nm, and the apex wavelengths of all syrups were 442 nm. Additionally, the intensities of the FDP showed a visible difference between the honey and syrup samples. Therefore, in this study, fluorescence spectra for each pure honey and pure sugar syrup were examined by excitation at 370 nm and data acquisition from 385 to 800 nm (Figure 2). The fluorescence emission spectra were clearly observable in the 385 nm to 700 nm range, with the FDP appearing at 740 nm. Spectral profiles were visibly different between honey and syrup samples, especially in the presence and size of the FDP.

In comparison to natural honey, the fluorescence emission spectra of the syrups produced a blueshift of about 26 nm. Detailed values of the five syrup types and honeys are shown in Table 1. The apexes of the fluorescence emission spectra from all five honeys were located at 470 nm, while the apexes of the syrups were centered at 450 nm. This shift is potentially attributable to differences in the array of conjugated compounds such as polyphenols, aromatic amino acids, Maillard reaction products and other small molecules present in the samples. Ghosh and colleagues investigated the fluorescence spectroscopic properties of honey and cane sugar syrup and found that all spectra from pure honey samples were characterized by two prominent features: a shoulder around 440 nm and a broad band around 510 nm. In contrast, a single prominent band around 430 nm was characteristic of spectra from cane sugar syrup. In addition, their analysis revealed that the major contributor to cane sugar syrup fluorescence is the reduced form of nicotinamide adenine dinucleotide, while the fluorescence of honey is predominantly caused by flavins [32].

Table 1. Apex wavelength and area of emission spectra and FDP area of honeys and syrups.

Honey	Fluorescence Emission Spectra		FDP
	Apex Wavelength (nm)	Area ± RSD	Area ± RSD
Medlar honey	471 ± 2 nm	3783 ± 169.3	1088 ± 30.4
Acacia honey	470 ± 3 nm	1897 ± 50.7	1221 ± 29.7
Linden honey	467 ± 4 nm	3324 ± 151.5	1006 ± 20.4
Chaste honey	465 ± 3 nm	3151 ± 80.1	935 ± 19.9
Rape honey	466 ± 2 nm	1615 ± 26.2	1080 ± 29.6
Average	468 ± 2.8 nm	2754 ± 95.6	1064 ± 26.0
Syrup			
Beet syrup	448 ± 4 nm	3280 ± 189.2	34 ± 5.1
Cassava syrup	438 ± 3 nm	573 ± 56.5	181 ± 16.8
Malt syrup	450 ± 3 nm	1664 ± 21.1	46 ± 4.9
Rice syrup	445 ± 2 nm	619 ± 30.8	103 ± 9.6
HFCS	427 ± 5 nm	2311 ± 230.0	53 ± 5.5
Average	442 ± 3.4 nm	1689 ± 105.2	83.4 ± 8.4

The areas of the emission spectra were substantially different between the different types of honey, possibly due to plant sources, geographic origins and climatic factors, all of which can influence the content of conjugated compounds. The same feature is true for the syrup spectral data. Furthermore, FDPs were highest at 740 nm for pure honey samples but exhibited a very low intensity in the syrup samples (values listed in Table 1). The FDP area for pure honey was consistently 10-fold greater than that of pure syrup. We propose that this difference between the FDP intensity of pure honey and that of syrup could be exploited to monitor adulteration by all types of syrups.

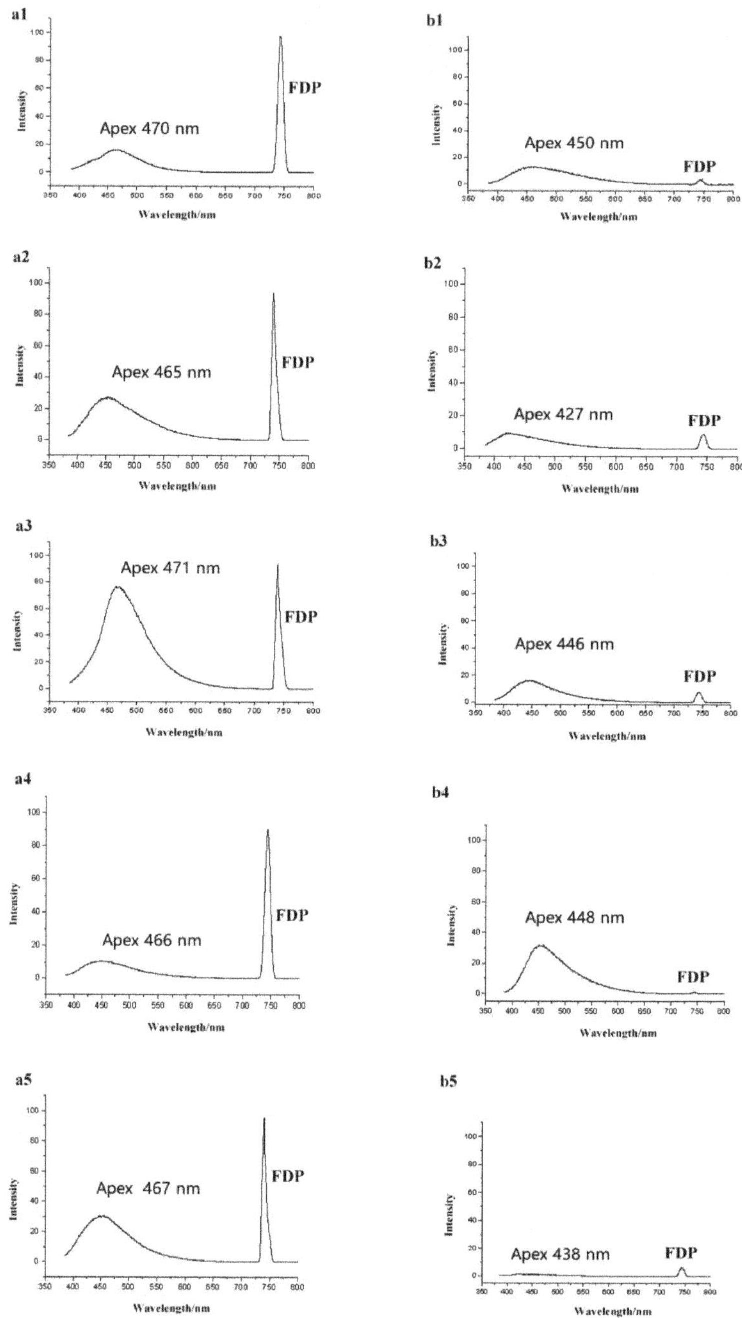

Figure 2. Typical fluorescence spectra of honey and syrup ((**a1**): acacia honey; (**a2**): chaste honey; (**a3**): medlar honey; (**a4**): rape honey; (**a5**): linden honey; (**b1**): maltose syrup; (**b2**): HFCS; (**b3**): rice syrup; (**b4**): beet syrup; (**b5**): cassava syrup).

3.3. The Fluorescence Spectra Profiles of Adulterated Honey

Previous studies have examined fluorescence spectra as a means of distinguishing between pure honey and sugar syrup as well as between honeys from different botanical and geographical origins [23]. In light of published articles describing the features of emission spectra, we found that the incorporation of the FDP intensity with these known spectra allowed us to develop profiles for the identification of sugar syrup adulterants in honey.

We blended beet syrup, cassava syrup, malt syrup, rice syrup and HFCS and incrementally added the mixture (0, 10%, 20%, 30% … 90%) to a blend of all five honeys to assess the correlation between the amount of added syrup and FDP intensities. The apex wavelengths and areas of the fluorescence emission spectra, as well as FDP intensities (Table 2), revealed that as the proportion of syrup increased, a significant blueshift (with the honey peak at 468 nm and the syrup peak at 442 nm) occurred, and the FDP areas decreased in a dose-dependent gradient. Thus, the FPD was positioned at 740 nm, and measurement of its intensity combined with the apex wavelength of the fluorescence emission spectra could be readily used to discriminate between pure and adulterated honey samples. As Figure S1 shows, the R^2 between the amounts of adulterated syrup and FDP intensities was 0.9873 ($p < 0.01$). Thus, the FDP intensity can strongly reflect the amount of syrup added.

Table 2. Fluorescence spectra information of artificially adulterated honey.

Items	Fluorescence Emission Spectra		FDP
	Apex (nm)	Area ± RSD	Area ± RSD
Honey	472 ± 5 nm	1699 ± 41.9	976 ± 35.8 a
10% syrup	468 ± 4 nm	2605 ± 42.8	813 ± 39.0 b
20% syrup	467 ± 5 nm	2530 ± 41.6	765 ± 33.2 bc
30% syrup	461 ± 4 nm	2375 ±39.3	713 ± 32.0 cd
40% syrup	460 ± 5 nm	2270 ± 38.3	623 ± 34.7 de
50% syrup	458 ± 4 nm	2076 ± 37.6	584 ± 28.9 ef
60% syrup	458 ± 6 nm	1873 ± 36.4	503 ± 25.6 fg
70% syrup	454 ± 5 nm	1776 ± 30.9	410 ± 27.5 gh
80% syrup	449 ± 3 nm	1519 ± 27.1	335 ± 25.5 h
90% syrup	446 ± 2 nm	1267 ± 21.6	200 ± 30.4 i

Note: lowercase letters indicate a significant difference ($p < 0.01$).

Commonly, the addition of 30% or higher syrup can make substantial profits. In this study, the minimum amount of syrup added was 10%, and compared with the honey sample, there was a significant difference ($p < 0.01$) (Table 2). We also found that a small deviation in the apex wavelength could be used as an auxiliary qualitative indicator for the relative purity of a given honey sample. Specifically, the apex wavelength shifted from 470 nm to 460 nm, and the FDP intensity decreased from 990 to 610 with the dilution of acacia honey to 30% syrup (Figure 3).

To verify the accuracy of the method, we chose acacia honey samples from different sources and different collection times to compare the intensities of the FDP. Among these acacia honey samples, there were no significant differences ($p > 0.05$) (for detailed information, see Table S1). Thus, the samples with the addition of at least 10% syrup could be identified by significance analysis ($p < 0.01$) compared with natural honey samples.

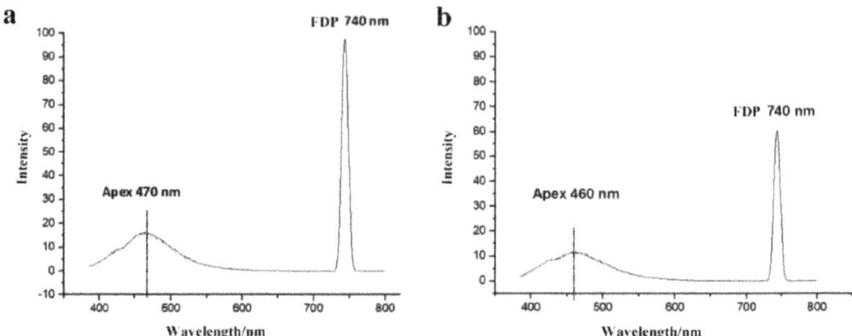

Figure 3. Typical fluorescence spectra of adulterated acacia honey with 30% syrup ((**a**): pure honey; (**b**): acacia honey adulterated with 30% syrup).

3.4. Application to Adulterated Market Samples

Acacia honey, from blossoms of *Robinia pseudoacacia*, is a transparent, pale-yellow honey, valued for its mild flavor compared to other honey varieties. In general, the yield of acacia honey is lower than for other honeys, and thus its market price is higher. Syrup adulteration of acacia honey is currently a serious problem [33,34]. Therefore, adulterant-positive samples of acacia honey were used to verify the effectiveness of our method (Table 3). Twenty samples were taken from local, small markets in Shandong and Henan.

Table 3. Results from four different methods.

Positive Sample Information	Total Numbers	Glucose, Fructose, Sucrose and Maltose Analysis	Stable Carbon Isotopic Ratio Analysis	TLC Method	Fluorescence Method
High-fructose corn syrup-adulterated honey	15	+(0) [a]	+(15)	+(4)	+(11)
Beet syrup-adulterated honey	2	+(0)	+(0)	+(0)	+(2)
Rice syrup-adulterated honey	2	+(0)	+(0)	+(0)	+(1)
Maltose syrup-adulterated honey	1	+(1)	+(0)	+(1)	+(1)

[a] Adulterated samples are indicated with +; parentheses indicate the number of samples.

As part of this study, we compared a method for glucose, fructose, sucrose and maltose analysis to effectively identify maltose syrup [5]. Similarly, we examined stable carbon isotope ratio analysis [29] for the identification of C4 plant sugar (corn syrup) in adulterated honey and found that, although it is effective with high accuracy for HFCS, C3 plant sugars (e.g., beet, rice and maltose syrup) cannot be readily distinguished in samples due to the similarity of their carbon isotope content to that of natural honey [3]. A third method, TLC [7,10], has also been used to target unhydrolyzed polysaccharides and oligosaccharides in some corn and maltose syrups. Each of these methods has unique advantages, but none of them is uniformly effective for the identification of all syrup types in adulterated honey.

The fluorescence-based method demonstrated in our study is a simple and relatively fast screening method to identify high-fructose corn syrup as well as beet, rice and maltose syrups that have been added to honey, with a reasonably low limit of detection at 10%. We found that five adulterated honey samples, one with rice syrup and four with HFCS, could not be identified (false negatives) because of the low syrup content, and thus, due to the detection limit, we conclude that the accuracy of the method is 75%. Among the fifteen positive samples, the apex wavelength ranged from 441 to 452 nm, and FDP intensities were below 200 (compared to the fluorescence spectra information of the syrups, see Table 1). These results indicate that these positive samples are almost pure syrup. In order to maximize profit, in some regions, samples can be found that are composed of almost 100% pure syrup. In these places, where effective market regulation is needed most, our method provides a fast and practical approach for the detection of adulterated honey with a high syrup content.

3.5. Interpretation of Differences in FDP between Honey and Syrup

Under 370 nm excitation, there was an FDP at 740 nm that could be used to distinguish between honey and syrup samples. The FDP arises in non-linear materials. When the monochromator was set to transmit 740 nm, a small fraction of 370 nm excitation light was also transmitted through the emission monochromator. Thus, the observed differences in the FDP at 740 nm were due to the differences in the 370 nm transmission through the samples. The difference in the FDP intensity between honey and syrup may be caused by differences in the amounts and types of macromolecular polymers in each sweetener, such as proteins or nucleotides. Different sources and processing techniques of honey and syrup are also likely to produce differences in the types and polymerization states of macromolecules, thus resulting in measurably large variation in the FDP between products. Interestingly, we found that jujube and buckwheat honeys are not clearly distinguishable from syrup using this method, due to their low FDP intensity, and we thus speculate that this method is currently unsuitable for dark honey. However, the underlying reasons and mechanisms require further study, and with some optimization, this technique may be modified for broader use.

4. Conclusions

We reported the development of a novel fluorescence screening method combining the FDP intensity with the apex wavelength of fluorescence emission spectra in order to identify syrup adulterants in honey. We demonstrated that the FDP intensity decreases significantly and the apex wavelength undergoes a distinct, dose-dependent blueshift with the addition of increasing amounts of syrup to pure honey. Using this method, 10% syrup and higher can be detected in adulterated honey. To validate this method, we analyzed 20 syrup-adulterated honey samples and successfully detected 11 out of 15 with HFCS, 1 out of 2 with rice syrup, 2 out of 2 with beet syrup and 1 (out of 1) with maltose syrup, thus demonstrating 75% accuracy. In comparison with other more sophisticated, labor-intensive and expensive spectroscopy methods, the analysis time was less than 1 min, and the instrument can be operated with minimal training and no subsequent statistical analyses to interpret the data. For most consumers, honey is purchased for its nutritional and medicinal benefits, and especially for its purity. This is also the first report, to our knowledge, on the application of the FDP to research on food adulteration. These experiments provide an exciting route toward wide adoption of this technique by honey producers and regulatory agencies.

Supplementary Materials: The following supporting information can be downloaded at: https://www.mdpi.com/article/10.3390/foods11152316/s1, Figure S1: The relationship of amount of syrup added and area of FDP. Table S1: Fluorescence Spectra Information of acacia honey Samples. Table S2: The comparison of fluorescence spectra information from 3 different manufacturers.

Author Contributions: Methodology, investigation, writing—original draft, funding acquisition: S.Y.; methodology, investigation, formal analysis: M.S.; methodology, investigation: X.W.; investigation, writing—review and editing, supervision: J.S.; conceptualization, writing—review and editing, supervision, funding acquisition: X.X. All authors have read and agreed to the published version of the manuscript.

Funding: This study was supported by the Agricultural Science and Technology Innovation Program under Grant, the project of the Bee Industry for Quality Improvement and the Shanxi Agricultural University Science and Technology Innovation Fund (grant number 2020BQ83).

Data Availability Statement: Data are contained within the article.

Acknowledgments: We are grateful to senior engineer Xinying Liu from the Shandong Bee Industry and Bee Products Quality Monitoring for providing the positive samples.

Conflicts of Interest: The authors declare no conflict of interest.

References

1. Osman, K.A.; Al-Doghairi, M.A.; Al-Rehiayani, S.; Helal, M.I.D. Mineral contents and physicochemical properties of natural honey produced in Al-Qassim region, Saudi Arabia. *J. Food Agric. Environ.* **2007**, *5*, 142–146.
2. Al, M.L.; Daniel, D.; Moise, A.; Bobis, O.; Laslo, L.; Bogdanov, S. Physico-chemical and bioactive properties of different floral origin honeys from Romania. *Food Chem.* **2009**, *112*, 863–867. [CrossRef]
3. Wu, L.; Du, B.; Vander Heyden, Y.; Chen, L.; Zhao, L.; Wang, M.; Xue, X. Recent advancements in detecting sugar-based adulterants in honey—A challenge. *TrAC Trends Anal. Chem.* **2017**, *86*, 25–38. [CrossRef]
4. Siddiqui, A.J.; Musharraf, S.G.; Choudhary, M.I. Application of analytical methods in authentication and adulteration of honey. *Food Chem.* **2017**, *217*, 687–698. [CrossRef]
5. Naila, A.; Flint, S.H.; Sulaiman, A.Z.; Ajit, A.; Weeds, Z. Classical and novel approaches to the analysis of honey and detection of adulterants. *Food Control* **2018**, *90*, 152–165. [CrossRef]
6. Tsagkaris, A.S.; Koulis, G.A.; Danezis, G.P.; Martakos, I.; Dasenaki, M.; Georgiou, C.A.; Thomaidis, N.S. Honey authenticity: Analytical techniques, state of the art and challenges. *RSC Adv.* **2021**, *11*, 11273–11294. [CrossRef]
7. Puscas, A.; Hosu, A.; Cimpoiu, C. Application of a newly developed and validated high-performance thin-layer chromatographic method to control honey adulteration. *J. Chromatogr. A* **2013**, *1272*, 132–135. [CrossRef] [PubMed]
8. Guler, A.; Kocaokutgen, H.; Garipoglu, A.V.; Onder, H.; Ekinci, D.; Biyik, S. Detection of adulterated honey produced by honeybee (*Apis mellifera* L.) colonies fed with different levels of commercial industrial sugar (C3 and C4 plants) syrups by the carbon isotope ratio analysis. *Food Chem.* **2014**, *155*, 155–160. [CrossRef] [PubMed]
9. Ruiz-Matute, A.I.; Soria, A.C.; Martínez-Castro, I.; Sanz, M.L. A new methodology based on GC-MS to detect honey adulteration with commercial syrups. *J. Agric. Food Chem.* **2007**, *55*, 7264–7269. [CrossRef] [PubMed]
10. Megherbi, M.; Herbreteau, B.; Faure, R.; Salvador, A. Polysaccharides as a Marker for Detection of Corn Sugar Syrup Addition in Honey. *J. Agric. Food Chem.* **2009**, *57*, 2105–2111. [CrossRef] [PubMed]
11. Xue, X.; Wang, Q.; Li, Y.; Wu, L.; Chen, L.; Zhao, J.; Liu, F. 2-acetylfuran-3-glucopyranoside as a novel marker for the detection of honey adulterated with rice syrup. *J. Agric. Food Chem.* **2013**, *61*, 7488–7493. [CrossRef] [PubMed]
12. Wang, S.; Guo, Q.; Wang, L.; Lin, L.; Shi, H.; Cao, H.; Cao, B. Detection of honey adulteration with starch syrup by high performance liquid chromatography. *Food Chem.* **2015**, *172*, 669–674. [CrossRef] [PubMed]
13. Sobrino-Gregorio, L.; Bataller, R.; Soto, J.; Escriche, I. Monitoring honey adulteration with sugar syrups using an automatic pulse voltammetric electronic tongue. *Food Control* **2018**, *91*, 254–260. [CrossRef]
14. Sobrino-Gregorio, L.; Vilanova, S.; Prohens, J.; Escriche, I. Detection of honey adulteration by conventional and real-time PCR. *Food Control* **2019**, *95*, 57–62. [CrossRef]
15. de Souza, R.R.; Fernandes, D.D.S.; Diniz, P. Honey authentication in terms of its adulteration with sugar syrups using UV-Vis spectroscopy and one-class classifiers. *Food Chem.* **2021**, *365*, 130467. [CrossRef] [PubMed]
16. Kelly, J.D.; Petisco, C.; Downey, G. Application of Fourier Transform Midinfrared Spectroscopy to the Discrimination between Irish Artisanal Honey and Such Honey Adulterated with Various Sugar Syrups. *J. Agric. Food Chem.* **2006**, *54*, 6166–6171. [CrossRef] [PubMed]
17. Chen, L.; Xue, X.; Ye, Z.; Zhou, J.; Chen, F.; Zhao, J. Determination of Chinese honey adulterated with high fructose corn syrup by near infrared spectroscopy. *Food Chem.* **2011**, *128*, 1110–1114. [CrossRef]
18. Li, S.; Yang, S.; Zhu, X.; Zhang, X.; Ling, G. Detection of honey adulteration by high fructose corn syrup and maltose syrup using Raman spectroscopy. *J. Food Compos. Anal.* **2012**, *28*, 69–74. [CrossRef]
19. Liu, T.; Ming, K.; Wang, W.; Qiao, N.; Qiu, S.; Yi, S.; Huang, X.; Luo, L. Discrimination of honey and syrup-based adulteration by mineral element chemometrics profiling. *Food Chem.* **2021**, *343*, 128455. [CrossRef]
20. Bertelli, D.; Lolli, M.; Papotti, G.; Bortolotti, L.; Serra, G.; Plessi, M. Detection of Honey Adulteration by Sugar Syrups Using One-Dimensional and Two-Dimensional High-Resolution Nuclear Magnetic Resonance. *J. Agric. Food Chem.* **2010**, *58*, 8495–8501. [CrossRef]
21. Guelpa, A.; Marini, F.; du Plessis, A.; Slabbert, R.; Manley, M. Verification of authenticity and fraud detection in South African honey using NIR spectroscopy. *Food Control* **2017**, *73*, 1388–1396. [CrossRef]
22. Lenhardt, L.; Bro, R.; Zeković, I.; Dramićanin, T.; Dramićanin, M.D. Fluorescence spectroscopy coupled with PARAFAC and PLS DA for characterization and classification of honey. *Food Chem.* **2015**, *175*, 284–291. [CrossRef] [PubMed]
23. Ruoff, K.; Karoui, R.; Dufour, E.; Luginbühl, W.; Bosset, J.O.; Bogdanov, S.; Amado, R. Authentication of the Botanical Origin of Honey by Front-Face Fluorescence Spectroscopy. A Preliminary Study. *J. Agric. Food Chem.* **2005**, *53*, 1343–1347. [CrossRef] [PubMed]
24. Ruoff, K.; Luginbühl, W.; Künzli, R.; Bogdanov, S.; Bosset, J.O.; von der Ohe, K.; von der Ohe, W.; Amadò, R. Authentication of the Botanical and Geographical Origin of Honey by Front-Face Fluorescence Spectroscopy. *J. Agric. Food Chem.* **2006**, *54*, 6858–6866. [CrossRef] [PubMed]
25. Sergiel, I.; Pohl, P.; Biesaga, M.; Mironczyk, A. Suitability of three-dimensional synchronous fluorescence spectroscopy for fingerprint analysis of honey samples with reference to their phenolic profiles. *Food Chem.* **2014**, *145*, 319–326. [CrossRef] [PubMed]
26. Mehretie, S.; Al Riza, D.F.; Yoshito, S.; Kondo, N. Classification of raw Ethiopian honeys using front face fluorescence spectra with multivariate analysis. *Food Control* **2018**, *84*, 83–88. [CrossRef]

27. Santana, J.E.G.; Coutinho, H.D.M.; da Costa, J.G.M.; Menezes, J.M.C.; Pereira Teixeira, R.N. Fluorescent characteristics of bee honey constituents: A brief review. *Food Chem.* **2021**, *362*, 130174. [CrossRef] [PubMed]
28. Valinger, D.; Longin, L.; Grbeš, F.; Benković, M.; Jurina, T.; Gajdoš Kljusurić, J.; Tušek, A.J. Detection of honey adulteration—The potential of UV-VIS and NIR spectroscopy coupled with multivariate analysis. *LWT-Food Sci. Technol.* **2021**, *145*, 111316. [CrossRef]
29. Cabañero, A.I.; Recio, J.L.; Rupérez, M. Liquid chromatography coupled to isotope ratio mass spectrometry: A new perspective on honey adulteration detection. *J. Agric. Food Chem.* **2006**, *54*, 9719–9727. [CrossRef] [PubMed]
30. Hao, S.; Li, J.; Liu, X.; Yuan, J.; Yuan, W.; Tian, Y.; Xuan, H. Authentication of acacia honey using fluorescence spectroscopy. *Food Control* **2021**, *130*, 108327. [CrossRef]
31. Kečkeš, S.; Gašić, U.; Veličković, T.Ć.; Milojković-Opsenica, D.; Natić, M.; Tešić, Ž. The determination of phenolic profiles of Serbian unifloral honeys using ultra-high-performance liquid chromatography/high resolution accurate mass spectrometry. *Food Chem.* **2013**, *138*, 32–40. [CrossRef]
32. Ghosh, N.; Verma, Y.; Majumder, S.K.; Gupta, P.K. A fluorescence spectroscopic study of honey and cane sugar syrup. *Food Sci. Technol. Res.* **2005**, *11*, 59–62. [CrossRef]
33. Kenjerić, D.; Mandić, M.L.; Primorac, L.; Bubalo, D.; Perl, A. Flavonoid profile of Robinia honeys produced in Croatia. *Food Chem.* **2007**, *102*, 683–690. [CrossRef]
34. Wang, J.; Xue, X.; Du, X.; Cheng, N.; Chen, L.; Zhao, J.; Zheng, J.; Cao, W. Identification of acacia honey adulteration with rape honey using liquid chromatography–electrochemical detection and chemometrics. *Food Anal. Method* **2014**, *7*, 2003–2012. [CrossRef]

Article

SDS-PAGE Protein and HPTLC Polyphenols Profiling as a Promising Tool for Authentication of Goldenrod Honey

Małgorzata Dżugan [1,*], Michał Miłek [1], Patrycja Kielar [2], Karolina Stępień [2], Ewelina Sidor [1,3] and Aleksandra Bocian [1,4]

1 Department of Chemistry and Food Toxicology, Institute of Food Technology and Nutrition, University of Rzeszow, Ćwiklińskiej 1a, 35-601 Rzeszow, Poland
2 Department of Biology, Institute of Biology and Biotechnology, University of Rzeszow, Zelwerowicza 4, 35-601 Rzeszow, Poland
3 Doctoral School, University of Rzeszow, Rejtana 16c, 35-959 Rzeszow, Poland
4 Department of Biotechnology and Bioinformatics, Faculty of Chemistry, Rzeszow University of Technology, Powstańców Warszawy 6, 35-959 Rzeszow, Poland
* Correspondence: mdzugan@ur.edu.pl; Tel.: +48-178721619

Abstract: The aim of the study was to use protein and polyphenolic profiles as fingerprints of goldenrod honey and to apply them for verification of the labeled variety. The markers for 10 honey samples were correlated with the standard physicochemical parameters and biological activity measured in vitro as antioxidant, antifungal and antibacterial activities. Honey proteins were examined regarding soluble protein, diastase and SDS-PAGE protein profile. The polyphenolic profile was obtained with the use of the HPTLC and the antioxidant activity was detected with standard colorimetric methods. The antimicrobial effect of representative honey samples of different chemical profiles was verified against *E. coli* and budding yeast. It was found that the SDS-PAGE technique allows for creating the protein fingerprint of the goldenrod honey variety which was consistent for 70% of tested samples. At the same time, the similarity of their polyphenolic profile was observed. Moreover, specific chemical composition resulted in higher bioactivity of honey against tested bacteria and yeast. The study confirmed the usefulness of both SDS-PAGE and HPTLC techniques in honey authentication, as an initial step for selection of samples which required pollen analysis.

Keywords: antibacterial activity; antifungal activity; authentication; goldenrod; honey; protein; polyphenols

Citation: Dżugan, M.; Miłek, M.; Kielar, P.; Stępień, K.; Sidor, E.; Bocian, A. SDS-PAGE Protein and HPTLC Polyphenols Profiling as a Promising Tool for Authentication of Goldenrod Honey. *Foods* **2022**, *11*, 2390. https://doi.org/10.3390/foods11162390

Academic Editors: Liming Wu and Qiangqiang Li

Received: 11 July 2022
Accepted: 5 August 2022
Published: 9 August 2022

Publisher's Note: MDPI stays neutral with regard to jurisdictional claims in published maps and institutional affiliations.

1. Introduction

Among nectar honeys, which differ in terms of properties and taste, goldenrod honey has recently become more and more popular. It is related to the growing supply of goldenrod as a honeyflow in Central and Eastern Europe, including Poland. On the other hand, goldenrod honey is still one of the less-studied varieties, and few literature data from recent years concern the chemical profiles and antioxidant properties of this honey [1,2]. Many beneficial properties of this honey are reported: antibiotic and supporting the urinary system, skin and circulatory system. However, due to the lack of confirmation of this varietal honey features in scientific literature, natural medicine attributes its healing effect to the properties of the goldenrod plant. Goldenrod (*Solidago* sp.) is a controversial plant: on the one hand, it has some phytotherapeutic effects (in diseases of the skin, respiratory system, circulation, urinary system, even in the treatment of certain cancers and depression [3–5], and on the other hand, popular species *S. gigantea* and *S. canadensis* are classified as invasive plants [6,7]. Only *Solidago virgaurea*, a native European species is listed in pharmacopoeias as an official herbal drug [5]. The most common species of *Solidago*, i.e., in south-eastern Poland, mainly *S. gigantea* and *S. canadensis*, are used as beeflows. Jasicka-Misiak et al. [1] refer to *S. virgaurea* as the source of nectar, but more recent research points to the remaining common species of goldenrod used by bees [2]. The availability of different species may be one of the reasons for the variability of goldenrod honeys, depending on their geographic origin.

Goldenrod honey owes its healing properties to the flowers from which it is obtained, i.e., *Solidago* spp. plants. Thanks to it, goldenrod honey contains flavonoids (including quercetin) as well as tannins and organic acids. It is rich in vitamins and bactericidal substances. It is characterized by a specific profile of volatile compounds among which germacrene D has been recognized as a marker compound of this honey variety [8]. In natural medicine, this type of honey is considered particularly valuable and recommended for its anti-inflammatory, diuretic, choleretic and diastolic effects on the genitourinary system; however, these features have been not objective of scientific research. This honey type has a characteristic, lemon-like, sour aftertaste and a pleasant aroma. It crystallizes relatively quickly, creating a creamy consistency. Goldenrod honey is available in shades of yellow to amber [2]. It becomes much brighter after crystallization. This honey variety does tend to crystallize fast in a matter of weeks [8]. Crystallization does not change the flavor or spoil the honey. Moreover, the specific feature of goldenrod honey is susceptibility to fermentation, which can occur during storage at room temperature.

Assessing the authenticity of honey is a serious problem that has gained much interest internationally because honey has frequently been subject to various fraudulent practices, including mislabeling of botanical and geographical origin and mixing with sugar syrups or honey of lower quality. Since now, melissopalynological analysis has been the only recognized analysis to confirm honey variety. However, this technique based on counting plant pollen which occurs in honey is expensive and time-consuming. To protect the health of consumers and reduce the unfair practice of honey adulteration, the different approaches used to assess the authenticity of honey, specifically by the application of advanced instrumental techniques, have been proposed. The protein and other nitrogen compounds have been rarely studied [8–13].

The aim of the study was to construct the fingerprints of goldenrod honey based on protein and polyphenolic profiles. For the first time, the SDS-PAGE protein analysis and HPTLC polyphenols profiling were used to verify the honey variety. An attempt was made to determine whether it is possible to identify the species of goldenrod that bees used for honey production.

2. Materials and Methods

2.1. Honey and Plant Material

Ten samples of honeys declared as goldenrod were purchased in the 2021 beekeeping season from apiaries located in the Podkarpackie Province (Poland).

The flowers of three species of goldenrod (*Solidago virgaurea, Solidago gigantea* and *Solidago canadensis*) were collected from natural habitats in August 2021 (50°045′ N 21°862′ E; 50°048′ N 21°869′ E; 50°039′N 21°894′ E). After botanical identification, the samples were dried in the dark, at a temperature not exceeding 40 °C. Voucher specimens were deposited in Department collection under the numbers SV/08/21, SG/08/21, and SC/08/21.

2.2. Extracts Preparation

Dried plant material was pulverized using pestle and mortar and weighted (5 g) into conical flasks. After flooding with 50 mL of 50% ethanol, it was extracted in an ultrasonic bath (Sonic 10, Polsonic, Warszawa, Poland) for 20 min, then filtered through a filter paper. Ethanol was evaporated using a centrifugal evaporator (VC 2–18 CDPlus, Martin Christ, Osterode am Harz, Germany) and then the aqueous residue was lyophilized using Alpha 1–2 LD plus freeze dryer (Martin Christ, Osterode am Harz, Germany).

For chromatographic analyses, honey samples were prepared using solid phase extraction (SPE). Twenty grams of each honey were dissolved in 100 mL of acidified distilled water (pH = 2) and passed through C-18 Sep Pack Cartridge (Waters Corporation, Milford, MA, USA), preconditioned with 10 mL of methanol and 10 mL of acidified (pH = 2) water.

After washing the sugars with acidified water, the polyphenols were eluted with methanol (2.5 mL).

2.3. Water Content

The determination of water content was done by the refractometric method, using an electronic refractometer for honey HI96800 (Hanna Instruments, Woonsocket, RI, USA). The value was determined in 21 ± 2 °C. with to the second decimal place. The determinations were made in triplicates.

2.4. pH and Free Acidity

To determine active acidity, a pH measurement of 20% solutions of honey extract in boiled, cold distilled water was performed using a CP-401 pH meter (Elmetron, Zabrze, Poland). To determine the free acidity, 50 mL of 20% appropriate honey extracts were titrated by 0.1 M NaOH to reach a pH of 8.3 measured by pH meter. The results were expressed in mEq/100 g.

2.5. Conductivity

To determine the electrical conductivity, 20% solutions of honey extracts in boiled, cold distilled water were used. The conductivity of each sample was determined by immersing the electrode in the test solution. Each sample was measured three times. The conductivity of each product solution was measured using a CP-401 conductometer (Elmetron, Zabrze, Poland) and the results (in mS/cm) were presented.

2.6. Color Analysis

The color of the honey was analyzed using a dedicated HI 96785 colorimeter (Hanna Instruments, Woonsocket, RI, USA). The results were expressed on the Pfund scale.

2.7. A-Amylase Assay

α-amylase was determined by a spectrophotometric method with the Phadebas Diastase test (Magle AB, Lund, Sweden) according to the manufacturer's instructions. The results of absorbance were measured at wavelength 620 nm against a blank (acetate buffer) using a Biomate 3 spectrophotometer (Thermo Scientific, Waltham, MA, USA). The results were calculated according to the formula attached in the manual and expressed as diastase number (DN).

2.8. Soluble Protein

The protein content in honey was determined using the Bradford method according to Latimer [14]. One thousand μL of Bradford reagent (G-250) was added to 20 μL of each honey sample. Samples were incubated for 5 min at room temperature and the absorbance at 595 nm was read using a Biomate 3 spectrophotometer (Thermo Scientific, Waltham, MA, USA). The results were calculated from a calibration curve 0–100 μg/sample (y = 0.0551x, $r^2 = 0.9991$). Bovine albumin has been used as the standard.

2.9. Total Phenolic and Flavonoids Content

Total phenolic and flavonoid content in honey and plant extracts samples were determined according to procedures described by us previously [13]. In the case of honeys, 5% solutions in distilled water were used, and in the case of plant extracts, crude extracts diluted properly were taken for analysis.

2.10. Antioxidant Capacity Assays

Antioxidant capacity using DPPH and FRAP methods was determined according to procedures described by us previously [13]. The CUPRAC method was applied according to Matłok et al. [15]. Briefly, 10 μL of each diluted sample was pipetted into microplate wells, then 40 μL of $CuCl_2$ (10 mM), 50 μL of neocuproine (7.5 mM), and 50 μL of ammonium acetate (1 M) were added. The absorbance was measured with a microplate reader (EPOCH 2, BioTek, Winooski, VT, USA) at 450 nm after 30 min incubation in the dark against a blank. The result was expressed in Trolox equivalents (μmol TE/100 g) from the standard curve (y = 0.026x, r^2 = 0.9973).

2.11. HPTLC Analyses

The analyses were performed with the use of the CAMAG (Muttenz, Switzerland) HPTLC chromatography set, consisting of a semi-automated sample applicator (Linomat 5), an automatic developing chamber (ADC 2), a derivatizer and a visualizer. The extract samples were applied on the HPTLC plate (HPTLC Silica Gel 60 F254 plates, 20 cm $\times$ 10 cm, Merck, Darmstadt, Germany) in a volume of 5 μL, as 8 mm wide bands. Two different systems for the mobile phase and derivatization reagent were used: A—mobile phase: ethyl acetate, acetic acid, formic acid, water (10:1.1:1.1:2.6) and derivatizing agent: Natural Product Reagent/Polyethylene glycol 400 (NP/PEG); B—mobile phase: chloroform, ethyl acetate, formic acid (5:4:1) and derivatizing agent: *p*-anisaldehyde reagent with heating in 110 °C (10 min). After derivatization, plates were photographed in visible light and UV (366 nm). The obtained images were analyzed using HPTLC software (Vision CATS, CAMAG, Muttenz, Switzerland).

2.12. Protein Profiling by SDS-PAGE

Ten samples of goldenrod honey and four reference honey samples (heather, honeydew, rape and multifloral) were prepared as in previous work with minor modifications [16]. One gram of raw honey was dissolved in 1 mL of deionized water containing 2% Nonidet P-40 substitute and 2% dithiothreitol. The samples were then mixed 2 to 1 with 4x concentrated standard Laemmli buffer and were incubated for 5 min at 95 °C. After cooling, 15 μL of the samples were applied to 15% denaturing gels (with 3% stacking gels). Electrophoresis was carried out on a Mini-Protean II apparatus (Bio-Rad Laboratories, Hercules, CA, USA) at 50 V for 15 min and then at 200 V for 1.5 h according to the standard method of Laemmli, with the BlueEasy Prestained Protein Marker (NIPPON Genetics EUROPE, Düren, Germany) as a molecular weight marker and Tris-Glycine-SDS buffer. After electrophoresis, gels were stained with Coomassie Brilliant Blue G-250. The staining was performed overnight and then gels were destained for 24 h with deionized water. Gels were scanned with Image Scanner III (GE Healthcare, Little Chalfont, UK) and processed by LabScan 6.0 (GE Healthcare, Little Chalfont, UK). Gels analysis was performed in ImageJ (1.52a) software to generate a graphical representation of each lane on the gel to assist in sample comparison. For better visualization, individual bands from the SDS-PAGE gel were contrasted with the profile obtained with the software.

2.13. Yeast Strain and Growth Conditions

The strain used in this study is a reference haploid wild-type yeast *Saccharomyces cerevisiae* strain BY4741 (*MATa his3Δ leu2Δ met15Δ ura3Δ*; Euroscarf, Germany). Yeast cells were grown in a standard rich liquid YPD medium (1% Difco yeast extract, 1% yeast bactopeptone, and 2% (*w/v*) glucose) on a rotary shaker at 150 rpm or on a solid YPD medium containing 2% agar. The experiments were carried out at an optimal temperature for yeast: 28 °C.

2.14. Kinetics of the Yeast Growth Assay

Yeast cells were grown in a liquid YPD medium (1% Difco yeast extract, 1% yeast bactopeptone, 2% (*w/v*) glucose) without (control) or with tested honeys or extracts on a rotary shaker at 150 rpm, or on a solid YPD medium containing 2% agar. The experiments were carried out at 28 °C. The growth was monitored using an Anthos 2010 type 17 550 microplate reader (Biochrom, Cambridge, UK) at 600 nm by measurements at 2 h intervals for 12 h. Each experiment was repeated at least three times.

2.15. Yeast Cell Viability

For determining cell death, standard staining with propidium iodide was used. Briefly, cells were washed twice in sterile water, suspended in PBS, and stained with 5 mg/mL propidium iodide (Sigma-Aldrich, Saint Louis, MO, USA) for 15 min in the dark at room temperature. Fluorescence pictures were taken using an Olympus BX-51 microscope equipped with a DP-72 digital camera and cell Sens Dimension software. Dead cells were red fluorescent (λex = 480 nm; λem = 520 nm). The data represent the mean values from three independent experiments. Statistical significance was assessed using Student's *t*-test ($p < 0.01$).

2.16. Bacteria Strain, Growth Condition, and Antibacterial Activity

The antibacterial activity of the honey and extracts samples was assessed by monitoring the cell growth of a wide-use Escherichia coli DH5α strain (Thermo Fisher Scientific, Hennigsdorf, Germany). Bacterial cells were grown in a standard LB medium (LB Broth Lennox; BioShop, Burlington, Canada) without (control) or with tested honeys or extracts on a rotary shaker at 180 rpm. The growth was monitored using an Anthos 2010 type 17 550 microplate reader at 600 nm by measurements at 2 h intervals for 12 h. Each experiment was repeated at least three times. The experiments were carried out at optimal for bacteria 37 °C. The optical density of the culture was monitored every 2 h, at 600 nm wavelength.

2.17. Statistical Analysis

All analyses were performed in triplicates. The results are presented as mean $\pm$ standard deviation. Significant differences between data obtained for tested samples were assessed using Tukey's test ($p = 0.05$) or Student's *t*-test ($p = 0.01$). The correlations between the results were expressed by the Pearson correlation coefficient. Principal component analysis (PCA) and cluster analysis were carried out to group the test samples according to the parameters tested. All calculations were performed using Statistica 13.1 software (StatSoft, Inc., Tulsa, OK, USA).

3. Results and Discussion

3.1. Physicochemical Parameters of Honey Quality

Firstly, the physicochemical parameters were analyzed for 10 goldenrod honeys to assess their quality regarding obligatory UE standards (Table S1). It can be concluded that the analyzed samples meet the requirements of the EU Directive of 2014 [17], which clearly indicates the quality requirements for varietal honeys. It was found that goldenrod honeys are characterized by a varied water content, which was on average 17.97% but did not exceed 19.55%, which fulfills guidelines (below 20%). Such moisture content is specific for Polish nectar honeys and has been reported earlier [18,19].

The mean pH value of the tested honeys was 4.27 and was higher than the results for goldenrod honeys obtained by Ratiu et al. [20], where the values ranged from 3.31 to 3.67. The inverse relationship was found by analyzing the acidity. Studied goldenrod honeys had lower values (13.95–32.55 mEq/kg) compared to foreign goldenrod honeys, where the parameter was in the range of 42.7–49 mEq/kg [20]. This feature can be specific for Polish nectar honey. Tomczyk et al. [19], examining Polish varietal honeys, found the lowest mean acidity values for acacia (16.1 mEq/kg), and the highest for multiflorous honey (37.0 mEq/kg).

Electrical conductivity can help distinguish nectar from honeydew honey. The EU Directive [17] indicates that the conductivity of nectar honeys cannot be lower than 0.2 mS/cm and not higher than 0.8 mS/cm. In the analyzed goldenrod honey samples, the average value of the conductivity was in the range of 0.139–0.592 mS/cm, and for four samples, the value of the tested parameter was below the limit, and the obtained mean value of the conductivity of 0.263 mS/cm is lower compared to the results obtained by Ratiu et al. [20], where the tested parameter for goldenrod honeys ranged from 0.331 to 0.669 mS/cm.

Tested honey samples were characterized by proper sugar profile (data not shown) and 5-hydroxymethylfurfural (HMF) content not exceeding obligatory limits (Table S1).

Tested samples were strongly diversified in the terms of α-amylase activity, determined as diastase number (DN). The values of DN were found from 11 to 22 (Table 1). However, all of the tested honeys are within the applicable limits amounted 8 DN [17]. As the diastase activity is a known indicator of the biological activity of honey and its overheating, samples no. 1, 3, 5 and 10 can be assumed as the most active honeys. The soluble protein content ranged from 22 to 89 mg/100 g, which are typical values for Polish light honeys—19.09–133.18 mg/100 g, for acacia, linden and multifloral honey [13].

Table 1. Protein content and diastase number (DN) of tested honey samples.

Sample	1	2	3	4	5	6	7	8	9	10
Soluble protein content [mg/100 g]	38.11 ± 7.97 [a]	22.68 ± 1.28 [a]	38.91 ± 8.35 [ab]	39.01 ± 8.98 [ab]	35.39 ± 9.08 [ac]	60.79 ± 11.55 [bd]	51.72 ± 14.25 [bc]	75.31 ± 8.98 [d]	57.16 ± 6.42 [bcd]	88.92 ± 10.27 [d]
Diastase number (DN)	25.96 ± 0.42 [a]	16.92 ± 1.15 [b]	24.40 ± 0.68 [a]	12.20 ± 0.85 [cd]	17.68 ± 1.25 [b]	10.68 ± 0.85 [c]	11.41 ± 0.00 [cd]	15.09 ± 0.42 [be]	13.62 ± 0.42 [de]	22.10 ± 0.85 [f]

Means sharing the same letter in a row are significantly different at $p = 0.05$.

Honey color is a parameter that is visually evaluated by beekeepers in Poland, however, we measured it by colorimeter (Figure 1). The tested samples of goldenrod honeys belonged to light varieties except for sample 10, which showed a slightly darker shade, and it was visible also in higher values of color parameters determined for this sample. Goldenrod honeys from three countries, studied by Czigle et al. [2] were classified as water white to light amber for Slovak and Hungarian samples, whereas honeys from Poland had a color in the range of 0.30 to 7.37 mm Pfund, which puts them in the water white group [2].

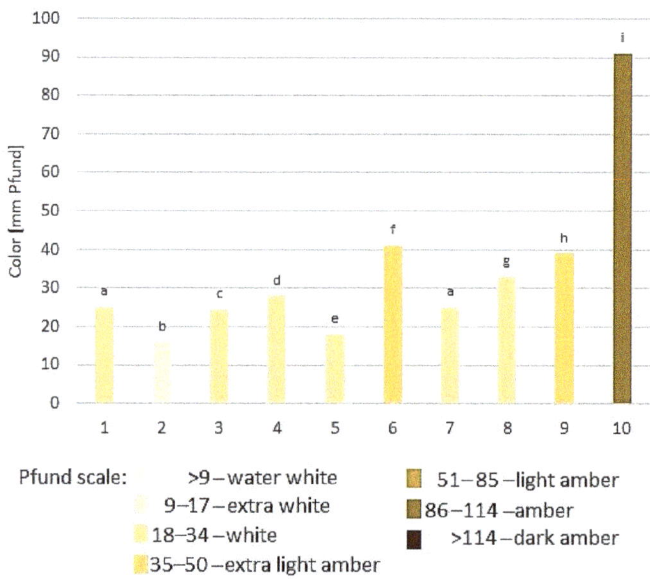

Figure 1. Color analysis of goldenrod honey samples (1–10). Different letters above the bars indicate significant differences ($p = 0.05$).

3.2. Protein Composition

The protein profiles of tested goldenrod honeys were obtained by SDS-PAGE analysis and compared to other varietal honeys (Figure 2). The presence of specific bands in the range between 50 and 75 kDa was found as specific for goldenrod honey, whereas all honeys contain protein fractions at around 60, 70 and 75 kDa.

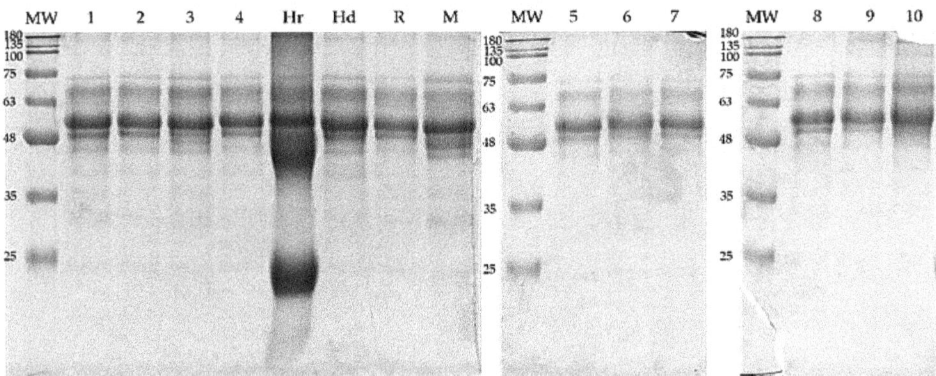

Figure 2. Protein profiles of analyzed goldenrod honey samples in comparison to other selected honeys obtained using SDS-PAGE. Abbreviations: 1–10—goldenrod honeys, Hr—heather honey, Hd—honeydew honey, R—rapeseed honey, M—multifloral honey, MW—BlueEasy Prestained Protein Marker.

The heather honey used as reference differs markedly from the rest with distinct fractions at 40 kDa and 25 kDa. All samples, except heather honey and test samples 6 and 10, have a band at a height of about 48 kDa. However, sample no. 3 as well as honeydew and multifloral honey have a specific arrangement of bands in the range of 40–48 kDa visible on the gels. These differences are best seen in profiles generated in ImageJ (Figure 3). Comparing the course of obtained graphs, the specific fingerprint for goldenrod honey can be pointed as a typical pattern as in the case of sample no. 1, 2, 4 and 8. Based on distinct profiles observed for samples 3, 6 and 10, it can be concluded that the variety is not pure, and the tested honey may originate from various plant sources and even contain an admixture of honeydew. We obtained similar profiles earlier for varietal honeys, although no clear peaks in the range of 40–48 kDa were observed in the case of the multifloral honey analyzed at that time [16]. Previously, the differences in the band pattern on SDS-PAGE gels between the different types of honey were demonstrated by Baroni et al. [9]. They also linked characteristic proteins present in plant pollen to those shown on gels for honey samples. Very similar profiles for varietal honeys were obtained by Muresan et al. [21], with repetitive proteins in the mass range of 45–85 kDa.

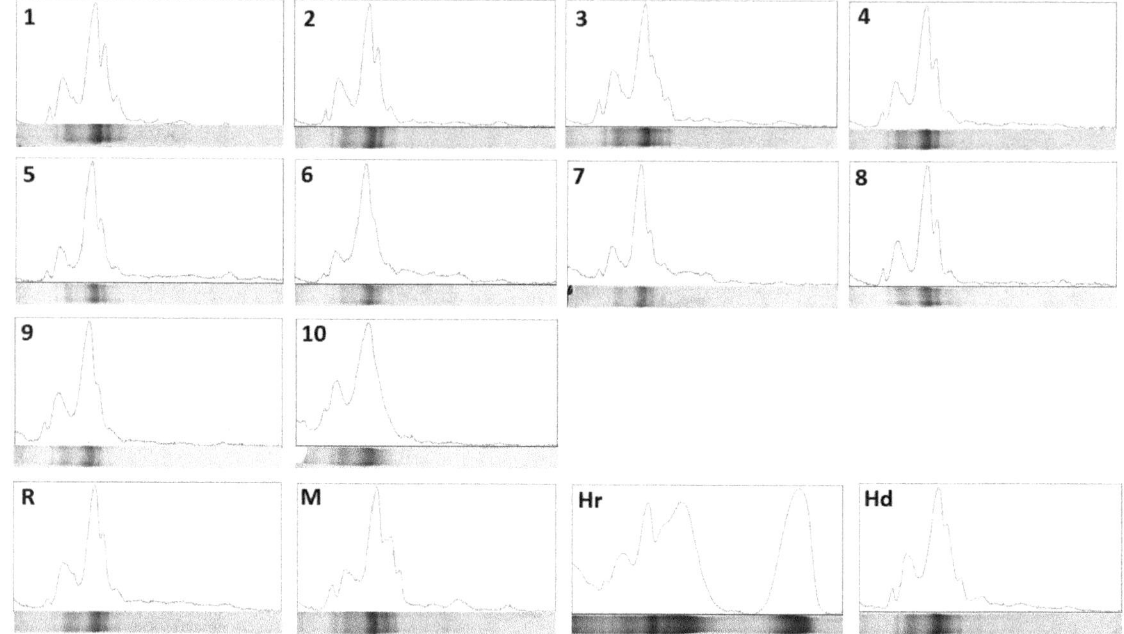

Figure 3. Detailed analysis of particular SDS-PAGE protein profiles obtained for analyzed goldenrod honeys (1–10) and four reference honeys (Hr—heather, Hd—honeydew, R—rapeseed, M—multifloral) in ImageJ software.

3.3. HPTLC Analysis

HPTLC phenolic profiles obtained with two mobile phase systems as well as two different derivatizing agents (Figure 4) are similar for most tested samples, allowing the specification of a typical profile for goldenrod honey and its authentication. The characteristic, repeating pattern of bands is visible, especially in the case of derivatization with *p*-anisaldehyde. In the case of samples 3, 6 and 10, the profiles are significantly different. In the case of sample 10, it is possible to infer an admixture of honeydew, which is manifested by characteristic additional pink bands at Rf approx. 0.48 and 0.6. This pattern is similar to the one we observed earlier for honeydew honey samples [16]. These pink bands are also present, although less intense, in sample 6. The obtained profiles seem to correlate with the higher content of phenolic compounds in the three above-mentioned samples and their stronger antioxidant properties which was particularly observed in sample 10 where the admixture of honeydew could explain the strongest antioxidant properties (see Section 3.4). In honey number 3, additional intense bands are visible. Figure 4A shows blue ones at Rf = 0.26 and a series of blue-green between 0.45 and 0.7. A similar series of bands are clearly visible in sample 6. Figure 4B shows an additional blue band at Rf = 0.11 in sample 3, observed by us earlier in the rape honey sample [16].

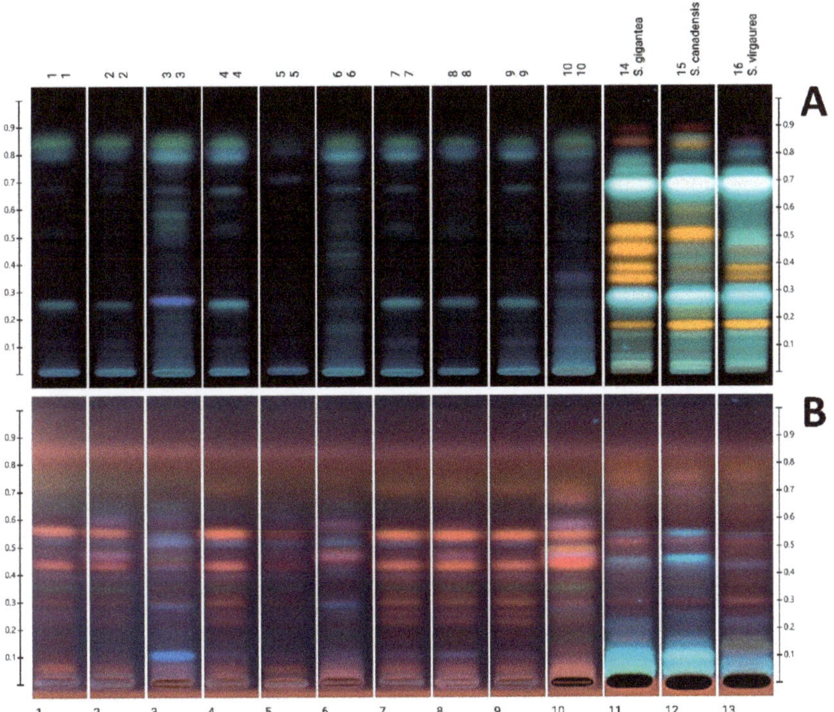

Figure 4. HPTLC plates images: (**A**)—developed with eluent system A, derivatized with NP/PEG reagent, (**B**)—developed with eluent system B, derivatized with *p*-anisaldehyde reagent, Track 1–10: goldenrod honey, track 11: *S. gigantea*, track 12: *S. canadensis*, track 13: *S. virgaurea*.

Since three of the samples (3, 6 and 10) differ from the others in terms of polyphenol profiles, we decided to evaluate the flower extract profiles of three species of goldenrod to check whether the differentiation of the honey profiles is not related to another species of *Solidago* as a honeybee flow. The extracts of the individual species differ in particular in the Rf region between 0.30 and 0.55 (Figure 4A), where the bands derived from flavonoid glycosides, mainly derivatives of quercetin and/or luteolin, were identified. *Solidago gigantea* flowers contain four distinct glycosides, whereas in *S. canadensis*, only one with the highest Rf was present in large amounts, and in *S. virgaurea*, two with lower Rf (0.34 and 0.38), were identified as quercitrin and hyperoside. Rutin was present in all three species (Rf = 0.18). In addition, in *S. canadensis*, extracts from the bands of flavonoid aglycones, quercetin and kaempferol (Rf = 0.82 and 0.84, respectively) were detected. Derivatization with *p*-anisaldehyde revealed an additional gray-green band in the *S. virgaurea* extract at Rf = 0.13, possibly from some phenolic acid. Based on the comparison of the tracks from honey and goldenrod flower extracts, it can be theorized that the two most intense bands present in all three species of flowers (Rf = 0.28 and 0.69, Figure 4A) are most likely derived from certain phenolic acids, and are also present in most of the analyzed honey samples.

Among the polyphenols mentioned as present in goldenrod honey are mainly phenolic acids: gallic, 4-hydroxybenzoic, *p*-coumaric and ferulic [1]. Flavonoids have also been identified, mainly chrysin, galanagin, pinocembrin, kaempferol, quercetin and luteolin [22]. In flowers of various species of goldenrod, flavonoids were identified: mainly quercetin glycosides, and also aglycones: quercetin, kaempferol andisoramnetin [23–25]. In addition, in *S. virgaurea* flowers, myricetin, naringenin, genistein and pinocembrin were identified [1]. Among phenolic acids, chlorogenic, caffeic and ferulic acids were mainly found in the flowers of *S. gigantea* and *S. canadensis* [23,25] and gallic, 3,4-dihydroxybenzoic and vanillic acids were found in the flowers of *S. virgaurea* [1]. The quoted literature data indicate a certain variability in the polyphenol profile between individual goldenrod species, with *S. gigantea* and *S. canadensis* being more similar to each other than to *S. virgaurea*. Similar polyphenol profiles were obtained by HPLC for *S. canadensis* and *S. gigantea* by Zekič et al. [26].

To confirm the richest polyphenol profile obtained for *S. gigantea*, the total content of phenols, flavonoids and antioxidant activity of the extracts of three species of goldenrod (flowers and leaves) were analyzed (Table S2). The obtained results also indicate that this species is the richest in phenolic compounds, which is reflected in its antioxidant capacity. Moreover, the data shows a high correlation (TPC vs. FRAP: r = 0.991, TPC vs. DPPH: r = 0.972, TPC vs. CUPRAC: r = 0.948). Literature data also indicate this goldenrod species as richer in polyphenols and antioxidants [27–29]. Based on provided literature data and own results, some species–specific differences were found; however, no obvious polyphenol markers were identified that could indicate which species of goldenrod is the source nectar for the analyzed honeys.

3.4. Antioxidant Properties of Goldenrod Honey

The goldenrod honeys were characterized in terms of the total content of phenols and flavonoids. The results obtained for ten honey samples are summarized in Table 2.

Table 2. Total phenolics and flavonoids content as well as antioxidant properties of goldenrod honeys.

Sample	TPC [mg GAE/100 g]	TFC [mg QE/100 g]	DPPH [μmol TE/100 g]	FRAP [μmol TE/100 g]	CUPRAC [μmol TE/100 g]
1	17.36 ± 1.69 [a]	0.31 ± 0.02 [a]	22.81 ± 12.62 [a]	19.74 ± 1.74 [a]	302.47 ± 23.49 [a]
2	18.06 ± 0.75 [a]	0.26 ± 0.01 [a]	24.66 ± 8.25 [a]	21.93 ± 2.31 [a]	379.37 ± 54.01 [ac]
3	31.75 ± 0.91 [b]	0.48 ± 0.02 [b]	29.27 ± 16.94 [a]	62.28 ± 2.74 [b]	625.45 ± 46.99 [b]
4	22.22 ± 1.89 [c]	0.33 ± 0.01 [ad]	25.58 ± 15.07 [a]	33.11 ± 0.38 [c]	333.23 ± 35.52 [ac]
5	15.77 ± 1.95 [a]	0.20 ± 0.03 [c]	15.44 ± 12.45 [a]	19.74 ± 3.29 [a]	640.83 ± 38.71 [b]
6	25.50 ± 1.20 [cd]	0.36 ± 0.01 [d]	31.34 ± 14.21 [a]	46.93 ± 1.00 [d]	435.76 ± 44.40 [c]
7	21.83 ± 0.69 [c]	0.34 ± 0.02 [ad]	21.43 ± 2.39 [a]	33.55 ± 0.66 [c]	317.85 ± 8.88 [ac]
8	18.65 ± 1.64 [ac]	0.30 ± 0.01 [a]	20.28 ± 4.17 [a]	23.46 ± 1.00 [a]	297.34 ± 49.44 [a]
9	27.78 ± 0.45 [d]	0.38 ± 0.02 [d]	20.74 ± 2.07 [a]	43.20 ± 1.00 [d]	451.14 ± 72.68 [c]
10	50.69 ± 0.69 [e]	0.77 ± 0.01 [e]	39.18 ± 2.88 [a]	96.93 ± 3.38 [e]	1168. 87 ± 30.76 [d]

GAE—gallic acid equivalents, QE—quercetin equivalents, TE—Trolox equivalents. Means sharing the same letter in a column are significantly different at $p = 0.05$.

The content of phenolic compounds ranged from 15.77 to 50.69 mg GAE/100 g, whereas sample 10 significantly differed from the others, for which TPC did not exceed 31.75 mg GAE/100 g. Flavonoids constituted a small percentage of phenolic compounds—on average between 0.2 and 0.4 mg QE/100 g, except for sample 10, containing 0.77 mg QE/100 g. According to literature data, goldenrod honey belongs to honeys with moderate phenolic content. For Polish goldenrod honey, this range was 11.29–21.03 [1], 28.4 to 96.6 [30] and even 250 mg GAE/100 g [31]. The content of flavonoids determined by Jasicka-Misiak et al. [1] was found at a similar level: 0.93–1.41 mg QE/100 g honey. It is well known that light honeys are characterized by a lower phenolic content and lower antioxidant activity than dark honeys [32,33]. The data for the antioxidant capacity expressed by the three methods show little differentiation of the tested samples. The content of phenols and flavonoids is strongly correlated (r = 0.986). The FRAP reducing power also strongly correlates with the content of polyphenols (r = 0.989 and 0.969 for TPC and TFC, respectively). There are three of them: 3, 6 and 10, and regardless of the method used, they achieved the highest results. Similar data for Polish honeys were obtained earlier. The FRAP reducing ability of Polish goldenrod honeys ranged from 60.57 to 235 μmol TE/100 g [30] and the antiradical activity (DPPH) was 44.9 μmol TE/100 g [31]. In the latest study comparing goldenrod honeys from Poland, Slovakia and Hungary, Polish honeys were the weakest among the compared [2]. This proves the high variability even within one honey variety, due to climatic conditions and the exact composition of the flora.

3.5. Statistical Analysis

To confirm the hypothesis resulting from the analysis of the presented results: samples 3, 6 and 10 are not typical goldenrod honeys, and a multivariate statistical analysis was performed. In the first step, principal components analysis (PCA) was used, by which the samples were grouped using the projection of cases onto the factor plane (PC 1 × PC 2) (Figure 5). Samples 3 and 10 clearly differ from the others; also sample 6 can be considered as different, located on the negative side of the PC 1 principal component. Among the honey samples considered typical for goldenrod, two groups can be distinguished: samples 4, 6, 7 and 9 as well as 1, 2 and 5.

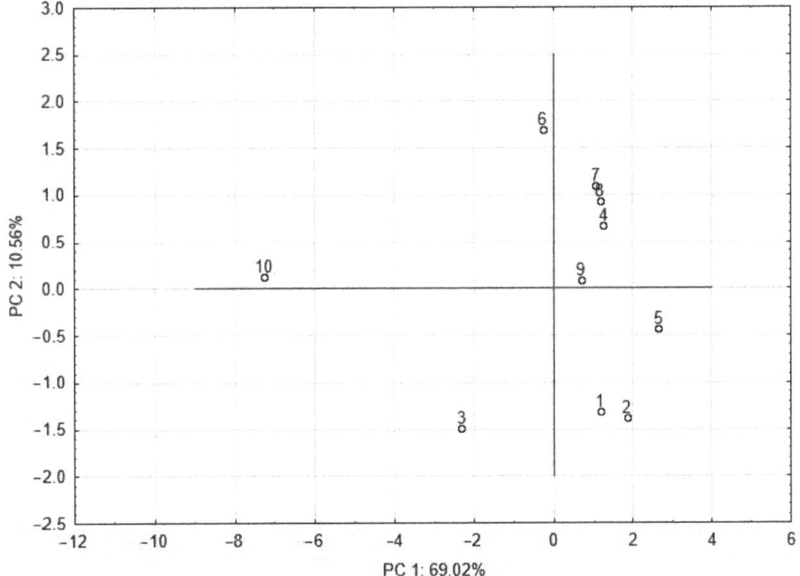

Figure 5. PCA score plot of tested goldenrod honey samples (1–10).

A cluster analysis was also performed, taking into account the tested honey parameters (Figure 6). The analysis showed the greatest differences for sample 10, which is confirmed by the highest bond distance. For samples 3 and 6, some similarities were demonstrated with other samples considered typical, 5 and 9, respectively, which may indicate the presence of admixtures to the base honey, which significantly changes its properties. The highest degree of similarity was shown by samples 1, 2, 4, 7 and 8, which was presented earlier with the use of SDS-PAGE protein and polyphenolic HPTLC profiles.

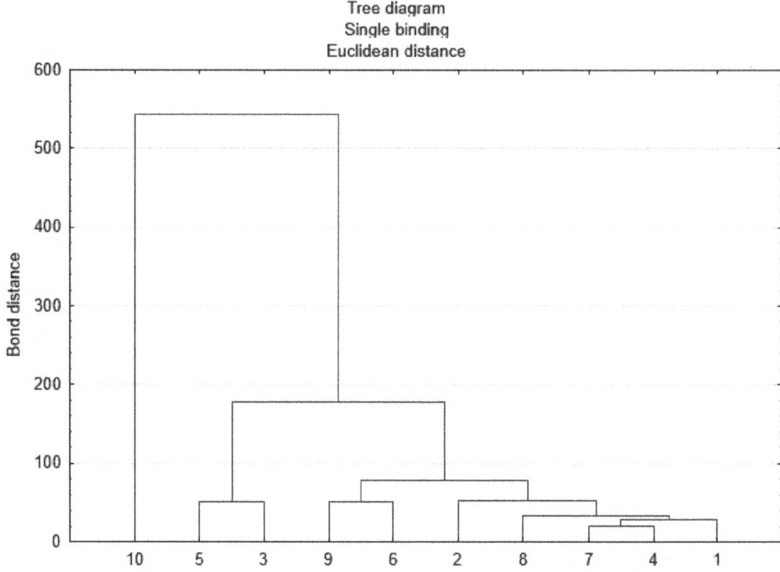

Figure 6. Tree diagram based on the average values of tested parameters for goldenrod honey samples (1–10).

3.6. Antimicrobial Activity

Various types of honey have been tested for their antibiotic or bacteriostatic activity for a long time, increasing our knowledge of the subject. Therefore, the next purpose of this study was to enhance general knowledge about goldenrod honey's bacteriostatic (against *Escherichia coli*) and fungicidal (against budding yeast) activities using four representative samples: 4 and 8 (with typical protein and HPTLC profiles) and 3 and 4 (with distinct features). First, we compared the growth rates of both yeast and bacteria cells. As can be seen in Figure 7A, 80% and 60% honeys completely stopped the growth of microbial cells, respectively. Then we diluted the solutions of the analyzed honeys in microbiological media in the ratio of 1: 5. As shown in Figure 7B, bacterial growth was much more restricted than yeast's. The most powerful inhibitory effect on yeast was found with 80% honey solutions 4 and 8. During the first 6 h, we found no significant differences in the effect of honey on bacteria. A slightly higher bacteriostatic effect was observed in the second part of the experiment with 80% and 60% of honey 3 (Figure 7B). Then, we diluted the honeys 1:10. Compared to the untreated honey control, yeast cell growth was only slightly inhibited (Figure 7C). In turn, this dilution significantly inhibited bacterial growth, especially in honey solution 3 (Figure 7C). In summary, we observe concentration-dependent inhibition of microorganism growth, especially bacteria.

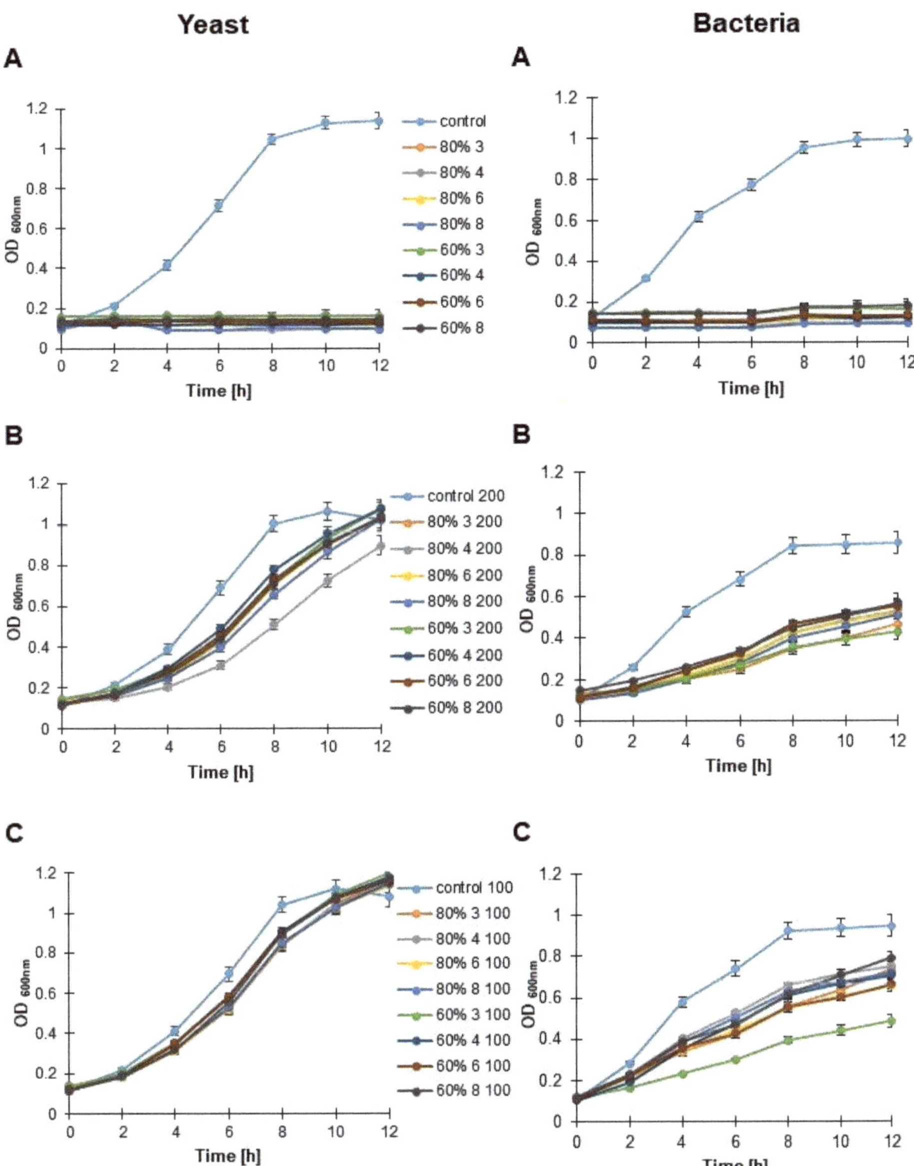

Figure 7. The effect of increasing doses of selected goldenrod honey samples (3, 4, 6 and 8) on budding yeast (**left**) and bacterial (**right**) growth. (**A**)—undiluted honey samples, (**B**)—samples in dilution 1:5 (200 mg/mL), (**C**)—samples in dilution 1:10 (100 mg/mL).

We then tested whether goldenrod honey and its 80% solution affected yeast viability. The dead cells were determined by fluorescently detecting the number of propidium iodide-labeled dead cells after incubation with honey and honey solution for 4 h at 28 °C. Figure 8 shows that the analyzed honeys do not cause 100% death of yeast cells, indicating that the arrest of growth is caused by the arrest of the cell cycle rather than the death of cells. Honey 8 displayed the highest fungicidal activity (7.67%). Meanwhile, no difference in survival was observed between honeys 3, 4 and 6. It is interesting to note that yeast cells treated with 80% honey solution survived at a rate of 44.67%. In conclusion, goldenrod honey does not cause cell death in all yeast species. It is necessary to conduct further research to identify a yeast cell population that is resistant to honey. The age of the cell and sensitivity of young cells possibly play a role here.

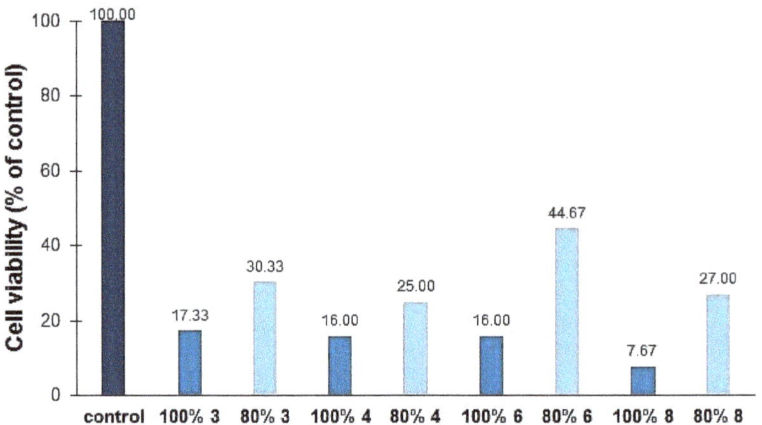

Figure 8. The fungicidal activity of selected goldenrod honey samples (no. 3, 4, 6 and 8).

It is well known that the components of honey show different activities against various microorganisms. Honey activity is dependent on the plant species, the weather conditions and the natural properties of the nectar [34]. The use of honey as a drug for the treatment of disease dates back to 2000 BC. Aristotle first described honey as "good for sore eyes and wounds" [35]. However, only a few decades ago, the first scientific reports confirmed the antibacterial and antifungal properties of honey. It has been shown many times that honey has strong antimicrobial properties against most fungi and bacteria that cause wound and surgical infections. The microorganisms were grown in both aerobic and anaerobic conditions. As was shown, growth inhibition was complete in the media containing 100%, partial in media containing 50% and no inhibition was produced by 20% of honey. Pure honey was therefore an ideal wound dressing agent, at least on a topical basis, for surgical infections, burns and wound infections [36].

Currently available agents are no longer effective against many dermatophytes. It seems that medicine, especially dermatology, has an urgent need for new effective antifungal agents suitable for the treatment of superficial skin infections. A recent study showed that Agastache honey had fungicidal activity against dermatophytes and yeast *Candida albicans* at 40% concentration [37]. Some previously report show that honey samples from different floral sources significantly inhibit the growth of few yeast species e.g., *Candida albicans*, *C. krusei*, *C. glabrata* and *Trichosoporon* spp. Inhibition of growth tests was found to depend on the type and concentration of honey and the pathogen tested [38]. In our previous study, we reported that honeys enriched with chokeberry fruits had anti-bacterial and anti-viral properties [39]. Here, we showed that goldenrod honey has antifungal properties that significantly inhibit yeast growth, but do not result in the death of the entire population. There are likely to be two factors contributing to this: the hydration of the honey and the goldenrod flowering in autumn. It is possible that some species of bacteria and yeast may ferment this honey because it has no antiseptic properties. Apparently, these data are crucial for beekeepers, for example, in the context of honey dehydration or honey fermentation in hives.

4. Conclusions

Protein and polyphenols profiling was used to verify the authenticity of 10 honey samples declared by beekeepers as goldenrod. The conclusions from the obtained varietal fingerprints (specific for 70% of the evaluated samples) were supported by standard physicochemical parameters and antioxidant activity of honey as well as PCA and cluster analysis. For selected samples, the antimicrobial effects against model bacteria and yeast strains were verified. HPTLC polyphenolic profiles comparison for honey and extracts of three *Solidago* species did not allow to determine of an exact botanical source of honey. It was found that SDS-PAGE and HPTLC assays allow for obtaining the protein and polyphenol fingerprints for goldenrod honey authentication. Final recognition of these methods requires confirmation for other honey varieties.

Supplementary Materials: The following supporting information can be downloaded at: https://www.mdpi.com/article/10.3390/foods11162390/s1, Table S1: Physicochemical parameters of goldenrod honeys; Table S2: Polyphenols content and antioxidant capacity of *Solidago* spp. extracts.

Author Contributions: Conceptualization, M.D. and M.M.; methodology, M.D., K.S. and A.B.; software, M.M.; validation, M.M., K.S. and A.B.; formal analysis, M.M.; investigation, M.M., P.K., K.S., E.S. and A.B.; resources, M.D.; data curation, E.S.; writing—original draft preparation, M.M., E.S.; writing—review and editing, M.D. and A.B; visualization, M.M.; supervision, M.D.; project administration, M.M.; funding acquisition, M.D. All authors have read and agreed to the published version of the manuscript.

Funding: This work was supported by the project financed under the program of the Minister of Science and Higher Education of Poland entitled "Regional Initiative of Excellence" in 2019–2022 (project no. 026/RID/2018/19).

Institutional Review Board Statement: Not applicable.

Informed Consent Statement: Not applicable.

Data Availability Statement: The data presented in this study are available in the article and Supplementary Material.

Conflicts of Interest: The authors declare no conflict of interest.

References

1. Jasicka-Misiak, I.; Makowicz, E.; Stanek, N. Chromatographic fingerprint, antioxidant activity, and colour characteristic of polish goldenrod (*Solidago virgaurea* L.) honey and flower. *Eur. Food Res. Technol.* **2018**, *244*, 1169–1184. [CrossRef]
2. Czigle, S.; Filep, R.; Balažová, E.; Szentgyörgyi, H.; Balázs, V.L.; Kocsis, M.; Purger, D.; Papp, N.; Farkas, Á. Antioxidant Capacity Determination of Hungarian-, Slovak-, and Polish-Origin Goldenrod Honeys. *Plants* **2022**, *11*, 792. [CrossRef] [PubMed]
3. Jarić, S.; Mačukanović-Jocić, M.; Djurdjević, L.; Mitrović, M.; Kostić, O.; Karadžić, B.; Pavlović, P. An ethnobotanical survey of traditionally used plants on Suva planina mountain (south-eastern Serbia). *J. Ethnopharmacol.* **2015**, *175*, 93–108. [CrossRef] [PubMed]
4. Jarić, S.; Popović, Z.; Mačukanović-Jocić, M.; Djurdjević, L.; Mijatović, M.; Karadžić, B.; Mitrović, M.; Pavlović, P. An ethnobotanical study on the usage of wild medicinal herbs from Kopaonik Mountain (Central Serbia). *J. Ethnopharmacol.* **2007**, *111*, 160–175. [CrossRef] [PubMed]
5. Fursenco, C.; Calalb, T.; Uncu, L.; Dinu, M.; Ancuceanu, R. *Solidago virgaurea* L.: A review of its ethnomedicinal uses, phytochemistry, and pharmacological activities. *Biomolecules* **2020**, *10*, 1619. [CrossRef] [PubMed]
6. Dudek, K.; Michlewicz, M.; Dudek, M.; Tryjanowski, P. Invasive Canadian goldenrod (*Solidago canadensis* L.) as a preferred foraging habitat for spiders. *Arthropod-Plant Interact.* **2016**, *10*, 377–381. [CrossRef]
7. Bobul'ská, L.; Demková, L.; Čerevková, A.; Renčo, M. Invasive goldenrod (Solidago gigantea) influences soil microbial activities in forest and grassland ecosystems in central Europe. *Diversity* **2019**, *11*, 134. [CrossRef]
8. Amtmann, M. The chemical relationship between the scent features of goldenrod (*Solidago canadensis* L.) flower and its unifloral honey. *J. Food Compos. Anal.* **2010**, *23*, 122–129. [CrossRef]
9. Baroni, M.V.; Chiabrando, G.A.; Costa, C.; Wunderlin, D.A. Assessment of the floral origin of honey by SDS-page immunoblot techniques. *J. Agric. Food Chem.* **2002**, *50*, 1362–1367. [CrossRef]
10. Nisbet, C.; Guler, A.; Ciftci, G.; Yarim, G.F. The Investigation of Protein Prophile of Different Botanic Origin Honey and Density Saccharose-Adulterated Honey by SDS-PAGE Method. *Kafkas Univ. Vet. Fak. Derg.* **2009**, *15*, 443–446.

11. Azevedo, M.S.; Valentim-Neto, P.A.; Seraglio, S.K.T.; da Luz, C.F.P.; Arisi, A.C.M.; Costa, A.C.O. Proteome comparison for discrimination between honeydew and floral honeys from botanical species Mimosa scabrella Bentham by principal component analysis. *J. Sci. Food Agric.* **2017**, *97*, 4515–4519. [CrossRef] [PubMed]

12. Borutinskaite, V.; Treigyte, G.; Čeksteryte, V.; Kurtinaitiene, B.; Navakauskiene, R. Proteomic identification and enzymatic activity of buckwheat (*Fagopyrum esculentum*) honey based on different assays. *J. Food Nutr. Res.* **2018**, *57*, 57–69.

13. Miłek, M.; Bocian, A.; Kleczyńska, E.; Sowa, P.; Dżugan, M. The comparison of physicochemical parameters, antioxidant activity and proteins for the raw local polish honeys and imported honey blends. *Molecules* **2021**, *26*, 2423. [CrossRef] [PubMed]

14. Latimer, G.W. *Official Methods of Analysis of AOAC International*; AOAC International: Gaithersburg, MD, USA, 2016.

15. Matłok, N.; Kapusta, I.; Piechowiak, T.; Zardzewiały, M.; Gorzelany, J.; Balawejder, M. Characterisation of some phytochemicals extracted from black elder (*Sambucus nigra* L.) flowers subjected to ozone treatment. *Molecules* **2021**, *26*, 5548. [CrossRef] [PubMed]

16. Tomczyk, M.; Bocian, A.; Sidor, E.; Miłek, M.; Zaguła, G.; Dżugan, M. The Use of HPTLC and SDS-PAGE Methods for Coniferous Honeydew Honey Fingerprinting Compiled with Mineral Content and Antioxidant Activity. *Molecules* **2022**, *27*, 720. [CrossRef] [PubMed]

17. European Parliament. Directive 2014/63/EU of the European Parliament and of the Council amending Council Directive2001/110/EC relating to honey. *Off. J. Eur. Communities* **2014**, *57*, L164/1.

18. Rybak-Chmielewska, H.; Szczęsna, T.; Waś, E.; Jaśkiewicz, K.; Teper, D. Characteristics of Polish Unifloral Honeys IV. Honeydew Honey, Mainly *Abies alba* L. *J. Apic. Sci.* **2013**, *57*, 51–59. [CrossRef]

19. Tomczyk, M.; Tarapatskyy, M.; Dżugan, M. The influence of geographical origin on honey composition studied by Polish and Slovak honeys. *Czech J. Food Sci.* **2019**, *37*, 232–238. [CrossRef]

20. Ratiu, I.A.; Al-Suod, H.; Bukowska, M.; Ligor, M.; Buszewski, B. Correlation study of honey regarding their physicochemical properties and sugars and cyclitols content. *Molecules* **2020**, *25*, 34. [CrossRef]

21. Muresan, C.I.; Cornea-Cipcigan, M.; Suharoschi, R.; Erler, S.; Mărgăoan, R. Honey botanical origin and honey-specific protein pattern: Characterization of some European honeys. *LWT* **2021**, *154*, 112883. [CrossRef]

22. Kečkeš, S.; Gašić, U.; Veličković, T.Ć.; Milojković-Opsenica, D.; Natić, M.; Tešić, Ž. The determination of phenolic profiles of Serbian unifloral honeys using ultra-high-performance liquid chromatography/high resolution accurate mass spectrometry. *Food Chem.* **2013**, *138*, 32–40. [CrossRef] [PubMed]

23. Sowa, P.; Marcinčáková, D.; Miłek, M.; Sidor, E.; Legáth, J.; Dzugan, M. Analysis of cytotoxicity of selected Asteraceae plant extracts in real time, their antioxidant properties and polyphenolic profile. *Molecules* **2020**, *25*, 5517. [CrossRef] [PubMed]

24. Avertseva, I.N.; Suleymanova, F.S.; Nesterova, O.V.; Reshetnyak, V.Y.; Matveenko, V.N.; Zhukov, P.A. Study of Polyphenolic Compounds in Extracts from Flowers and Leaves of Canadian Goldenrod and Dwarf Goldenrod (*Solidago canadensis* L. and *Solidago nana* Nitt.). *Mosc. Univ. Chem. Bull.* **2020**, *75*, 47–51. [CrossRef]

25. Shelepova, O. Phytochemistry and Inflorescences Morphometry of Invasive *Solidago* L. (Goldenrods) Species-Valuable Late Autumn Mellifers. *Agrobiodiversity Improv. Nutr. Health Life Qual.* **2021**, *5*, 209–214. [CrossRef]

26. Zekič, J.; Vovk, I.; Glavnik, V. Extraction and analyses of flavonoids and phenolic acids from canadian goldenrod and giant goldenrod. *Forests* **2021**, *12*, 40. [CrossRef]

27. Woźniak, D.; Ślusarczyk, S.; Domaradzki, K.; Dryś, A.; Matkowski, A. Comparison of Polyphenol Profile and Antimutagenic and Antioxidant Activities in Two Species Used as Source of Solidaginis herba-Goldenrod. *Chem. Biodivers.* **2018**, *15*, e1800023. [CrossRef]

28. Mietlińska, K.; Przybyt, M.; Kalemba, D. Polish plants as raw materials for cosmetic purposes. *Biotechnol. Food Sci.* **2019**, *83*, 95–106. [CrossRef]

29. Marksa, M.; Zymone, K.; Ivanauskas, L.; Radušienė, J.; Pukalskas, A.; Raudone, L. Antioxidant profiles of leaves and inflorescences of native, invasive and hybrid Solidago species. *Ind. Crop. Prod.* **2020**, *145*, 112123. [CrossRef]

30. Dżugan, M.; Tomczyk, M.; Sowa, P.; Grabek-Lejko, D. Antioxidant activity as biomarker of honey variety. *Molecules* **2018**, *23*, 2069. [CrossRef]

31. Gośliński, M.; Nowak, D.; Kłębukowska, L. Antioxidant properties and antimicrobial activity of manuka honey versus Polish honeys. *J. Food Sci. Technol.* **2020**, *57*, 1269–1277. [CrossRef]

32. Wesołowska, M.; Dżugan, M. Aktywność i stabilność termiczna diastazy występującej w Podkarpackich miodach odmianowych. *Zywn. Nauk. Technol. Jakosc/Food. Sci. Technol. Qual.* **2017**, *24*, 103–112. [CrossRef]

33. Halagarda, M.; Groth, S.; Popek, S.; Rohn, S.; Pedan, V. Antioxidant activity and phenolic profile of selected organic and conventional honeys from Poland. *Antioxidants* **2020**, *9*, 44. [CrossRef] [PubMed]

34. Almasaudi, S.B.; Al-Nahari, A.A.M.; Abd El-Ghany, E.S.M.; Barbour, E.; Al Muhayawi, S.M.; Al-Jaouni, S.; Azhar, E.; Qari, M.; Qari, Y.A.; & Harakeh, S. Antimicrobial effect of different types of honey on Staphylococcus aureus. *Saudi J. Biol. Sci.* **2016**, *24*, 1255–1261. [CrossRef] [PubMed]

35. Mandal, M.D.; Mandal, S. Honey: Its medicinal property and antibacterial activity. *Asian Pac. J. Trop. Biomed.* **2011**, *1*, 154–160. [CrossRef]

36. Efem, S.E.E.; Iwara, C.I. The antimicrobial spectrum of honey and its clinical significance. *Infection* **1992**, *20*, 227–229. [CrossRef]

37. Anand, S.; Deighton, M.; Livanos, G.; Pang, E.C.K.; Mantri, N. Agastache honey has superior antifungal activity in comparison with important commercial honeys. *Sci. Rep.* **2019**, *9*, 18197. [CrossRef]

38. Koc, A.N.; Silici, S.; Ercal, B.D.; Kasap, F.; Hörmet-Öz, H.T.; Mavus-Buldu, H. Antifungal Activity of Turkish Honey against Candida spp. and Trichosporon spp: An in vitro evaluation. *Med. Mycol.* **2009**, *47*, 707–712. [CrossRef]
39. Miłek, M.; Grabek-Lejko, D.; Stępień, K.; Sidor, E.; Mołoń, M.; Dżugan, M. The enrichment of honey with Aronia melanocarpa fruits enhances its in vitro and in vivo antioxidant potential and intensifies its antibacterial and antiviral properties. *Food Funct.* **2021**, *12*, 8920–8931. [CrossRef]

 *foods*

Article

Characterization of Romanian Bee Pollen—An Important Nutritional Source

Mircea Oroian *, Florina Dranca and Florin Ursachi

Faculty of Food Engineering, Stefan cel Mare University of Suceava, 720229 Suceava, Romania
* Correspondence: m.oroian@fia.usv.ro; Tel.: +40-744-524-872

Abstract: Bee pollen represents an important bee product, which is produced by mixing flower pollens with nectar honey and bee's salivary substances. It represents an important source of phenolic compounds which can have great importance for importance for prophylaxis of diseases, particularly to prevent cardiovascular and neurodegenerative disorders, those having direct correlation with oxidative damage. The aim of this study was to characterize 24 bee pollen samples in terms of physicochemical parameters, organic acids, total phenolic content, total flavonoid content, individual phenolics compounds, fatty acids, and amino acids from the Nort East region of Romania, which have not been studied until now. The bee pollen can be considered as a high protein source (the mean concentration was 22.31% d.m.) with a high energy value (390.66 kcal/100 g). The total phenolic content ranged between 4.64 and 17.93 mg GAE/g, while the total flavonoid content ranged between 4.90 and 20.45 mg QE/g. The high protein content was observed in *Robinia pseudoacacia*, the high content of lipids was observed in *Robinia pseudoacacia* pollen, the high fructose content in *Prunus* spp. pollen while the high F/G ratio was observed in *Pinaceae* spp. pollen. The high TPC was observed in *Prunus* spp. pollen, the high TFC was observed in *Robinia pseudoacacia* pollen, the high free amino acid content was observed in *Pinaceae* spp. pollen, and the high content of PUFA was reported in *Taraxacum* spp. pollen. A total of 16 amino acids (eight essential and eight non-essential amino acids) were quantified in the bee pollen samples analyzed. The total content of the amino acids determined for the bee pollen samples varied between 11.31 µg/mg and 45.99 µg/mg. Our results can indicate that the bee pollen is a rich source of protein, fatty acids, amino acids and bioactive compounds.

Keywords: bee pollen; physicochemical parameters; fatty acids; amino acids

Citation: Oroian, M.; Dranca, F.; Ursachi, F. Characterization of Romanian Bee Pollen—An Important Nutritional Source. *Foods* **2022**, *11*, 2633. https://doi.org/10.3390/foods11172633

Academic Editors: Liming Wu, Qiangqiang Li and Olga Escuredo

Received: 8 July 2022
Accepted: 25 August 2022
Published: 30 August 2022

Publisher's Note: MDPI stays neutral with regard to jurisdictional claims in published maps and institutional affiliations.

1. Introduction

Bee pollen is collected by honeybees (*Apis* spp.) and is stored and used as food for all development stages in the hive [1]. Bee pollen is a honey bee derivate that is produced by mixing flower pollens with nectar (and/or) honey and bee's salivary substances. The main compounds found in the bee pollen are: proteins (10–40% in dry weight), carbohydrates (13–55% in dry weight), lipids (1–13% in dry weight), dietary fibers (0.3–20% in dry weight), phenolic compounds (up to 2.5% in dry weight), fatty acids, minerals, amino acids, carotenoids and vitamins [2]. The phenolic compounds (e.g., flavonoids and phenolic acids) presented in the bee pollen are of a great interest for pharmaceutical industry due to their great importance for prophylaxis of diseases, particularly to prevent cardiovascular and neurodegenerative disorders, those having direct correlation with oxidative damage [3]. Given its unique composition, bee pollen is consumed as a food supplement and scientists considered that it is an important functional food [2,4,5] and was reported to have strong health properties such as antioxidant, antiallergen, anti-inflammatory, antiulcer, immune-stimulating, antimicrobial and anticarcinogenic [6,7]. The chemical composition of bee pollen is influenced by different factors such as: floral source, geographical origin, and harvesting technique [8–10]. Carbohydrates represent 13–55% in dry weight of the bee pollen depending on botanical and geographical origin of the product; the main carbohydrates present are fructose, glucose and sucrose (more than 90% of the total carbohydrates content) [11]. Proteins represent a high percentage of bee pollen (10–40% in dry weight), but the amino acids define much better the biological value of the bee pollen; they play an important role in human nutrition (e.g., in metabolism, reduce excessive body fat, modulates gene expression, enhances skeletal muscle) [12–15]. The essential amino acids presented in the bee pollen represent 34.59% to 48.49% of the total content of amino acids; the main amino acids presented are aspartic acids, leucine, glutamic acid, proline and lysine [16]. The bee pollen was reported to have a high antioxidant activity mainly due to the polyphenols which generate a high free radical scavenging potential [10,14,17–20]. Among the phenolic compounds reported to be determined in the bee pollen were: kaempferol, caffeic acid, quercetin, isoquercetrin, galangin and chrysisn; the glycosides of isorhamnetin, quercetin and kaempferol are the predominant flavonoids in bee pollen [3,21,22]. Bee pollen is a rich source of oil (1–13% in dry weight) and in consequence an important source of fatty acids for hive development; they are important not only for their role as a structural component for cell membranes and energy, but also for their role for bees health [12,23–27]. The main fatty acids presented in the bee pollen are saturated fatty acids (e.g., myristic acid (C14:0), palmitic acid (C16:0), and stearic acid (C18:0)) followed by unsaturated fatty acids (e.g., oleic acid (C18:1), linoleic acid (C18:2), and alpha-linolenic acid (C18:3)) [28]. There were reported two polyunsaturated fatty acids (linoleic acid (omega-6 fatty acid) and alpha-linolenic acid (omega-3 fatty acid)) in the bee pollen, which cannot be biosynthesized by humans and bees [25].

In the last year there has been a high interest from humans and the scientific community for using natural sources as alternatives for synthetic drugs, however the knowledge regarding the bee pollen from Romania is little, and the consumption of it is not high. To gather the two demands, it is necessary to classify the pollen according to the botanical origin and to deeply characterize it from physicochemical point of view in order to achieve its characteristics. Bee pollen can be used as a food supplement due to its positive effects as an antioxidant or antimicrobial effects; moreover, the low content of sugars and saturated fatty acids make the bee pollen a perfect component for food diets [7,29]. The composition of bee pollen is correlated to the botanical origin, geographical origin (the climatic and pedological factors influence the chemical composition) and the processing techniques used [29,30].

Nowadays, the bee pollen is considered a functional food product due to its high nutritional value and chemical composition (e.g., vitamins, carotenoids, phenolic compounds) [5]. The bee pollen extracts can be used for complementary treatment of different diseases (e.g., benign prostatic hyperplasia, vasomotor symptoms), but the most important compounds that may pose a high pharmacological activity are phenolic acids, fatty acids, flavonoids, carotenoids [31]. In the market, the bee pollen can be found as capsules, granules, pellets and powders [32], and the daily recommended dose for an adult is from 20 to 40 g [5].

The aim of this study is to characterize 24 samples of bee pollen in terms of physicochemical parameters (pH, free acidity, protein content, oil content, moisture content), fatty acids, amino acids, carbohydrates, organic acids, individual phenolics compounds and FT-IR spectra. Until now there have been no other studies related to the characterization of bee pollen samples from the North-East part of Romania.

2. Materials and Methods

2.1. Bee Pollen Samples

A total of 24 samples of bee pollen from the North East part of Romania were collected in 2020 from spring to autumn, and the samples were dried. The samples were kept at −20 °C prior to analysis.

2.2. Materials

H_2SO_4 (37%), NaOH (>99%), metaphosphoric acid (>99%), gluconic acid (>99%), formic acid (>99%), butyric acid propionic acid (>99%), lactic acid(>99%), acetic acid (100%), rosmarinic acid (>99%), *p*-coumaric acid(>99%), chlorogenic acid (>99%), vanillic acid (>99%), caffeic acid (>99%), *p*-hydroxybenzoic acid (>99%), protocatechiuc acid (>99%), gallic acid (>99%), kaempferol (>99%), quercetin (>95%), luteolin (>99%), myricetin (>99%), methanol (>99%), fructose (>99%), glucose (>99%), sucrose (>99%), melesitose (>99%), raffinose (>99%), AlCl3 (>99%), sodium carbonate (>99%), trichloroacetic acid (>99%) were purchased from Sigma Aldrich (Germany). Fatty acids methyl esters (FAME) mix was purchased from Restek, (Bellefonte, PA, USA, 35077). Alanine, sarcosine, glycine, α-aminobutyric acid, valine, β-amino isobutyric acid, internal standard, leucine, allo-isoleucine, isoleucine, threonine, serine, proline, asparagine, thioproline, aspartic acid, methionine, 3-hydroxyproline/4-hydroxyproline, phenylalanine, glutamic acid, α-aminoadipic acid, α-aminopimelic acid, glutamine, ornithine, glycyl-prolizine—2 isomers, proline-hydroxyproline, histidine, lysine, tyrosine, tryptophan, cystathionine and cystine were purchased from Phenomenex (Torrance, CA, USA).

2.3. Methods

2.3.1. Palynological Analysis

The palynological analysis was conducted as follows: 2 g of bee pollen was prepared by dissolving and washing it in H_2SO_4 (5‰), and placed on the slide [33]. The slide was examined using a Primostar 3 KMAT Carl Zeiss microscope at 400× magnification. The pollen frequencies were determined based on the Louveaux et al. (1978) methodology, and were divided into four categories as: predominant pollen (>45% of the total pollen detected); secondary pollen (16–45%); minor important pollen (3–15%); minor pollen (<3%) [33].

2.3.2. Determination of Routine Physicochemical Parameters: Moisture Content, Water Activity, pH and Free Acidity

- Moisture content

A total of 1 g of sample was weighed and heated at 103 °C for 2 h for start, weighted after cooling and heated again until constant weight was obtained [18].

- Water activity

Water activity was measured using a water activity meter AquaLab Lite (Decagon, USA).

- pH and free acidity

2 g of bee pollen was mixed with 5 mL of water, after homogenization the solution was filtered and titrated with NaOH 0.05 M [17]. The pH was determined on the same solution using a Metler Toledo pH meter Seven compact S210. All the measurements were carried out in triplicate.

2.3.3. Proximate Composition of Bee Pollen

The nutritional value of the bee pollen involved the determination of total lipid (AOAC 920.85), total protein (AOAC 978.04) and ash (AOAC 920.85) [34]. All the measurements were carried out in triplicate. Carbohydrates were determined by difference. The energetic values of the bee pollen were determined as:

$$\text{Energy (kcal)} = 4 \times (\text{g protein} + \text{g carbohydrates}) + 9 \times (\text{g lipid}) \tag{1}$$

2.3.4. Determination of Organic Acids

A total of 0.5 g of bee pollen was mixed with 2.5 mL 4% metaphosphoric acid (*w/v*), after homogenization the solution was centrifuged for 10 min at 5000 rpm. The supernatant was filtered using a 0.45 μm filter prior analysis. The determination was carried out on a high performance liquid chromatograph Shimadzu (Kyoto, Japan) with diode array detector. The separation of the organic acids (gluconic acid, formic acid, butyric acid propionic acid, lactic acid and acetic acid, respectively) was made using a Phenomenex Kinetex® 5 μm C18 100 Å HPLC Column 250 × 4.6 mm [35]. The mobile phase consisted of a mixture of 0.5% metaphosphoric acid and acetonitrile (50/50, *v/v*), and the flow rate was set at of 0.8 mL·min^{-1}. The injection volume was set at 10 μL. The detector was set at 210 nm. The organic acids were expressed as mg/kg dry mater. All the measurements were carried out in triplicate.

2.3.5. Determination of Free Sugars

A total of 1g of bee pollen was mixed with 20 mL of methanol and filled to 50 mL with water. The solution was centrifuged for 10 min at 5000 rpm; the supernatant was transferred into a 50 mL flask and filled with water. The solution was filtered using a 0.45 μm filter prior analysis. The separation of the free sugars (fructose, glucose, sucrose, melesitose and raffinose) was made on Schimadzu HPLC instrument (Kyoto, Japan) with RID (refractive index detector). A Phenomenex Luna® Omega 3 μm SUGAR 100 Å HPLC Column 150 × 4.6 mm (Torrance, CA, USA) was used for the separation. The mobile phases were acetonitrile and water in a mixture of 80:20 (*v/v*). The flow rate was set at 1.3 mL·min^{-1}; column and detector temperature was 30 °C and the sample volume injection was 10 μL. The free sugars were expressed as % reported to dry mater. All the measurements were carried out in triplicate.

2.3.6. Bee Pollen Extracts for the Determination of Phenolic Compounds

The extraction procedure was carried out as follows: 0.1 g of bee pollen was mixed with 25 mL of 80% methanol and ultrasonicated for 30 min at 50 °C. After the heating, the solution was transferred into a centrifuge tube and centrifuged for 5 min at 5000 rpm. The supernatant was collected, filled up to 50 mL with 80% methanol and kept at 4 °C for TPC, TFC and individual phenolics compounds determination.

2.3.7. Determination of Total Phenolic Content (TPC) and Total Flavonoid Content (TFC)

Total phenolic content (TPC) determination: 0.1 mL of bee pollen extract was mixed with 1.9 mL water, 0.1 mL of Folin Ciocalateau reagent, the mixture was homogenized for 2 min and after it 0.8 mL of 5% sodium carbonate was added. The mixture was kept at 40 °C for 20 min and cooled down in an ice bath for stopping the reaction. The total phenolic content was determined at 750 nm, based on a gallic acid calibration curve expressing the results as gallic acid equivalent (mg GAE/g) dry mater [35]. All the measurements were carried out in triplicate.

Total flavonoid content (TFC) determination: 0.2 mL of bee pollen extract was mixed with 2 mL of methanol and 0.1 mL of 5% AlCl$_3$ (prepared in methanol). The solution was left for 30 min at room temperature and its absorbance was measured at 425 nm. The concentration was determined based on a quercetin calibration curve expressing the results as mg quercetin equivalent/g (mg QE/g) dry mater [35]. All the measurement were carried out in triplicate.

2.3.8. Determination of Individual Phenolic Compounds

The phenolic acids (rosmarinic acid, *p*-coumaric acid, chlorogenic acid, vanillic acid, caffeic acid, *p*-hydroxybenzoic acid, protocatechiuc acid, gallic acid) and flavonoids (kaempferol, quercetin, luteolin and myricetin) were determined from the methanolic extract using a high performance liquid chromatograph Shimadzu (Kyoto, Japan) with diode array detector. The separation was carried out on a Zorbax SP-C18 column, with 150 mm length, 4.6 mm i.d. 5 μm-diameter particle was used for the separation [35]. The separation of the compounds were realized on a system with 0.1% acetic acid (mobile phase A) and acetonitrile (mobile phase B) based on the elution range described by Palacios et al. [36] as: min 0—A 100%, min 6.66—B 5%, min 66.66—B 40% and min 74—B 80%. The flow rate was set at 1 mL·min^{-1}, and the injection volume was 10 μL. Gallic acid, vanillic acid, protocatechuic acid and *p*-hydroxybenzoic acid were determined at 280 nm, and chlorogenic acid, *p*-coumaric acid, caffeic acid, rosmarinic acid, myricetin, quercetin, luteolin and kaempherol were determined at 320 nm, respectively. The phenolics compounds were expressed as mg/kg dry mater. All the measurements were carried out in triplicate.

2.3.9. Determination of Total Free Amino Acids

For the extraction and identification of free amino acids, 1.75 ± 0.1 g of sample was mixed with 15 mL of 15% trichloroacetic acid (TCA). The pH of the mixture was adjusted to 2.2 (isoelectric precipitation point of the proteins) and the extract was further diluted to 25 mL with 15% trichloroacetic acid. Then, the supernatant was collected and filtered using 0.45 μm microfilters. A total of 100 μL of filtered supernatant was subjected to the determination of organic components, using the EZfaast GC-MS kit (Phenomenex, Torrance, CA, USA), following the protocol given by the manufacturer. Identification and separation of free amino acids was performed using a gas chromatograph coupled with a mass spectrometer (MS) equipped with a Zebron TM ZB-AAA column (10 m × 0.25 mm, film thickness: 0.25 μm). Injection: split 1:15, carrier gas: helium 1.1 mL/min, oven program: 30 °C/min from 110 °C to 320 °C. The MS parameters: source temperatrure 240 °C, scan range 45–450 *m/z*, sampling rate 3.5 scans/s. The identification of each amino acid was performed by calculating the area of each "peak" and comparing it with a standard consisting of 33 amino acids (alanine, sarcosine, glycine, α-aminobutyric acid, valine, β-amino isobutyric acid, internal standard, leucine, allo-isoleucine, isoleucine, threonine, serine, proline, asparagine, thioproline, aspartic acid, methionine, 3-hydroxyproline/4-hydroxyproline, phenylalanine, glutamic acid, α-aminoadipic acid, α-aminopimelic acid, glutamine, ornithine, glycyl-prolizine—2 isomers, proline-hydroxyproline, histidine, lysine, tyrosine, tryptophan, cystathionine and cystine). The results are expressed as μg/mg bee pollen; all the determinations were carried out in triplicate.

2.3.10. Fatty Acids Determination Using GC-MS

Prior the analysis, the bee pollen was extracted for 48 h with n-hexane at room temperature [37]. The bee pollen oil (0.1 g) was mixed with 1 mL of n-hexane and 1 mL of 15% BF$_3$ in methanol. The mixture was thermostated for 15 min at 60 °C in a water bath. The mixture was cooled to 20 °C and mixed with 5 mL of NaCl saturated solution, after mixing the solution was centrifuged for 5 min at 3000 rpm. The supernatant was filtered with a 0.45 μm nylon filter and kept at −20 °C prior analysis. The separation of the fatty acids methyl esters was carried out on SUPELCOWAX 10 column (60 m × 0.25 mm i.d., 0.25 μm film thickness; Supelco Inc., Bellefonte, PA, USA) using a Shimadzu GC-MS instrument (GC MS-QP 2010 Plus, Shimadzu, Japan) equipped with an AOC-01 auto-injector that was used to perform the gas chromatographic-mass spectrometric analyses. The initial oven temperature was 140 °C and was increased to 220 °C at a rate of 7 °C/min and then held at this temperature for 23 min. The flow rate of the carrier gas (He) and the split ratio were 0.8 mL/min and 1:24, respectively. Identification of FAMEs was done by comparing their retention times to those of known standards (37 component FAME Mix, Restek, Bellefonte, PA, USA, 35077) and the resulting mass spectra to the ones from our database (NIST MS

Search 2.0) [17]. The results are expressed as µg/g bee pollen; all the determinations were carried out in triplicate.

2.3.11. Determination of FT-IR

The bee pollen spectra in the wave number range of 4000–650 cm^{-1} was recorded using a Nicolet i-20 spectrophotometer (Thermo Scientific, Waltham, MA, USA) with ATR module in absorbance mode. The bee pollen spectra were analysed using Spectra Gryph–spectroscopy software (Version 1.2.11). Each sample was ground into powder and filtered with 200 mesh, pressed and analysed directly on the ATR module. Each spectra was the media of 32 scans at a resolution of 4 cm^{-1}. All the measurements were carried out in triplicate.

2.4. Statistical Analysis

The results were submitted to analysis of variance (ANOVA) using Statgraphics Centurion XIX software (trial version, Statgraphics Technologies, Inc., The Plains, VA, USA). Tukey (HSD)/Analysis of the differences between the categories with a confidence interval of 95% was used. Principal component analysis (PCA) was performed using Unscrambler X software version 10.1 (Camo, Norway).

3. Results and Discussion

3.1. Botanical Origin of the Bee Pollen

The melissopalynological analysis of the bee pollen is presented in Table 1 and covers the identified plant families, species and genera presented. According to the analysis, neither one sample has reached 80% of a one single pollen to be considered as monofloral. From the 24 samples analyzed, 20 samples had a pollen which represented more than 45% of the pollen variability (five samples with *Helianthus annuus*, five samples with *Robinia pseudoacacia*, two samples with *Pinaceae* spp., two samples with *Quercus* spp., two samples with *Prunus* spp. and one sample with *Zea mays*, *Tillia* spp. *Crataegus monogyna Taraxacum* spp., respectively), while in the case of four samples neither one pollen type has reached the 45% level. Secondary pollens were found: *Robinia pseudoacacia*, *Tilia* spp. and *Helianthus annuus*. Minor pollens were observed *Fagus sylvatica*, *Corylus* spp., *Taraxacum* spp., *Vicia* spp., *Salicaceae* spp., *Sophora* spp., *Poaceae* spp., *Trifolium* spp., *Asteraceae* spp., *Urtisaceae* spp., *Pinaceae* spp., *Cucumber* spp., *Castaneae* spp., *Oleaceae* spp., *Allium* spp., *Plantago* spp., *Myrcia* spp., *Fireweed* spp., *Mimosa* spp., *Cucumber* spp. Pollen botanical origin from the pollen pellets may vary according to the region of collection, and vegetation available for bees at the collecting moment.

Table 1. Palynological analysis: Plant species giving the predominant, secondary, important minor and minor pollen in the analyzed bee samples.

Sample Code	Dominant Pollen (>45%)	Secondary Pollen (16–44%)	Important Minor Pollen (4–15%)	Minor Pollen (<3%)
S1	*Helianthus annuus*	-	*Robinia pseudoacacia* *Zea mays* *Tilia* spp. *Fagus sylvatica*	*Quercus* spp. *Betulus pendula*
S2	*Pinaceae* spp.	-	*Corylus* spp. *Taraxacum* spp. *Vicia* spp. *Helianthus annuus*	*Carduus* spp. *Ambrosia* spp.

Table 1. *Cont.*

Sample Code	Dominant Pollen (>45%)	Secondary Pollen (16–44%)	Important Minor Pollen (4–15%)	Minor Pollen (<3%)
S3	*Helianthus annuus*	-	*Robinia pseudoacacia* *Tilia* spp.	*Conicum* spp. *Vicia* spp. *Pruunus spinosa*
S4	*Tillia* spp.	-	*Allium* spp. *Helianthus annuus* *Asteraceae* spp.	*Brassicaceae* spp. *Fagopyrum* spp.
S5	-	-	*Helianthus annuus* *Taraxacum* spp. *Quercus* spp. *Zea mays*	*Prunus* spp.
S6	-	-	*Trifolium* spp. *Robinia pseudoacacia* Helianthus annuus *Poaceae* spp. *Vicia* spp.	*Fabaceae* spp. *Conicum* spp. *Fagus* spp. *Ulmu* spp.
S7	*Robinia pseudoacacia*	-	*Asteraceae* spp. *Urtisaceae* spp.	*Rosaceae* spp. *Prunus* spp.
S8	*Quercus* spp.	-	*Pinaceae* spp.	*Castaneae* spp.
S9	*Robinia pseudoacacia*	-	*Cucumber* spp. *Castaneae* spp. *Oleaceae* spp.	*Rosaceae* spp. *Prunus* spp.
S10	-	-	*Trifolium* spp. *Robinia pseudoacacia* *Urticaceae* spp. *Castanea* spp.	*Quercus* spp.
S11	*Zea mays*	-	*Trifolium* spp.	*Quercus* spp.
S12	*Helianthus annuus*	*Robinia pseudoacacia*	-	*Vicia* spp. *Pruunus spinosa* *Tilia* spp.
S13	*Crataegus monogyna*	-	*Helianthus annuus* *Taraxacum* spp *Quercus* spp.	*Robinia pseudoacacia*
S14	*Helianthus annuus*	*Tilia* spp.	*Asteraceae* spp.	*Taraxacum* spp.
S15	*Taraxacum* spp.	-	*Plantago* spp. *Quercus* spp.	
S16	*Quercus* spp.	-	*Taraxacum* spp. *Castaneae* spp.	*Ambrosia* spp. *Helianthus annuus*
S17	*Robinia pseudoacacia*	*Helianthus annuus*	*Taraxacum* spp. *Castaneae* spp.	*Tilia* spp.
S18	*Robinia pseudoacacia*	*Helianthus annuus*	*Tilia* spp.	*Taraxacum* spp. *Castaneae* spp.
S19	*Prunus* spp.	-	*Taraxacum* spp. *Quercus* spp.	*Robinia pseudoacacia* *Helianthus annuus* *Teucrium* spp.

Table 1. *Cont.*

Sample Code	Dominant Pollen (>45%)	Secondary Pollen (16–44%)	Important Minor Pollen (4–15%)	Minor Pollen (<3%)
S20	*Pinaceae* spp.	-	*Fagaceae* spp.	*Prunus* spp. *Asteraceae* spp. *Gramineae* spp.
S21	-	-	*Zea mays* *Myrcia* spp. *Helianthus annuus* *Fireweed* spp. *Mimosa* spp.	*Salix* spp. *Taraxacum* spp.
S22	*Prunus* spp.		*Cucumber* spp.	*Quercus* spp.
S23	*Robinia pseudoacacia*	-	*Salicaceae* spp. *Sophora* spp. *Poaceae* spp.	*Humulus* spp. *Salicaceae* spp. *Allium* spp.
S24	*Helianthus annuus*	-	*Robinia pseudoacacia* *Tilia* spp.	*Quercus* spp. *Vicia* spp. *Pruunus spinosa*

3.2. Routine Physicochemical Parameters: Moisture Content, Water Activity, pH and Free Aciditiy

In Table 2 are presented the physicochemical properties of bee pollen samples analyzed. The stability and the shelf life of bee pollen is influenced strongly by the pH and the titratable acidity; these two parameters are a good indicator of the dynamic microbial activity [14]. The bee pollen pH ranged between 3.90 and 5.84 with a mean of 4.60 ($p < 0.05$); the values are in agreement with those reported in the case of pollen from Tuscany, Portugal, Greece and India [8,27,38]. The free acidity ranged between 124.58 and 306.90 meq/kg with a mean of 215.51 meq/kg ($p < 0.05$); the magnitude of free acidity confirmed the acidic nature of the bee pollen. The free acidity of the samples were in agreement with those reported for Colombian bee pollen [39] and Brazilian bee pollen [40]. Moisture content ranged between 2.96% and 11.90% ($p > 0.05$), with an average of 5.00%, which confirms the dry characteristics labeled by the beekeepers on the products; the moisture content was much lower than those reported for bee pollen from Colombia, Italy and Spain [22], but in agreement with the levels determined in Brazilian bee pollen [40]. According to the literature, the moisture content of dry bee pollen should be between 6 to 8% to ensure the bee pollen quality and stability [29]. The water activity is correlated to the stability and shelf life of a product; a high water activity stimulates the growth of microorganisms (e.g., molds, yeasts) and can cause pollen toxicity due to mycotoxins formation [8]. The water activity ranged between 0.17 and 0.55 ($p < 0.05$), with an average of 0.29, in agreement with those reported for bee pollen from Portugal, Spain and Colombia [22,41].

Table 2. Physicochemical parameters (moisture content, water activity, pH and free acidity), proximate composition and free sugars of bee pollen samples (mean values and standard deviation in brackets).

Parameter	*Crataegus monogyna*	*Helianthus annuus*	*Pinaceae* spp.	Polyfloral	*Prunus* spp.	*Quercus* spp.	*Robinia pseudoacacia*	*Taraxacum* spp.	*Tillia* spp.	*Zea mays*	F-Value
pH	4.57 (0.07) [a,b,c]	4.36 (0.27) [a,b]	4.12 (0.30) [a]	4.85 (0.23) [a,b,c]	5.19 (0.69) [c]	4.69 (0.14) [a,b,c]	4.24 (0.23) [a,b]	4.96 (0.07) [b,c]	4.85 (0.07) [a,b,c]	4.53 (0.06) [a,b,c]	6.4 ***
Free acidity (meq/kg)	233.4 (3.3) [a,b]	187.7 (33.0) [a]	272.79 (26.90) [b]	179.6 (16.5) [a]	199.3 (87.8) [a,b]	252.0 (59.8) [b]	249.13 (24.41) [b]	214.2 (3.0) [a,b]	194.0 (2.7) [a,b]	127.5 (1.8) [a]	5.2 ***
Moisture content (%)	4.92 (0.07) [a,b]	5.07 (1.62) [a,b]	8.84 (3.40) [b]	4.74 (1.12) [a]	4.86 (2.05) [a,b]	4.89 (1.16) [a,b]	4.34 (0.57) [a]	3.32 (0.05) [a]	4.50 (0.06) [a]	2.90 (0.04) [a]	3.8 **
a_w	0.27 (0.01) [a]	0.30 (0.10) [a,b]	0.48 (0.08) [b]	0.27 (0.06) [a]	0.24 (0.08) [a]	0.29 (0.09) [a,b]	0.28 (0.05) [a]	0.17 (0.01) [a]	0.28 (0.01) [a]	0.20 (0.01) [a]	4.2 ***
Protein. d.m. (%)	22.66 (0.32) [b,c,d]	18.03 (1.05) [a,b]	23.09 (2.48) [c,d]	23.06 (1.35) [c,d]	24.54 (1.40) [c,d]	26.86 (0.95) [d]	23.13 (2.82) [c,d]	23.71 (0.34) [c,d]	21.14 (0.30) [b,c]	15.58 (0.22) [a]	13.7 ***
Lipids d.m. (%)	2.62 (0.04) [a,b]	3.72 (1.11) [a,b]	6.05 (1.36) [c]	4.31 (1.01) [a,b]	5.25 (0.43) [b,c]	3.28 (0.87) [a,b]	6.02 (0.89) [c]	3.51 (0.05) [a,b]	3.30 (0.05) [a,b]	2.22 (0.03) [a]	8.7 ***
Ash (%)	3.23 (0.05) [b,c]	2.57 (0.14) [a]	3.16 (0.45) [b]	3.30 (0.23) [b,c]	3.50 (0.13) [b,c]	3.83 (0.18) [e]	3.32 (0.40) [b,c]	3.44 (0.05) [b,c]	3.03 (0.04) [a,b]	2.27 (0.03) [a]	12.4 ***
Energy kcal/100 g	377.0 (5.3) [a]	384.8 (8.0) [a]	380.30 (18.77) [a]	386.2 (5.6) [a]	389.9 (7.5) [a,b]	378.1 (4.4) [a]	396.49 (7.49) [b]	387.0 (5.5) [a,b]	382.9 (5.4) [a]	386.6 (5.3) [a,b]	2.6 *
Fructose d.m. (%)	18.82 (0.27) [a,b]	19.49 (1.54) [a,b,c]	20.00 (1.11) [a,b,c]	19.59 (0.64) [a,b,c]	21.31 (3.03) [c]	20.08 (0.25) [a,b,c]	18.46 (0.98) [a]	19.44 (0.28) [a,b,c]	21.44 (0.31) [c]	20.68 (0.30) [b,c]	2.4 *
Glucose d.m. (%)	12.78 (0.18) [a,b]	14.50 (2.75) [a,b]	9.49 (1.03) [a]	14.91 (1.78) [a,b]	16.48 (5.47) [b]	14.52 (0.21) [a,b]	11.53 (1.25) [a,b]	17.19 (0.25) [b]	16.19 (0.23) [b]	17.40 (0.25) [b]	4.9 ***
Sucrose d.m. (%)	0.13 (0.01) [a]	0.48 (0.33) [a]	0.73 (0.11) [a,b]	0.79 (0.22) [a,b]	0.43 (0.29) [a]	0.34 (0.39) [a]	1.23 (0.43) [b]	0.23 (0.01) [a]	1.45 (0.02) [b]	0.84 (0.01) [a,b]	7.2 ***
Turanose d.m. (%)	0 (0) [a]	0.10 (0.13) [a]	0 (0) [a]	0.08 (0.14) [a]	0 (0) [a]	0.07 (0.09) [a]	0.05 (0.01)	0 (0) [a]	0(0) [a]	0(0) [a]	0.6 ns
Maltose d.m. (%)	0.21 (0.01) [a]	0.30 (0.32) [a]	0.77 (0.07)	0.61 (0.14) [a]	0.45 (0.09) [a]	1.06 (0.97) [a]	0.58 (0.25)	0.46 (0.01) [a]	0.92 (0.01) [a]	0.50 (0.01) [a]	1.3 ns
Trehalose d.m. (%)	0.37 (0.01) [a]	1.14 (0.81) [a]	1.21 (0.36)	1.13 (0.04) [a]	1.93 (1.27) [a]	1.16 (0.60) [a]	0.85 (0.23)	0.96 (0.01) [a]	1.03 (0.01) [a]	1.03 (0.01) [a]	1.5 ns
Melesitose d.m. (%)	2.99 (0.04) [a]	2.13 (0.31) [a]	0.90 (0.84)	1.21 (1.30) [a]	2.41 (0.73) [a]	1.86 (2.14) [a]	1.56 (0.89)	2.29 (0.03) [a]	2.60 (0.04) [a]	1.53 (0.02) [a]	1.7 ns
Raffinose d.m. (%)	0.16 (0.01) [a,b]	0.09 (0.09) [a]	0.29 (0.19) [a,b,c,d]	0.22 (0.14) [a,b,c]	0.20 (0.20) [a,b,c]	0.56 (0.57) [d]	0.07 (0.06) [a]	0.44 (0.01) [b,c,d]	0.50 (0.01) [c,d]	0.06 (0.01) [a]	3.4 **
F/G	1.46 (0.02) [b,c]	1.37 (0.23) [a,b,c]	2.11 (0.34) [d]	1.31 (0.11) [a,b,c]	1.35 (0.26) [a,b,c]	1.37 (0.02) [a,b,c]	1.60 (0.15) [c]	1.12 (0.02) [a]	1.31 (0.02) [a,b,c]	1.18 (0.02) [a,b]	8.4 ***

d.m.—dry matter. [a-e] different letters in the same column indicate differences between samples ($p < 0.05$). ns-not significant ($p > 0.05$), *—$p < 0.05$, **—$p < 0.01$, ***—$p < 0.001$.

3.3. Proximate Composition of Bee Pollen

Protein content ranged between 15.74 and 27.92% ($p < 0.05$) with an average of 22.31%, which confirms the important role of bee pollen for human nutrition. The level of proteins was in agreement with those reported from Brazil (12.28% to 27.07.%) [40], Spain (15.19–20.23%), Colombia (21.6%) and Italy (19.5%) [22]. Gardana et al. observed a lower content in terms of proteins (12.3%) for bee pollen from Spain [22]. The great variability of protein level in the 24 samples analyzed might be influenced by the floral sources, geographical origin and/or storage conditions.

Lipids are considered the third group of substances present in the bee pollen, after carbohydrates and proteins, and are vital for the generation of royal jelly. The lipids ranged between 2.24 and 7.30% ($p < 0.05$), with a mean concentration of 4.49%. The principal compounds are triglycerides, carotenoids and sterols [42].

Ash represents 2.29 to 4.02% ($p < 0.05$) of the bee pollen with a mean of 3.18%, in agreement with the literature [17]. In the case of energy value of the bee pollen it was 368.19 to 407.68 kcal/100 g ($p < 0.05$), with a mean of 390.66 kcal/100 g; this fact confirms the importance of this bee product for human nutrition.

3.4. Free Sugars of Bee Pollen

The carbohydrates represent the main compounds present in the bee pollen; the main compound determined was fructose followed by glucose, melesitose, trehalose, sucrose, maltose, raffinose and turanose in the bee pollen [4]. Fructose and glucose were in the same range as those reported for bee pollen from Colombia, Italy and Spain, while sucrose was much lower (5.1–6.2% for Colombia, Italy and Spain, and for the samples analysed in this study) [22]. In other studies, others sugars were reported such as arabinose, melibiose, isomaltose, melesitose, ribose, turanose and trehalose but they do not represent more than 1% of them [43]. The fructose/glucose ratio was between 1.13 and 2.43 ($p < 0.05$) with a mean ratio of 1.46. The Spanish bee pollen was reported to have a fructose/glucose ratio between 1.13 and 1.53 [44], the Polish bee pollen was reported to have a fructose/glucose ratio between 1.03 to 2.51 [30], while the Brazilian bee pollen was reported to have a fructose/glucose ratio between 1.01 to 2.24 [40]; one possible reason of the differences between the samples may attributed to the harvesting seasons [30,40].

3.5. Organic Acids of Bee Pollen

In this study the presence of gluconic acids, formic acid, lactic acid, acetic acid, succinic acid, propionic acid and butyric acid was investigated, and the results are presented in Table 3. As can be observed from the Table 3, the major organic acid was gluconic acid, followed by lactic acid, acetic acid and propionic acid. Formic acid, succinic acid and butyric acid were not present in any of the samples. Similar levels from gluconic acid, lactic acid and acetic acid were reported by Kalaycıoglu et al. [45] in the case of bee pollen from Turkey. The lactic acid presented in the bee pollen is probably the result of the fermentation process during the fermentation of carbohydrates using lactic acid bacteria present in the bees' stomach [46]. The organic acids present preservation potential from foods and are promoted as a new generation instead of antibiotics, so the bee pollen can be considered a potential preservation agent [45,46]. The absence of formic acid and butyric acid represents a good indicator that the bee pollen is not contaminated with undesired microorganisms.

Table 3. Organic acids, total phenolic content, total flavonoid content and individual phenolics compounds of bee pollen (mean values and standard deviation in brackets).

	Crataegus monogyna	*Helianthus annuus*	*Pinaceae* spp.	Polyfloral	*Prunus* spp.	*Quercus* spp.	*Robinia pseudoacacia*	*Taraxacum* spp.	*Tillia* spp.	*Zea mays*	F-Value
Gluconic acid (g/kg)	33.78 (0.48) [a,b,c]	21.68 (1.05) [a,b]	34.47 (0.60) [b,c]	30.00 (5.04) [a,b,c]	29.37 (13.89) [a,b,c]	22.22 (4.48) [a,b]	36.33 (7.98) [c]	24.88 (0.35) [a,b,c]	26.24 (0.37) [a,b,c]	14.03 (0.20) [a]	4.9 ***
Lactic acid (g/kg)	0.67 (0.01) [a,b]	0.61 (0.16) [a]	1.10 (0.01) [b]	0.69 (0.13) [a,b]	0.74 (0.41) [a,b]	0.67 (0.10) [a,b]	0.76 (0.13) [a,b]	0.54 (0.01) [a]	0.77 (0.01) [a,b]	0.47 (0.01) [a]	3.8 **
Acetic acid (g/kg)	0.26 (0.01) [a]	0.61 (0.23) [a,b]	1.20 (0.01) [c]	0.44 (0.09) [a,b]	0.71 (0.52) [b]	0.28 (0.14) [a,b]	0.49 (0.41) [a,b]	0.28 (0.01) [a,b]	0.58 (0.01) [a,b]	0.28 (0.01) [a,b]	10.4 ***
Propionic acid (g/kg)	0.43 (0.01) [a]	0.11 (0.23) [a]	0 (0) [a]	0.05 (0.09) [a]	0.45 (0.52) [a]	0.29 (0.14) [a]	0.37 (0.31) [a]	0.04 (0.01) [a]	0.26 (0.01) [a]	0.13 (0.01) [a]	1.6 ns
TPC (GAE mg/g)	8.73 (0.12) [a,b,c]	7.56 (3.02) [a,b]	12.39 (0.17) [a,b,c,d]	13.53 (2.16) [b,c,d]	15.74 (2.33) [d]	15.52 (1.17) [d]	14.11 (3.08) [c,d]	16.45 (0.24) [d]	14.83 (0.21) [c,d]	7.10 (0.10) [a]	9.3 ***
TFC (QE mg/g)	8.55 (0.12) [a,b,c]	5.95 (1.04) [a]	11.03 (0.16) [b,c,d]	13.97 (1.57) [d,e,f]	17.37 (3.33) [f]	15.68 (0.42) [e,f]	18.81 (2.19) [c,d,e]	16.39 (0.23) [e,f]	14.79 (0.21) [d,e,f]	6.28 (0.09) [a,b]	29.1 ***
P-A	0 (0) [a]	0 (0) [a]	88.93 (0) [a]	0 (0) [a]	0 (0) [a]	0 (0) [a]	0 (0) [a]	0 (0) [a]	0 (0) [a]	0 (0) [a]	3.8 ***
p-H-A	0 (0) [b]	0 (0) [b]	21.02 (0) [a]	0 (0) [b]	0 (0) [b]	0 (0) [b]	0 (0) [b]	0 (0) [b]	0 (0) [b]	0 (0) [b]	3.9 ***
V-A (mg/kg)	0 (0) [a]	21.19 (1.20) [b]	0 (0) [a]	0 (0) [a]	0 (0) [a]	0 (0) [a]	0 (0) [a]	0 (0) [a]	0 (0) [a]	0 (0) [a]	4.1 ***
C-A	0 (0) [a]	0.78 (0.10) [a]	0 (0) [a]	3.82 (1.08) [a]	0 (0) [a]	0 (0) [a]	3.63 (4.90) [a]	0 (0) [a]	0 (0) [a]	0 (0) [a]	0.8 ns
p-C-A (mg/kg)	2.92 (0.04) [a]	24.75 (17.36) [a]	139.79 (3.82) [a,b]	79.26 (66.99) [a]	92.74 (49.07) [a]	27.78 (0.33) [a]	238.97 (67.93) [b]	23.36 (0.33) [a]	168.70 (2.41) [a,b]	18.58 (0.27) [a]	9.5 ***
R-A	0 (0) [a]	2.01 (1.25) [a,b]	0 (0) [a]	11.65 (12.61) [a,b]	3.63 (4.19) [a,b]	26.08 (14.77) [b]	14.92 (12.89) [a,b]	14.50 (0.21) [a,b]	22.75 (0.32) [a,b]	85.14 (1.22) [c]	16.8 ***
Myricetin (mg/kg)	397.49 (5.68) [b,c,d,e]	33.36 (21.85) [a,b]	209.11 (4.54) [a,b,c,d]	558.08 (160.75) [d,e]	183.92 (22.48) [a,b,c]	712.13 (211.67) [e]	256.93.16 (206.45) [a,b,c,d]	284.01 (4.06) [a,b,c,d]	439.01 (6.27) [c,d,e]	0 (0) [a]	12.7 ***
Luteolin (mg/kg)	0 (0) [a]	13.45 (10.73) [a,b,c]	10.79 (0.31) [a,b]	9.75 (8.05) [a,b]	3.08 (3.56) [a]	0 (0) [a]	26.22 (17.50) [b,c]	0 (0) [a]	33.46 (0.48) [c]	0 (0) [a]	3.1 **
Quercitin (mg/kg)	126.38 (1.81) [a]	22.11 (40.39) [a]	28.14 (0.80) [a]	172.43 (125.23) [a]	757.22 (454.34) [c]	686.07 (35.27) [b,c]	105.063 (100.80) [a]	296.34 (4.23) [a,b]	381.56 (5.45) [a,b,c]	71.36 (1.02) [a]	14.2 ***
Kaempferol (mg/kg)	126.45 (1.81) [a,b]	52.22 (26.18) [a]	186.77 (2.45) [a,b]	269.20 (148.03) [a,b]	283.42 (56.03) [a,b]	269.41 (4.81) [a,b]	363.19 (181.96) [b,c]	652.38 (9.32) [c]	406.47 (5.81) [b,c]	323.18 (4.62) [a,b]	8.6 ***

TPC—total phenolic content, TFC—total flavone content, p-a—protocatechuic acid, p-h-a—p-hydroxybenzoic acid, C-a—chlorogenic acid, p-c-A—p-coumaric acid, R-a—rosmarinic acid, V-A vanillic acid. [a–f] different letters in the same column indicate differences between samples ($p < 0.05$). ns-not significant ($p > 0.05$), **—$p < 0.01$, ***—$p < 0.001$.

3.6. TPC, TFC and Individual Phenolics Compounds

In the Table 3 the total phenolic content, total flavonoid content and individual phenolics compounds of bee pollen are presented. The TPC ranged between 4.64 and 17.93 GAE mg/g ($p < 0.05$) while the TFC ranged between 4.93 and 20.45 QE mg/g. The significant differences ($p < 0.05$) were found between TPC and TFC of pollen extracts from different sources which might be due to variation in the botanical origins as well as different climatic conditions. The values of TPC were in the same range as those reported for Indian bee pollen (9.79–35.63 GAE mg/g) [10] and Brazilian pollen (6.50 to 29.20 mg GAE/g) [47]. Regarding the TFC, the values are in agreement with those reported for Brazilian bee pollen (0.30–17.50 mg QE/g) [47] and Indian bee pollen (9.72–15.62 GAE mg/g) [10]. From the phenolics compounds studied, gallic acid and caffeic acid were not reported in any samples studied. The protocatechiuc acid and *p*-hydroxybenzoic acid were only observed in one sample (S2—sample with more than 45% of the pollen from *Pinaceae* spp.), vanillic acid was reported in three samples (S1, S3 and S24) and chlorogenic acid in five samples (S1, S12, S18, S21 and S23). The other phenolics studied were more present in the pollen samples. The major compound found was quercetin (S19—sample with more than 45% of the pollen from *Prunus* spp.), followed by myricetin (S16—sample with more than 45% of the pollen from *Quercus* spp.) and kaempferol (S23—sample with more than 45% of the pollen from *Robinia pseudoacacia*). The Brazilian bee pollen [48] had a similar concentration of chlorogenic acid and vanillic acid; from 56 samples just two samples contained gallic acid and three caffeic acid, so their findings are similar to ours and we can conclude that bee pollen is not a source of this phenolic acids; the *p*-coumaric acid, quercetin and kaempferol were in a much lower concentration than those reported in this study which may be attributed to the different botanical or geographical origin of the samples. Thakur and Nanda [10] reported that flavonoids are influenced by the botanical origin of the bee pollen (catechin: 0.94–19.10 mg/100 g; rutin: 4.81–24.83 mg/100 g; quercetin: 3.14–15.94 mg/100 g; luteolin: 1.06–5.86 mg/100 g; kaempferol: 0.12–9.35 mg/100 g; and apigenin: 0.46–3.02 mg/100 g), quercetin and kaempferol were the major flavonoids reported by them but in lower concentration than in this study.

3.7. Total Free Amino Acid Composition

As the data presented in Table 4 states, 16 amino acids (eight essential amino acids and eight non-essential amino acids) were quantified in the bee pollen samples analyzed, and there was observed a significant difference in terms of amino acids concentration between the samples ($p < 0.05$). The total content of the amino acids determined for the bee pollen samples varied between 11.31 µg/mg (sample 21) and 45.99 µg/mg (sample 4). These values were comparable to those reported for the total free amino acid content of commercial bee pollen from Colombia (25.3 ± 1.0 mg/g), Italy (29.4 ± 0.7 mg/g) and Spain (30.8 ± 0.2 mg/g) [22], and higher than the total amino acids content of bee pollen from floral sources such as sunflower (12.20 g/100 g) and rape (12.25 g/100 g) [49]. From the data presented, in the samples analyzed in our study, the most abundant essential amino acids were histidine (values of 0.29–2.30 µg/mg), lysine (0.14–0.74 µg/mg) and phenylalanine (0.12–0.43 µg/mg). Histidine was found to predominate in the bee pollen collected during autumn, while high levels of lysine and phenylalanine were determined in bee pollen collected during winter [50]. The distribution of these essential amino acids in the analyzed bee pollen was therefore in accordance with the period when the samples were collected. Leucine, isoleucine and tryptophan were detected in low amounts, and similar findings were reported for monofloral bee pollen of *Geranium* botanical origin [15]. Glutamic acid was the main amino acid in all bee pollen samples, with values that varied between 0.34 and 18.77 µg/mg, followed by aspartic acid (0.01–14.08 µg/mg) and proline (2.79–7.19 µg/mg). Glutamic acid, aspartic acid and proline were also reported as the major amino acids in different bee pollen varieties (coconut, coriander, rapeseed and multifloral) from India [27]. The variation of the amino acid content was found to be influenced by both botanical origin and processing and storage conditions. In regard to the processing and storage conditions, previous studies reported that glutamic acid is the most abundant amino acid in bee pollen that is freshly collected, while proline is the free amino acid that is found in high amounts in well dried and stored bee pollen [22]. In the case of our study, the high content of glutamic acid was well correlated to the fact that the bee pollen samples analyzed in this study were freshly collected.

3.8. Fatty Acids Composition

For the bee pollen samples analyzed in this study, 19 fatty acids were quantified by the GC-MS method. The total content of fatty acids of the bee pollen samples varied between 81.69 µg/g (sample 10) and 645.72 µg/g (sample 24) (Table 4). The unsaturated fatty acids were predominant (UFA; 62.65–927.50 µg/g), of which levels of 7.51–88.69 µg/g were determined for monounsaturated fatty acids (MUFAs) and 43.86–838.82 µg/g for polyunsaturated fatty acids (PUFAs). The main MUFAs were oleic acid (C18:1 (Z)-octadec-9-enoic acid) and 11-eicosenoic acid (C20:1 (cis-11) (Z)-icos-11-enoic acid), which were also reported as prevalent in *Brassica napus* pollen from India [27]. Of the PUFAs, γ-linoleic acid (C18:3 (all-cis-6,9,12) octadeca-6,9,12-trienoic acid) and linoleic acid (C18:2 (all-cis-9,12) (9Z,12Z)-octadeca-9,12-dienoic acid) were determined in high levels in all pollen samples. By comparison, in 18 bee pollen samples from Turkey and Romania, Mărgăoan et al. [51] determined a higher content of α-linoleic acid than linoleic acid. In our study, α-linoleic acid was quantified only in samples 3 and 4. Palmitic acid (C16:0 hexadecanoic acid) and stearic acid (C18:0 octadecanoic acid) were determined as the main saturated fatty acids (SFA); these two fatty acids were also reported as the predominant saturated fatty acids in the bee pollen from 11 different floral sources from Taiwan [25] and in commercial bee pollen samples from Colombia, Italy and Spain [22]. C15:0 pentadecanoic acid and C17:0 heptadecanoic acid were determined in lower amounts and were found in less than half of the bee pollen samples analyzed in this study. When studying the fatty acids profile of bee pollen, the ratio between UFA and SFA is of great importance. It was considered that a value of the UFA/SFA ratio higher than 1 is characteristic of bee pollen with considerable nutritional value, while a value below 1 indicates degradation of unsaturated fatty acids due to storage and dehydration process [50]. For the 24 bee pollen samples analyzed in our study, the UFA/SFA ratio varied between 1.86 and 5.78 and was comparable with the values of 2.2–6.7 reported for the bee pollen from India [27] and the 1.9–2.2 UFA/SFA ratio calculated for commercial bee pollen from Colombia, Italy and Spain [22].

Table 4. Amino acids profile and fatty acids composition of bee pollen (mean values and standard deviation in brackets).

	Crataegus monogyna	*Helianthus annuus*	*Pinaceae* spp.	Polyfloral	*Prunus* spp.	*Quercus* spp.	*Robinia pseudoacacia*	*Taraxacum* spp.	*Tillia* spp.	*Zea mays*	F-Value
Amino acids, µg/mg bee pollen											
Valine	0.18 (0.01) [a]	0.14 (0.07) [a]	0.12 (0.02) [a]	0.19 (0.08) [a]	0.22 (0.03) [a]	0.24 (0.09) [a,b]	0.20 (0.03) [a]	0.41 (0.01) [b]	0.21 (0.01) [a]	0.18 (0.01) [a]	5.1 ***
Leucine	0.08 (0.01) [a,b]	0.08 (0.01) [a,b]	0.08 (0.03) [a,b]	0.09 (0.01) [a,b]	0.06 (0.03) [a]	0.08 (0.01) [a,b]	0.11 (0.02) [b]	0.09 (0.01) [a,b]	0.09 (0.01) [a,b]	0.09 (0.01) [a,b]	3.5 **
Isoleucine	0.12 (0.01) [a]	0.16 (0.06) [a]	0.13 (0.03) [a]	0.17 (0.04) [a]	0.16 (0.05) [a]	0.20 (0.04) [a]	0.16 (0.03) [a]	0.24 (0.01) [a]	0.21 (0.01) [a]	0.16 (0.01) [a]	1.8 ns
Threonine	0 (0) [a]	0.10 (0.03) [b]	0.12 (0.04) [b,c]	0.21 (0.05) [d,e]	0.16 (0.06) [b,c,d]	0.20 (0.03) [c,d,e]	0.15 (0.02) [b,c,d]	0.27 (0.01) [e]	0.09 (0.01) [b]	0.22 (0.01) [d,e]	15.2 ***
Phenylalanine	0.25 (0.01) [a,b]	0.23 (0.08) [a,b]	0.18 (0.05) [a]	0.29 (0.09) [a,b]	0.21 (0.08) [a]	0.28 (0.02) [a,b]	0.19 (0.04) [a]	0.40 (0.01) [b]	0.30 (0.01) [a,b]	0.21 (0.01) [a]	3.2 **
Histidine	1.18 (0.02) [a,b,c]	1.28 (0.71) [a,b,c]	0.45 (0.18) [a]	0.97 (0.13) [a,b,c]	0.84 (0.44) [a,b,c]	1.83 (0.52) [c]	0.62 (0.26) [a,b]	1.64 (0.02) [b,c]	1.09 (0.02) [a,b,c]	0.50 (0.01) [a]	4.8 ***
Lysine	0.40 (0.01) [a,b,c]	0.35 (0.23) [a,b,c]	0.01 (0.00) [a]	0.32 (0.20) [a,b,c]	0.10 (0.11) [a,b]	0.47 (0.20) [b,c]	0.29 (0.23) [a,b,c]	0.25 (0.01) [a,b,c]	0.73 (0.01) [c]	0.46 (0.01) [a,b,c]	3.3 **
Tryptophan	0.18 (0.01) [a]	0.12 (0.07) [a]	0.11 (0.01) [a]	0.18 (0.02) [a]	0.14 (0.01) [a]	0.16 (0.01) [a]	0.16 (0.04) [a]	0.18 (0.01) [a]	0.17 (0.01) [a]	0.16 (0.01) [a]	1.5 ns
Alanine	0.31 (0.01) [a,b]	0.26 (0.07) [a,b]	0.21 (0.03) [a]	0.26 (0.04) [a,b]	0.22 (0.01) [a]	0.31 (0.04) [a,b]	0.38 (0.07) [b]	0.26 (0.01) [a,b]	0.31 (0.01) [ab]	0.32 (0.01) [a,b]	5.6 ***
Sarcosine	0.04 (0.01) [a]	0.03 (0.01) [a]	0.03 (0.00) [a]	0.03 (0.01) [a]	0.03 (0.01) [a]	0.03 (0.01) [a]	0.03 (0.01) [a]	0.04 (0.01) [a]	0.04 (0.01) [a]	0.04 (0.01) [a]	1.6 ns
Glycine	0.08 (0.01) [a]	0.05 (0.03) [a]	0.06 (0.03) [a]	0.07 (0.02) [a]	0.03 (0.01) [a]	0.06 (0.02) [a]	0.06 (0.02) [a]	0.04 (0.01) [a]	0.04 (0.01) [a]	0.04 (0.01) [a]	1.9 ns
Serine	1.32 (0.02) [d]	0.71 (0.29) [a,b,c]	0.33 (0.11) [a]	0.76 (0.16) [a,b,c]	0.49 (0.28) [a,b]	0.85 (0.10) [a,b,c,d]	0.85 (0.22) [a,b,c]	0.67 (0.01) [a,b,c]	0.97 (0.01) [b,c,d]	1.12 (0.02) [c,d]	5.4 ***
Proline	7.12 (0.10) [c]	3.66 (0.46) [a,b]	3.32 (0.64) [a,b]	3.74 (0.37) [a,b]	3.78 (0.86) [a,b]	4.92 (0.27) [b]	4.92 (0.94) [a,b]	4.88 (0.07) [b]	4.25 (0.06) [a,b]	3.20 (0.05) [a]	8.7 ***
Asparagine	1.34 (0.02) [a,b,c]	0.74 (0.53) [a,b]	0.50 (0.01) [a]	1.22 (0.50) [a,b,c]	1.03 (0.29) [a,b,c]	1.64 (0.03) [b,c]	1.64 (0.18) [a]	1.71 (0.01) [c]	0.99 (0.01) [a,b,c]	0.77 (0.01) [a,b]	5.9 ***
Aspartic acid	6.94 (0.10) [a,b,c]	5.62 (4.92) [a,b,c]	10.77 (0.86) [b,c]	3.32 (3.55) [a,b,c]	1.02 (0.89) [a]	2.57 (2.51) [a,b]	2.57 (2.99) [a,b,c]	3.08 (0.04) [a,b,c]	11.64 (0.17) [c]	0.38 (0.01) [a]	4.2 **
Glutamic acid	11.50 (0.16) [a,b,c]	8.35 (6.17) [a,b,c]	14.82 (1.49) [b,c]	5.12 (5.08) [a,b]	1.31 (0.96) [a]	4.20 (2.16) [a,b]	4.20 (4.01) [a,b]	3.06 (0.04) [a]	17.83 (0.25) [c]	3.88 (0.06) [a,b]	4.9 ***
Tyrosine	0.25 (0.01) [a,b]	0.21 (0.05) [a,b]	0.18 (0.01) [a]	0.21 (0.04) [a,b]	0.15 (0.01) [a]	0.20 (0.05) [a,b]	0.15 (0.03) [a,b]	0.18 (0.01) [a,b]	0.22 (0.01) [b]	0.16 (0.01) [a,b]	2.5 *
Total AA content	31.26 (0.45) [b,c]	22.11 (9.49) [a,b,c]	31.40 (2.24) [b,c]	17.13 (8.14) [a,b]	9.94 (1.28) [a]	18.23 (4.67) [a,b]	8.84 (7.22) [a,b]	17.38 (0.25) [a,b]	28.41 (0.56) [c]	11.87 (0.17) [a]	5.4 ***
Fatty acids, µg/g bee pollen											
(C6:0)	0.04 (0.01) [a]	0.29 (0.21) [a,b]	0.75 (0.35) [d]	0.44 (0.30) [b,c,d]	0.42 (0.35) [a,b,c,d]	0.09 (0.03) [a,b]	0.75 (0.23) [c,d]	0.09 (0.01) [a,b]	0.52 (0.01) [c,d]	0.11 (0.01) [a,b]	3.8 **
(C8:0)	0.29 (0.01) [a]	1.38 (1.27) [a,b]	2.73 (1.56) [b]	1.65 (1.05) [a,b]	1.18 (0.31) [a,b]	0.33 (0.19) [a,b]	1.05 (0.39) [ab]	0.23 (0.01) [a]	1.81 (0.03) [a,b]	0.30 (0.01) [a]	3.1 **
(C10:0)	0.26 (0.01) [a]	0.99 (0.71) [a]	1.47 (1.08) [a]	0.66 (0.42) [a]	0.57 (0.36) [a]	0.44 (0.20) [a]	1.09 (0.61) [a]	0.26 (0.01) [a]	0.76 (0.01) [a]	0.45 (0.01) [a]	1.5 ns
(C12:0)	0.33 (0.01) [a]	3.64 (2.92) [a]	4.42 (3.99) [a]	2.43 (1.51) [a]	1.81 (0.53) [a]	0.85 (0.21) [a]	2.37 (0.84) [a]	0.74 (0.01) [a]	3.09 (0.04) [a]	1.29 (0.02) [a]	1.9 ns

Table 4. *Cont.*

	Crataegus monogyna	*Helianthus annuus*	*Pinaceae* spp.	Polyfloral	*Prunus* spp.	*Quercus* spp.	*Robinia pseudoacacia*	*Taraxacum* spp.	*Tillia* spp.	*Zea mays*	F-Value
(C14:0)	0.61 (0.01) [a]	4.12 (5.43) [a]	5.46 (3.86) [a]	2.63 (1.16) [a]	3.71 (2.10) [a]	0.93 (0.07) [a]	6.56 (1.83) [a]	0.49 (0.01) [a]	3.31 (0.05) [a]	0.55 (0.01) [a]	1.5 ns
(C15:0)	0 (0) [a]	0.12 (0.16) [a]	0.24 (0.28) [a]	0.23 (0.19) [a]	0 (0) [a]	0.16 (0.08) [a]	0.08 (0.03) [a]	0 (0) [a]	0.23 (0.01) [a]	0 (0) [a]	1.2 ns
(C16:0)	9.14 (0.13) [a]	48.96 (50.24) [a]	149.07 (137.91) [a]	46.56 (39.24) [a]	115.20 (25.43) [a]	13.57 (2.94) [a]	120.34 (57.35) [a]	12.98 (0.19) [a]	23.85 (0.34) [a]	12.78 (0.18) [a]	2.4 ns
(C16:1 [cis-9])	0.59 (0.01) [a]	1.77 (1.95) [a]	10.81 (11.80) [b]	0.68 (0.38) [a]	1.24 (0.09) [a]	0.25 (0.10) [a]	1.62 (0.82) [a]	0.33 (0.01) [a]	0.21 (0.01) [a]	0.32 (0.01) [a]	3.5 **
(C17:0)											
(C17:1 [cis-10])	0 (0) [a]	1.95 (3.87) [a]	1.60 (1.05) [a]	2.34 (1.11) [a]	0 (0) [a]	5.18 (4.99) [a]	3.21 (3.06) [a]	0 (0) [a]	2.58 (0.34) [a]	0 (0) [a]	0.8 ns
(C18:0)	0 (0) [a]	0.02 (0.05) [a]	0 (0) [a]	1.00 (0.16) [a]	0 (0) [a]	1.43 (0.65) [a]	2.09 (1.22) [a]	0 (0) [a]	0 (0) [a]	0 (0) [a]	1.6 ns
(C18:1 [trans-9]) + (C18:1 [cis-9])	10.07 (0.14) [a,b]	19.58 (9.55) [a,b]	32.28 (22.78) [b]	14.76 (7.53) [a,b]	24.78 (15.43) [a,b]	7.03 (3.73) [a]	33.11 (11.29) [a,b]	14.34 (0.20) [ab]	9.15 (0.13) [a,b]	13.85 (0.20) [a,b]	2.6 *
(C18:2 [trans-9,12])	15.84 (0.23) [a]	21.62 (17.91) [a]	33.39 (24.03) [a]	19.59 (12.89) [a]	35.24 (0.09) [a]	8.52 (3.58) [a]	42.93 (13.75) [a]	18.74 (0.27) [a]	10.56 (0.15) [a]	9.65 (0.14) [a]	2.3 ns
(C18:2 [cis-9,12])	0 (0) [a]	2.82 (3.74) [a]	9.83 (11.35) [a]	29.06 (18.53) [a]	0 (0) [a]	0 (0) [a]	0 (0) [a]	0 (0) [a]	14.75 (0.21) [a]	0 (0) [a]	1.7 ns
(C18:3 [cis-6,9,12])	15.36 (0.22) [a]	44.26 (35.23) [a,b]	93.06 (64.14) [a,b]	46.94 (38.53) [a,b]	121.73 (14.53) [b]	23.31 (1.15) [a]	117.91 (32.98) [ab]	28.00 (0.40) [a]	21.49 (0.31) [a]	24.08 (0.34) [a]	5.0 ***
(C20:1 [cis-11])	19.39 (0.28) [a]	85.87 (74.06) [a]	283.61 (253.73) [a]	90.81 (83.28) [a]	207.81 (21.02) [a]	35.64 (6.99) [a]	232.90 (98.51) [a]	34.74 (0.50) [a]	28.13 (0.40) [a]	36.38 (0.52) [a]	2.7 ns
(C18:3 [cis-9,12,15])	2.35 (0.03) [a]	7.33 (3.13) [a,b,c,d]	7.52 (5.84) [c,d]	3.83 (1.59) [a,b,c,d]	2.59 (1.04) [a,b]	2.39 (0.64) [a]	10.24 (3.23) [d]	3.16 (0.05) [a,b,c]	3.28 (0.05) [a,b,c]	3.25 (0.05) [a,b,c]	3.6 **
(C20:3 [cis-11,14,17])	0 (0) [a]	7.44 (5.68) [a]	0 (0) [a]	0 (0) [a]	0 (0) [a]	0 (0) [a]	0 (0) [a]	0 (0) [a]	0 (0) [a]	2.10 (0.01) [a]	0.8 ns
(C22:2 [cis-13,16])	0 (0) [a]	0.95 (2.00) [a]	2.79 (2.22) [a]	0 (0) [a]	0 (0) [a]	5.57 (0.01) [a]	0 (0) [a]	0 (0) [a]	0 (0) [a]	0 (0) [a]	1.9 ns
Total FA content	12.74 (0.18) [a]	19.27 (12.47) [a]	106.36 (83.32) [b]	36.54 (29.33) [a,b]	92.96 (32.59) [a,b]	34.21 (4.99) [a]	18.67 (9.98) [a,b]	65.01 (0.33) [a,b]	23.15 (0.12) [a]	15.61 (0.26) [a]	3.8 **
MUFA	87.01 (1.24) [a]	272.35 (200.87) [a,b]	745.39 (585.93) [b]	300.15 (221.09) [a,b]	609.33 (111.84) [a,b]	238.04 (9.36) [a,b]	118.79 (35.87) [a,b]	641.26 (1.96) [a,b]	137.23 (1.94) [a,b]	136.66 (1.73) [a,b]	3.1 *

Table 4. *Cont.*

	Crataegus monogyna	*Helianthus annuus*	*Pinaceae* spp.	Polyfloral	*Prunus* spp.	*Quercus* spp.	*Robinia pseudoacacia*	*Taraxacum* spp.	*Tillia* spp.	*Zea mays*	F-Value
PUFA	18.78 (0.27) [a]	30.71 (19.30) [a]	51.71 (41.68) [a]	24.10 (6.45) [a]	39.08 (9.94) [a]	15.62 (4.31) [a]	11.17 (6.75) [a]	54.79 (0.32) [a]	22.23 (0.20) [a]	18.10 (0.19) [a]	2.5 *
UFA	47.49 (0.68) [a]	160.60 (116.14) [a,b]	495.65 (386.63) [b]	203.35 (160.92) [a,b]	422.49 (67.82) [a,b]	160.88 (10.88) [a,b]	77.62 (58.37) [a,b]	415.82 (1.23) [a,b]	85.89 (1.10) [a,b]	81.27 (1.12) [a,b]	3.3 **
SFA	66.27 (0.95) [a]	191.31 (135.42) [a,b]	547.36 (428.31) [b]	227.45 (167.06) [a,b]	461.57 (77.76) [a,b]	176.50 (6.61) [a,b]	88.78 (73.60) [a,b]	470.61 (1.54) [a,b]	108.12 (1.30) [a,b]	99.37 (1.31) [a,b]	3.2 **
UFA/SFA	20.73 (0.30) [a]	81.02 (66.27) [a]	198.02 (157.62) [a]	71.70 (54.96) [a]	147.77 (34.19) [a]	61.54 (14.13) [a]	28.58 (7.75) [a]	168.56 (0.42) [a]	29.12 (0.65) [a]	37.30 (0.42) [a]	2.7 *
(C6:0)	3.17 (0.05) [a]	2.51 (0.64) [a]	2.78 (0.64)	3.22 (0.54) [a]	3.09 (0.39) [a]	2.84 (0.14) [a]	3.87 (0.61) [a]	3.50 (0.05) [a]	3.67 (0.03) [a]	2.82 (0.04) [a]	1.8 ns

AA—amino acids; FA—fatty acids; MUFA—monounsaturated fatty acids; PUFA—polyunsaturated fatty acids; UFA—unsaturated fatty acids; SFA—saturated fatty acids. [a–e] different letters in the same row indicate differences between samples ($p < 0.05$). ns-not significant ($p > 0.05$), *—$p < 0.05$, **—$p < 0.01$, ***—$p < 0.001$.

3.9. FTIR-ATR Spectroscopy of Pollen

The FTIR-ATR spectra of 24 pollen samples, recorded in absorbance mode in the mid-infrared region, is presented in Figure 1. The broad band around 3290 cm^{-1} that was observed in all pollen samples corresponded to O–H stretching vibration due to the presence of water [52,53]. As it was previously reported that the moisture content of fresh pollen varies between 21 and 30% [9], the presence of a broad band in this spectral region was expected. Between 3000 and 2850 cm^{-1}, two peaks were identified in all the samples analyzed: a peak around 2920 cm^{-1} and one around 2850 cm^{-1}, both assigned to C–H stretching, and mainly CH$_2$ and CH$_3$ vibrations of lipids, proteins and carbohydrates [52]. For one pollen sample (sample 4) a small peak was found at 2360 cm^{-1} and was also attributed to C–H stretching vibrations of lipids [54]. Furthermore, another signal corresponding to lipids, and namely the peak at 1740 cm^{-1}, which is characteristic of stretching vibrations of C=O groups, was prominent in some pollen samples; previous research found that this signal shows large variation within pollen samples of related plant species [55]. In the spectra of all pollen samples, peaks were observed at 1650 cm^{-1} and 1540 cm^{-1} that were attributed to stretching vibrations of amide I and II [56]. Characteristic of all samples was also the peak at 1414 cm^{-1} that was assigned to asymmetric in-plane bending of the –CH$_3$ group [53]. The peak in the region between 1350 and 1200 cm^{-1} was assigned to amide III, and more precisely an in-phase combination of N–H deformation vibrations and C–N stretching vibrations [56]. All samples had high absorption peaks around 1030 cm^{-1} that corresponded to stretching vibrations of saccharides and proteins, and were also reported for crude pollen and defatted pollen samples in our previous study on the extraction of polyphenols from crude pollen [17]. In the spectral range between 1200 and 500 cm^{-1}, which is considered the fingerprint region of pollen, the peaks observed for the analyzed samples were due to C–O and C–C stretching vibrations, and their variation among pollen samples indicated differences in the saccharide, protein and lipid composition. This overlapped the region of 1500–800 cm^{-1}, where characteristic signals were attributed to C–O and C–C stretching vibrations of flavonoids and phenolic compounds [57].

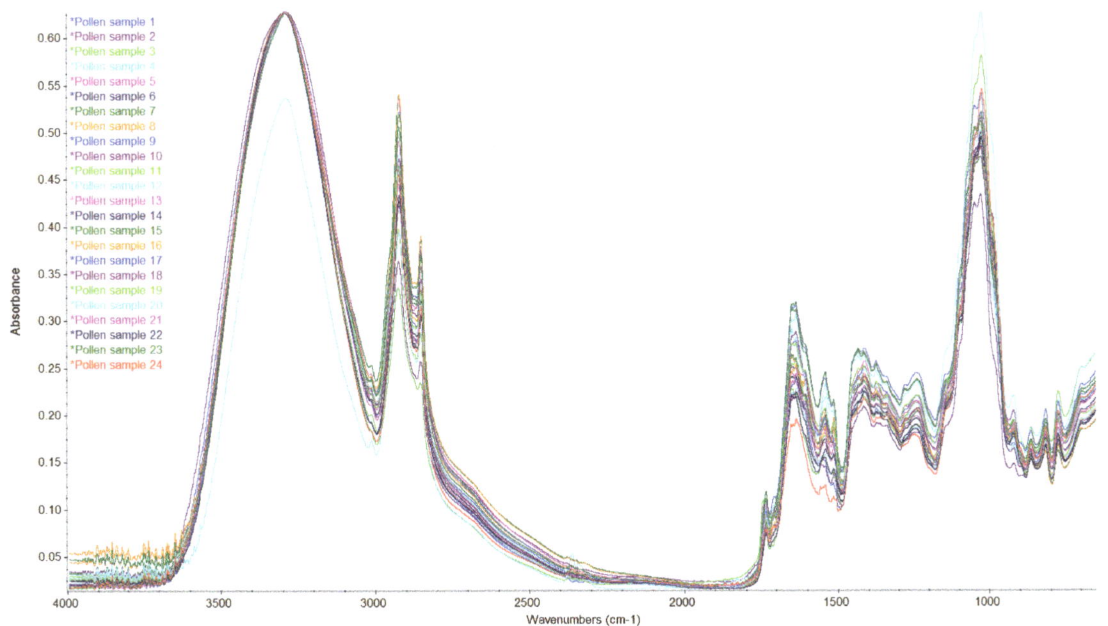

Figure 1. Bee pollen FT-IR spectra in the region 4000–650 cm^{-1}.

3.10. Principal Component Analysis

The PCA was conducted based on the analysis discussed in order to discriminate the bee pollen samples based on their botanical origin. The first two principal components (PC1 and PC2) explained 77% of the data variance (PC1 explained 50% of the data variance, while PC2 explained 27% of the data variance). The PC1 is influenced strongly by lipids and negatively by raffinose and C18:2, the PC2 is influenced positively by quercetin while myricetin influences it negatively. Propionic acid, C17:1, TFC and asparagine do not influence the projection of the scores due to their closeness to the origins of the two PC. As can be seen in Figure 2A, the samples with high percentage of *Helianthus annus, Robinia pseudoacacia, Pinaceae* spp., *Quercus* spp. and *Prunus* spp. formed clusters which confirms that the analysis carried out is useful for their discrimination. Regarding the polyfloral bee pollen it can be observed that the samples are near on the other one but they include in their region the bee pollen from *Crataegus monogyna, Tilia* spp. and *Taraxacum*. The *Quercus* spp. and *Crataegus monogyna* exhibited a high myricetin content, while *Prunus* spp. exhibited a high quercetin content. The *Helianthus annus* samples exhibited a high free acidity, C18:0, F/G, lactic acid, lipids and C20:1. The polyfloral samples exhibited a high by raffinose and C18:2, myricetin and raffinose (Figure 2B). The *Robinia pseudoacacia* pollen samples were associated with C15:0, chlorogenic acid, turanose, maltose, C17:0, and *p*-hydroxibenzoic acid. A high positive correlation between moisture content and water activity (r = 0.898), moisture content and F/G (r = 0.818) was observed. The lipids content was positively correlated with C18:1 [trans-9]) + (C18:1 [cis-9] (r = 0.812), C18:2 [cis-9,12] (r = 0.898), C18:3 [cis-6,9,12] (r = 0.852), C22:2 [cis-13,16] (r = 0.772), MUFA (r = 0.796), PUFA (r = 0.870), UFA (r = 0.869) and SFA (r = 0.825). The TPC and TFC were correlated positively with quercetin (r = 0.579, r = 0.705) and kaempferol (r = 0.705, r = 0.679). The protein content was correlated with proline (r = 0.309) and asparagine (r = 0.302). The energetic value was correlated negatively with moisture content (r = −0.527), and positively with lipids (r = 0.587), sucrose (r = 0.328), MUFA (r = 0.620) and PUFA (r = 0.591).

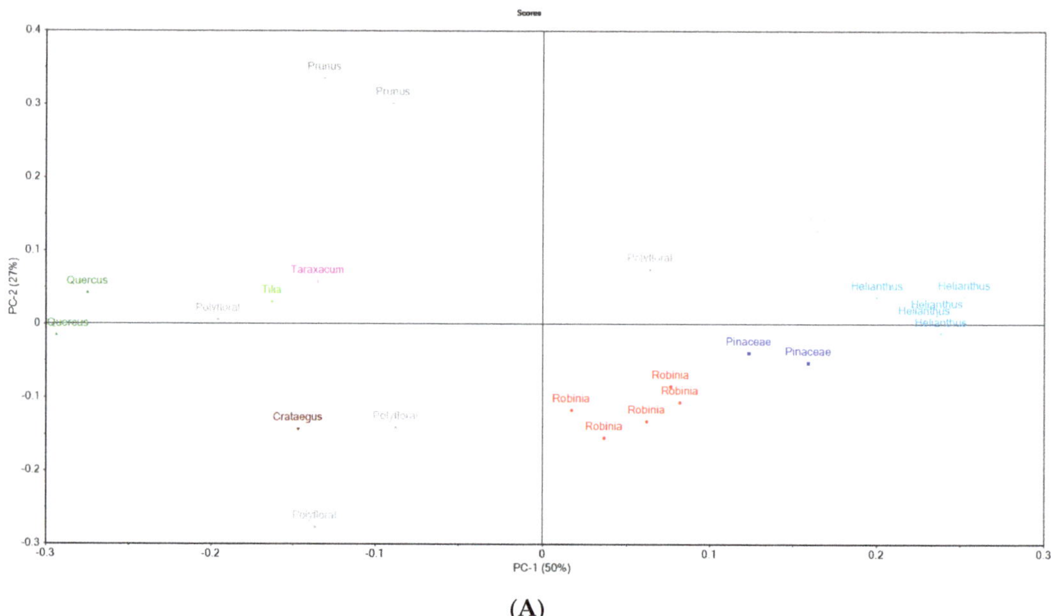

(A)

Figure 2. *Cont.*

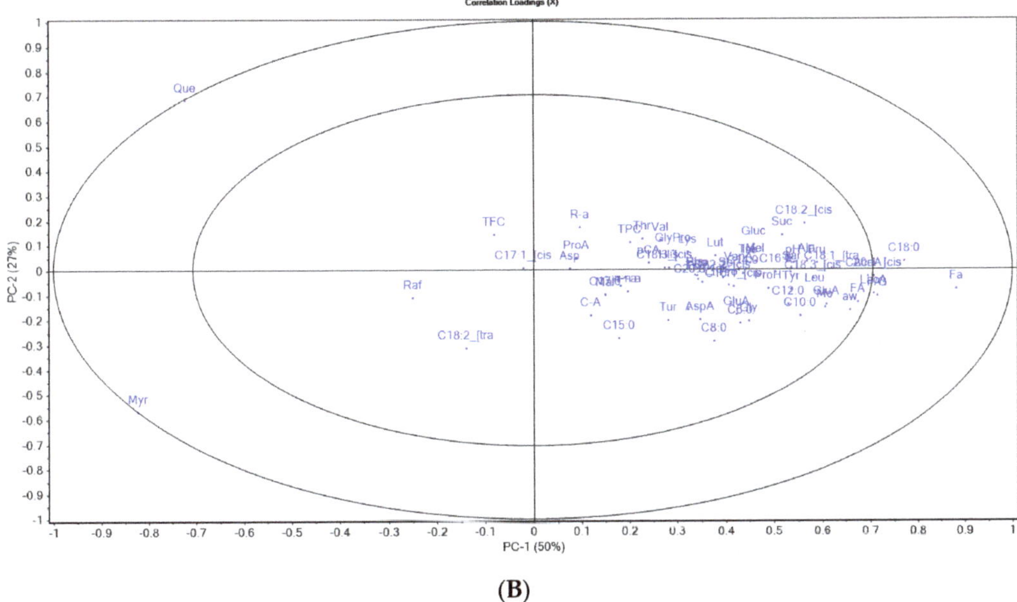

(B)

Figure 2. Principal component analysis scores (**A**) and loadings (**B**) of the bee pollen based on their physicochemical, organic acids, sugars, TPC, TFC, phenolics compounds, amino acids, fatty acids: a—scores, b—loadings, Fa-free acidity, Mo—moisture content, Fa-lipids, p-protein content, Fru-fructose, Glu-glucose, Suc-sucrose, Tur-turanose, Mal-maltose, Tre-trehalose, Mel-melesitose, Raf-raffinose, GluA-gluconic acid, LacA-lactic acid, AceA-acetic acid, ProA- propionic acid, TPC-total phenolic content, TFC-total flavone content, p-a—protocatecuic acid, 4-h-a—p-hydroxibenzoic acid, C-a—chlorogenic acid, p-c-A—p-coumaric acid, R-a—rosmarinic acid, V-A vanillic acid, Ala-alanine, Sar-Sarcosine, Gly-glicine, Val-valine, Leu-leucine, Iso—isoleucine, Thr-threonine, Ser-Serine, Pro-Proline, Asp-asparagine, AspA-aspartic acid, Phe-phenylalanine, GluA-glumatic acid, His-histidine, Lys-lysine, Tyr-tyrosine, Tryp-tryptophan.

4. Conclusions

In this study, we established that bee pollen is a rich source of protein, polyphenols, fatty acids, organic acids and amino acids. The organic acids (gluconic, lactic, acetic and propionic acids) provide antimicrobial properties for foods and are promoted as a new generation alternative to antibiotics, so bee pollen can be considered a potential preservation agent. The high protein content was observed in *Robinia pseudoacacia*, the high content of lipids was observed in *Robinia pseudoacacia* pollen, the high fructose content in *Prunus* spp. pollen while the high F/G ratio was observed in *Pinaceae* spp. pollen. The high TPC was observed in *Prunus* spp. pollen, the high TFC was observed in *Robinia pseudoacacia* pollen, the high free amino acid content was observed in *Pinaceae* spp. pollen, and the high content of PUFA was reported in *Taraxacum* spp. pollen. A total of 16 amino acids (eight essential amino acids and eight non-essential amino acids) were quantified in the bee pollen samples analyzed. Predominant were the unsaturated fatty acids (UFA; 62.65–927.50 µg/g), of which levels of 7.51–88.69 µg/g were determined for monounsaturated fatty acids (MUFAs) and 43.86–838.82 µg/g for polyunsaturated fatty acids (PUFAs). According to the data obtained, bee pollen can be considered a complex matrix with a high potential as food supplement or source of bioactive compounds for the pharmaceutical industry.

Author Contributions: Conceptualization, F.U.; data curation, F.D.; formal analysis, M.O.; funding acquisition, M.O.; investigation, M.O. and F.U.; methodology, F.D. and F.U.; project administration, M.O.; resources, F.D.; software, F.D.; validation, M.O. and F.U.; writing—original draft, F.D.; writing—review and editing, M.O. All authors have read and agreed to the published version of the manuscript.

Funding: This work was funded by Ministry of Research, Innovation and Digitalization within Program 1—Development of national research and development system, Subprogram 1.2—Institutional Performance—RDI excellence funding projects, under contract no. 10PFE/2021.

Data Availability Statement: Data is contained within the article.

Conflicts of Interest: The authors declare no conflict of interest.

References

1. Di Chiacchio, I.M.; Paiva, I.M.; de Abreu, D.J.M.; Carvalho, E.E.N.; Martínez, P.J.; Carvalho, S.M.; Mulero, V.; Murgas, L.D.S. Bee pollen as a dietary supplement for fish: Effect on the reproductive performance of zebrafish and the immunological response of their offspring. *Fish Shellfish Immunol.* **2021**, *119*, 300–307. [CrossRef] [PubMed]
2. Bayram, N.E.; Gercek, Y.C.; Çelik, S.; Mayda, N.; Kostić, A.; Dramićanin, A.M.; Özkök, A. Phenolic and free amino acid profiles of bee bread and bee pollen with the same botanical origin—Similarities and differences. *Arab. J. Chem.* **2021**, *14*, 103004. [CrossRef]
3. Alimoglu, G.; Guzelmeric, E.; Yuksel, P.I.; Celik, C.; Deniz, I.; Yesilada, E. Monofloral and polyfloral bee pollens: Comparative evaluation of their phenolics and bioactivity profiles. *LWT* **2021**, *142*, 110973. [CrossRef]
4. Kieliszek, M.; Piwowarek, K.; Kot, A.M.; Błażejak, S.; Chlebowska-Śmigiel, A.; Wolska, I. Pollen and bee bread as new health-oriented products: A review. *Trends Food Sci. Technol.* **2018**, *71*, 170–180. [CrossRef]
5. Kostić, A.; Milinčić, D.D.; Barać, M.B.; Shariati, M.A.; Tešić, Ž.L.; Pešić, M.B. The application of pollen as a functional food and feed ingredient—The present and perspectives. *Biomolecules* **2020**, *10*, 84. [CrossRef]
6. Denisow, B.; Denisow-Pietrzyk, M. Biological and therapeutic properties of bee pollen: A review. *J. Sci. Food Agric.* **2016**, *96*, 4303–4309. [CrossRef]
7. Mărgăoan, R.; Stranţ, M.; Varadi, A.; Topal, E.; Yücel, B.; Cornea-Cipcigan, M.; Campos, M.G.; Vodnar, D.C. Bee collected pollen and bee bread: Bioactive constituents and health benefits. *Antioxidants* **2019**, *8*, 568. [CrossRef]
8. Feas, X.; Vazquez-Tato, M.P.; Estevinho, L.; Seijas, J.A.; Iglesias, A. Organic bee pollen: Botanical origin, nutritional value, bioactive compounds, antioxidant activity and microbiological quality. *Molecules* **2012**, *17*, 8359–8377. [CrossRef]
9. Mayda, N.; Özkök, A.; Ecem Bayram, N.; Gerçek, Y.C.; Sorkun, K. Bee bread and bee pollen of different plant sources: Determination of phenolic content, antioxidant activity, fatty acid and element profiles. *J. Food Meas. Charact.* **2020**, *14*, 1795–1809. [CrossRef]
10. Thakur, M.; Nanda, V. Screening of Indian bee pollen based on antioxidant properties and polyphenolic composition using UHPLC-DAD-MS/MS: A multivariate analysis and ANN based approach. *Food Res. Int.* **2021**, *140*, 110041. [CrossRef]
11. Bertoncelj, J.; Polak, T.; Pucihar, T.; Lilek, N.; Kandolf Borovšak, A.; Korošec, M. Carbohydrate composition of Slovenian bee pollens. *Int. J. Food Sci. Technol.* **2018**, *53*, 1880–1888. [CrossRef]
12. Ma, L.; Wang, Y.; Hang, X.; Wang, H.; Yang, W.; Xu, B. Nutritional effect of alpha-linolenic acid on honey bee colony development (*Apis mellifera* L.). *J. Apic. Sci.* **2015**, *59*, 63–72. [CrossRef]
13. Rzepecka-Stojko, A.; Stojko, J.; Kurek-Górecka, A.; Górecki, M.; Kabała-Dzik, A.; Kubina, R.; Moździerz, A.; Buszman, E.; Iriti, M. Polyphenols from Bee Pollen: Structure, absorption, metabolism and biological activity. *Molecules* **2015**, *20*, 21732–21749. [CrossRef] [PubMed]
14. Thakur, M.; Nanda, V. Composition and functionality of bee pollen: A review. *Trends Food Sci. Technol.* **2020**, *98*, 82–106. [CrossRef]
15. Themelis, T.; Gotti, R.; Orlandini, S.; Gatti, R. Quantitative amino acids profile of monofloral bee pollens by microwave hydrolysis and fluorimetric high performance liquid chromatography. *J. Pharm. Biomed. Anal.* **2019**, *173*, 144–153. [CrossRef]
16. Nicolson, S.W.; Human, H. Chemical composition of the "low quality" pollen of sunflower (*Helianthus annuus*, Asteraceae). *Apidologie* **2013**, *44*, 144–152. [CrossRef]
17. Oroian, M.; Ursachi, F.; Dranca, F. Ultrasound-assisted extraction of polyphenols from crude pollen. *Antioxidants* **2020**, *9*, 322. [CrossRef]
18. Urcan, A.C.; Criste, A.D.; Dezmirean, D.S.; Bobiş, O.; Bonta, V.; Dulf, F.V.; Mărgăoan, R.; Cornea-Cipcigan, M.; Campos, M.G. Botanical origin approach for a better understanding of chemical and nutritional composition of beebread as an important value-added food supplement. *LWT* **2021**, *142*, 111068. [CrossRef]
19. Fatrcová-Šramková, K.; Nôžková, J.; Máriássyová, M.; Kačániová, M. Biologically active antimicrobial and antioxidant substances in the *Helianthus annuus* L. bee pollen. *J. Environ. Sci. Health-Part B Pestic. Food Contam. Agric. Wastes* **2016**, *51*, 176–181. [CrossRef]
20. Kostić, A.; Milinčić, D.D.; Gašić, U.M.; Nedić, N.; Stanojević, S.P.; Tešić, Ž.L.; Pešić, M.B. Polyphenolic profile and antioxidant properties of bee-collected pollen from sunflower (*Helianthus annuus* L.) plant. *LWT* **2019**, *112*, 108244. [CrossRef]
21. Han, L.; Liu, X.; Yang, N.; Li, J.; Cai, B.; Cheng, S. Simultaneous chromatographic fingerprinting and quantitative analysis of flavonoids in Pollen Typhae by high-performance capillary electrophoresis. *Acta Pharm. Sin. B* **2012**, *2*, 602–609. [CrossRef]

22. Gardana, C.; Del Bo, C.; Quicazán, M.C.; Corrrea, A.R.; Simonetti, P. Nutrients, phytochemicals and botanical origin of commercial bee pollen from different geographical areas. *J. Food Compos. Anal.* **2018**, *29*, 29–38. [CrossRef]
23. Arien, Y.; Dag, A.; Yona, S.; Tietel, Z.; Lapidot Cohen, T.; Shafir, S. Effect of diet lipids and omega-6:3 ratio on honey bee brood development, adult survival and body composition. *J. Insect Physiol.* **2020**, *124*, 104074. [CrossRef] [PubMed]
24. Arien, Y.; Dag, A.; Zarchin, S.; Masci, T.; Shafir, S. Omega-3 deficiency impairs honey bee learning. *Proc. Natl. Acad. Sci. USA* **2015**, *112*, 15761–15766. [CrossRef]
25. Hsu, P.S.; Wu, T.H.; Huang, M.Y.; Wang, D.Y.; Wu, M.C. Nutritive value of 11 bee pollen samples from major floral sources in taiwan. *Foods* **2021**, *10*, 2229. [CrossRef]
26. Kostić, A.; Mačukanović-Jocić, M.P.; Špirović Trifunović, B.D.; Vukašinović, I.; Pavlović, V.B.; Pešić, M.B. Fatty acids of maize pollen—Quantification, nutritional and morphological evaluation. *J. Cereal Sci.* **2017**, *77*, 180–185. [CrossRef]
27. Thakur, M.; Na, V. Assessment of physico-chemical properties, fatty acid, amino acid and mineral profile of bee pollen from India with a multivariate perspective. *J. Food Nutr. Res.* **2018**, *57*, 328–340.
28. Manning, R. Fatty acids in pollen: A review of their importance for honey bees. *Bee World* **2001**, *82*, 60–75.
29. Campos, M.G.R.; Bogdanov, S.; de Almeida-Muradian, L.B.; Szczesna, T.; Mancebo, Y.; Frigerio, C.; Ferreira, F. Pollen composition and standardisation of analytical methods. *J. Apic. Res.* **2008**, *47*, 154–161.
30. Szczesna, T.; Rybak-Chmielewska, H.; Chmielewski, W. Sugar composition of pollen loads harvested at different periods of the beekeeping season. *J. Apic. Sci.* **2002**, *46*, 107–115.
31. Antonelli, M.; Donelli, D.; Firenzuoli, F. Therapeutic efficacy of orally administered pollen for nonallergic diseases: An umbrella review. *Phytother. Res.* **2019**, *33*, 2938–2947.
32. Li, Q.Q.; Wang, K.; Marcucci, M.C.; Sawaya, A.C.H.F.; Hu, L.; Xue, X.F.; Wu, L.M.; Hu, F.L. Nutrient-rich bee pollen: A treasure trove of active natural metabolites. *J. Funct. Foods* **2018**, *49*, 472–484.
33. Louveaux, J.; Maurizio, A.; Vorwohl, G. Methods of Melissopalynology. *Bee World* **1978**, *59*, 139–157. [CrossRef]
34. AOAC International. *Official Methods of Analysis of AOAC International*, 20th ed.; AOAC: Gaithersbg, MD, USA, 2016.
35. Pauliuc, D.; Dranca, F.; Oroian, M. Antioxidant activity, total phenolic content, individual phenolics and physicochemical parameters suitability for Romanian honey authentication. *Foods* **2020**, *9*, 306. [CrossRef]
36. Palacios, I.; Lozano, M.; Moro, C.; D'Arrigo, M.; Rostagno, M.A.; Martínez, J.A.; García-Lafuente, A.; Guillamón, E.; Villares, A. Antioxidant properties of phenolic compounds occurring in edible mushrooms. *Food Chem.* **2011**, *128*, 674–678. [CrossRef]
37. Dranca, F.; Ursachi, F.; Oroian, M. Bee bread: Physicochemical characterization and phenolic content extraction optimization. *Foods* **2020**, *9*, 1358. [CrossRef]
38. Karabagias, I.K.; Karabagias, V.K.; Gatzias, I.; Riganakos, K.A. Bio-functional properties of bee pollen: The case of "bee pollen yoghurt". *Coatings* **2018**, *8*, 423. [CrossRef]
39. Fuenmayor, B.C.; Zuluaga, D.C.; Díaz, M.C.; Quicazán de C, M.; Cosio, M.; Mannino, S. Evaluation of the physicochemical and functional properties of Colombian bee pollen. *Rev. MVZ Córdoba* **2014**, *19*, 4003–4014. [CrossRef]
40. Martins, M.C.T.; Morgano, M.A.; Vicente, E.; Baggio, S.R.; Rodriguez-Amaya, D.B. Physicochemical composition of bee pollen from eleven Brazilian states. *J. Apic. Sci.* **2011**, *55*, 107–116.
41. Nogueira, C.; Iglesias, A.; Feás, X.; Estevinho, L.M. Commercial bee pollen with different geographical origins: A comprehensive approach. *Int. J. Mol. Sci.* **2012**, *13*, 11173–11187. [CrossRef]
42. Mărgăoan, R.; Al Mărghitaş, L.; Dezmirean, D.S.; Dulf, F.V.; Bunea, A.; Socaci, S.A.; Bobiş, O. Predominant and secondary pollen botanical origins influence the carotenoid and fatty acid profile in fresh honeybee-collected pollen. *J. Agric. Food Chem.* **2014**, *62*, 6306–6316. [CrossRef] [PubMed]
43. Liolios, V.; Tananaki, C.; Dimou, M.; Kanelis, D.; Goras, G.; Karazafiris, E.; Thrasyvoulou, A. Ranking pollen from bee plants according to their protein contribution to honey bees. *J. Apic. Res.* **2015**, *54*, 582–592. [CrossRef]
44. Bonvehí, J.S.; Jordà, R.E. Nutrient Composition and Microbiological Quality of Honeybee-Collected Pollen in Spain. *J. Agric. Food Chem.* **1997**, *45*, 725–732. [CrossRef]
45. Kalaycıoğlu, Z.; Kaygusuz, H.; Döker, S.; Kolaylı, S.; Erim, F.B. Characterization of Turkish honeybee pollens by principal component analysis based on their individual organic acids, sugars, minerals, and antioxidant activities. *LWT-Food Sci. Technol.* **2017**, *84*, 402–408. [CrossRef]
46. Olofsson, T.C.; Vásquez, A. Detection and identification of a novel lactic acid bacterial flora within the honey stomach of the honeybee *Apis mellifera*. *Curr. Microbiol.* **2008**, *57*, 356–363. [CrossRef]
47. De-Melo, A.A.M.; Estevinho, L.M.; Moreira, M.M.; Delerue-Matos, C.; de Freitas, A.d.S.; Barth, O.M.; de Almeida-Muradian, L.B. Phenolic profile by HPLC-MS, biological potential, and nutritional value of a promising food: Monofloral bee pollen. *J. Food Biochem.* **2018**, *42*, e12536. [CrossRef]
48. De-Melo, A.A.M.; Estevinho, L.M.; Moreira, M.M.; Delerue-Matos, C.; de Freitas, A.d.S.; Barth, O.M.; de Almeida-Muradian, L.B. A multivariate approach based on physicochemical parameters and biological potential for the botanical and geographical discrimination of Brazilian bee pollen. *Food Biosci.* **2018**, *25*, 91–110. [CrossRef]
49. Taha, E.K.A.; Al-Kahtani, S.; Taha, R. Protein content and amino acids composition of bee-pollens from major floral sources in Al-Ahsa, eastern Saudi Arabia. *Saudi J. Biol. Sci.* **2019**, *26*, 232–237. [CrossRef]
50. Al-Kahtani, S.N.; Taha, E.K.A.; Farag, S.A.; Taha, R.A.; Abdou, E.A.; Mahfouz, H.M. Harvest Season Significantly Influences the Fatty Acid Composition of Bee Pollen. *Biology* **2021**, *10*, 495. [CrossRef]

51. Mărgăoan, R.; Özkök, A.; Keskin, Ş.; Mayda, N.; Urcan, A.C.; Cornea-Cipcigan, M. Bee collected pollen as a value-added product rich in bioactive compounds and unsaturated fatty acids: A comparative study from Turkey and Romania. *LWT* **2021**, *149*, 111925. [CrossRef]
52. Depciuch, J.; Kasprzyk, I.; Drzymała, E.; Parlinska-Wojtan, M. Identification of birch pollen species using FTIR spectroscopy. *Aerobiologia* **2018**, *34*, 525–538. [CrossRef] [PubMed]
53. Anjos, O.; Santos, A.J.A.; Dias, T.; Estevinho, L.M. Application of FTIR-ATR spectroscopy on the bee pollen characterization. *J. Apic. Res.* **2017**, *56*, 210–218. [CrossRef]
54. Bouknana, D.; Hammouti, B.; Jodeh, S.; Sbaa, M.; Lgaz, H. Extracts of olive inflorescence flower pre-anthesis, at anthesis and grain pollen as eco-friendly corrosion inhibitor for steel in 1M HCl medium. *Anal. Bioanal. Electrochem.* **2018**, *10*, 751–777.
55. Kendel, A.; Zimmermann, B. Chemical analysis of pollen by FT-Raman and FTIR spectroscopies. *Front. Plant Sci.* **2020**, *11*, 352. [CrossRef] [PubMed]
56. Kasprzyk, I.; Depciuch, J.; Grabek-Lejko, D.; Parlinska-Wojtan, M. FTIR-ATR spectroscopy of pollen and honey as a tool for unifloral honey authentication. The case study of rape honey. *Food Control* **2018**, *84*, 33–40. [CrossRef]
57. Subari, N.; Mohamad Saleh, J.; Md Shakaff, A.; Zakaria, A. A Hybrid Sensing Approach for Pure and Adulterated Honey Classification. *Sensors* **2012**, *12*, 14022–14040. [CrossRef]

 *foods*

Article

Allergenicity Alleviation of Bee Pollen by Enzymatic Hydrolysis: Regulation in Mice Allergic Mediators, Metabolism, and Gut Microbiota

Yuxiao Tao [1,†], Enning Zhou [1,†], Fukai Li [2], Lifeng Meng [1], Qiangqiang Li [1,*] and Liming Wu [1]

1   Institute of Apicultural Research, Chinese Academy of Agricultural Sciences (CAAS), Beijing 100093, China
2   Key Laboratory of Agro-Product Quality and Safety, Institute of Quality Standards and Testing Technology for Agro-Products, Chinese Academy of Agricultural Sciences (CAAS), Beijing 100081, China
*   Correspondence: qiangqiangli1991@163.com; Tel.: +86-132-6949-5300
†   These authors contributed equally to this work.

Abstract: Bee pollen as a nutrient-rich functional food has been considered for use as an adjuvant for chronic disease therapy. However, bee pollen can trigger food-borne allergies, causing a great concern to food safety. Our previous study demonstrated that the combined use of cellulase, pectinase and papain can hydrolyze allergens into peptides and amino acids, resulting in reduced allergenicity of bee pollen based on in vitro assays. Herein, we aimed to further explore the mechanisms behind allergenicity alleviation of enzyme-treated bee pollen through a BALB/c mouse model. Results showed that the enzyme-treated bee pollen could mitigate mice scratching frequency, ameliorate histopathological injury, decrease serum IgE level, and regulate bioamine production. Moreover, enzyme-treated bee pollen can modulate metabolic pathways and gut microbiota composition in mice, further supporting the alleviatory allergenicity of enzyme-treated bee pollen. The findings could provide a foundation for further development and utilization of hypoallergenic bee pollen products.

Keywords: bee pollen; enzyme-treatment; allergenicity alleviation; metabolism; gut microbiota

Citation: Tao, Y.; Zhou, E.; Li, F.; Meng, L.; Li, Q.; Wu, L. Allergenicity Alleviation of Bee Pollen by Enzymatic Hydrolysis: Regulation in Mice Allergic Mediators, Metabolism, and Gut Microbiota. *Foods* **2022**, *11*, 3454. https://doi.org/10.3390/foods11213454

Academic Editor: M. Carmen Seijo

Received: 28 September 2022
Accepted: 28 October 2022
Published: 31 October 2022

Publisher's Note: MDPI stays neutral with regard to jurisdictional claims in published maps and institutional affiliations.

1. Introduction

Bee pollen as a natural food source is regarded as an excellent nutritional supplement for human consumption. It contains a variety of nutrients including proteins, carbohydrates, lipids, polyphenols, and many other nutrients [1]. Bee pollen also displays a variety of beneficial health properties, such as antioxidant, antibacterial, hepatoprotective, and cardioprotective activities [2,3]. By 2024, the global bee pollen market is expected to reach a value of USD 720 million according to Marketwatch.com. Global bee pollen consumption shows an upward trend with increasing numbers of consumers regarding bee pollen as a nutritional supplement. However, bee pollen consumption can cause a number of clinical allergic symptoms in certain individuals with allergies [4,5]. In addition, bee pollen can induce cross-allergic reactions when consumed with other foods [6–8]. The potential allergenicity of bee pollen has become one of the key issues limiting the development and utilization of bee pollen. Developing an efficient approach for reducing the allergenicity of bee pollen is thus necessary for expanding its further utilization.

Enzyme treatment is a critical food processing technique that can increase the nutritional value and reduce the allergenicity of foods. Furthermore, this technique is low cost and highly efficient [9]. Currently, α-chymotrypsin, trypsin and flavourzyme are widely used to reduce the allergenicity of certain allergenic foods, such as peanuts [10,11], soybean [11,12], and wheat [13]. Enzyme treatment has also been applied to break down the pollen wall and release nutrients contained within [14]. In our previous study, cellulase, pectinase and papain were combined and used for allergen degradation into small peptides and amino acids, resulting in decreased bee pollen allergenicity based on in vitro assays [15]. In this study, we aim to further clarify the mechanisms of allergenicity alleviation of enzyme-treated bee pollen on the changes in serum allergic mediators, metabolic pathways, and gut microbiota composition.

Immunoglobulin (Ig) E-mediated type I hypersensitivity accounts for the vast majority of food allergies [16]. As a high-IgE response strain, the BALB/c mouse is suitable for IgE-mediated food allergy research, such as egg- [17,18], milk- [19], fish- [20] and peanut-induced [21,22] hypersensitive reactions. Recently, BALB/c mice were adopted to study the pathogenesis of allergy syndrome caused by oral pollen [23], providing the basis for the selection of an in vivo model in our study. Additionally, food allergies can activate the release of cytokines such as interleukin(IL)-4, IL-5, IL-13, and other mediators that can induce the production of IgE antibody in B cells [16]. Food allergies can also cause changes in host metabolism and gut microbiota. Some studies have applied metabolomics to investigate IgE-mediated food allergies [24–26]. A close association between food allergies and the dysbiosis of gut microbiota was also proposed in numerous studies [27–29]. The immune system can be influenced by host metabolic disorders and gut microbiota dysbiosis, although the related mechanism remains unclear [30].

To further clarify the mechanisms behind allergenicity alleviation of enzyme-treated bee pollen, the BALB/c mouse model was used to investigate the regulatory effects of enzyme-treated bee pollen on the production of serum allergic mediators, changes in metabolic pathways, and gut microbiota composition. Bee pollen samples were treated with a combination of cellulase, pectinase and papain, and mice were fed enzyme-treated and non-enzyme-treated bee pollen, respectively. The scratching behavior and histopathological injury were evaluated to identify the allergic state in mice. Subsequently, the mice serum was collected for allergic mediator and metabolite assays. Mice fecal DNA was extracted for gut microbiota composition analysis. Our findings might provide a basis for further development and utilization of hypoallergenic bee pollen products.

2. Materials and Methods

2.1. Reagents and Apparatus

Imject^{TM} Alum Adjuvant (No. 77161) was obtained from Thermo Scientific Inc. (Pittsburgh, PA, USA). Ovalbumin (OVA) was obtained from Sigma-Aldrich Inc. (Saint Louis, MO, USA). BCA protein assay kit was obtained from Beyotime Biotechnology Co., Ltd. (Shanghai, China). Goat anti-mouse IgE antibody and HRP-labeled donkey anti-goat IgG antibody were obtained from Abcam Inc. (Cambridge, UK). DAB peroxidase substrate kit was purchased from Solarbio Co., Ltd. (China). The HPLC-grade acetonitrile, methanol, ammonium formate and formic acid were obtained from Fisher Scientific (Pittsburgh, PA, USA). Ultrapure water was collected from Millipore Milli-Q system (Bedford, MA, USA). Cellulase (400 U/mg), pectinase (500 U/mg) and papain (800 U/mg) were obtained from Yuanye Bio-Technology Co., Ltd. (Shanghai, China). Ultrafiltration centrifugal tube (15 mL, 10 kDa) was purchased from Millipore Inc. (Bedford, MA, USA). Other reagents were purchased from Sigma-Aldrich Inc. (Saint Louis, MO, USA).

2.2. Enzymatic Treatment of Bee Pollen

Bee pollen samples (composed of more than 92% *Brassica campestris* pollen according to palynological counting) were collected from the beekeeping base of the Institute of Apicultural Research (IAR), Chinese Academy of Agricultural Sciences (CAAS). The sample was lyophilized after grinding into powder, and then sterilized by irradiation at 7 kGy. Cellulase, pectinase and papain were used for enzymatically treating bee pollen as described in our previous study [15]. In brief, 10 mL of Millipore water and 5 g of bee pollen powder were combined and vortexed for 5 min. For two kinds of enzyme-treated bee pollen (2E-BP) groups, 3000 U cellulase and 3000 U pectinase were added into the sample. For three kinds of enzyme-treated bee pollen (3E-BP) groups, 3000 U cellulase, 3000 U pectinase and 3000 U papain were added into the sample. All samples were adjusted to pH 4.0 with vitamin C solution, and then incubated at 45 °C for 24 h. A vacuum freeze dryer was used for sample lyophilization, and samples were stored at $-80\,^{\circ}$C for further study.

2.3. Protein Sample Preparation

Bee pollen protein was extracted using water and 5% NaCl solution under ultrasonication for 45 min, respectively. The mixture was centrifuged at 3,500 g for 15 min. The supernatant was then filtered with a nylon membrane (22 µm). The filtered solution was subsequently ultrafiltered using a Millipore ultrafiltration tube (15 mL, 10 kDa) at 3500 g for 30 min. The protein concentration of the ultrafiltration retentate was detected using a BCA protein assay kit.

2.4. Animal Experiment

The animal experiments were conducted in Beijing Animal Experimental Center. The Animal Ethics Committee of IAR, CAAS (Beijing, China) provided approval for the animal experimentation. The registration number is CAAS-IAR-ER0046. Female BALB/c mice (6-week-old, 16–20 g) were kept in cages, housed in a 12-h light–dark cycle at 20–24 $^{\circ}$C and a relative humidity of $50 \pm 5\%$, and supplied with filtered pathogen-free air, standard AIN-93 laboratory diet and sterile water. After acclimation for one week, all mice were randomly divided into five groups (n = seven per group), and named as CK, OVA, BP, 2E-BP and 3E-BP. The mice administration procedure is shown in Figure 1A. Briefly, mice in all groups were given a standard diet and water. For the CK group, mice were intraperitoneally injected with a saline solution every seven days for 28 days; but for the OVA group (as a positive control), mice were injected intraperitoneally with 0.2 mL of 0.1 mg/mL OVA solution (containing 1% Alum Adjuvant) every seven days for 21 days, and finally injected with 0.2 mL of 0.5 mg/mL OVA solution (containing 1% Alum Adjuvant) on the 28th day; for BP, 2E-BP and 3E-BP groups, mice were injected intraperitoneally with 0.2 mL of 0.5 mg/mL BP, 2E-BP and 3E-BP solution (containing 1% Alum Adjuvant) every seven days for 21 days, respectively, and finally injected with 0.2 mL of 2.5 mg/mL BP, 2E-BP and 3E-BP solution (containing 1% Alum Adjuvant) on the 28th day, respectively. The frequency of scratching behavior was observed and recorded 15 min after final injection. Mice were sacrificed after intraperitoneal injection with an excitation dose of samples for 30 min on the 28th day. The blood was extracted from eyes, and then centrifuged at 3000 rpm for 10 min for serum collection.

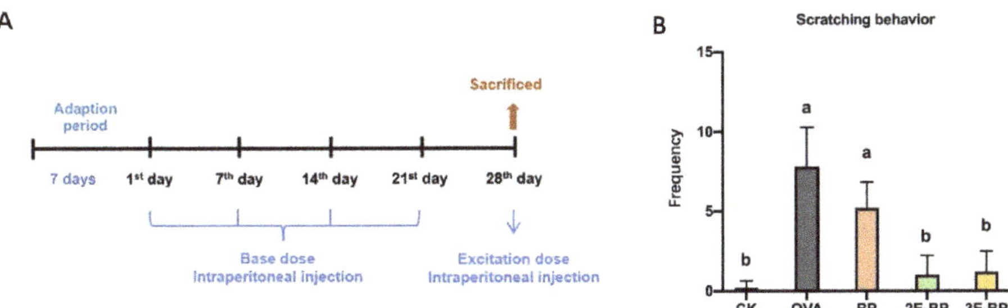

Figure 1. BALB/c mice administration schedule. (**A**) Scratching frequency of mice in different treatment groups. (**B**) Different letters indicate a significant difference among different groups ($p < 0.05$).

2.5. Histopathological Testing

The formalin-fixed spleen tissue was embedded in paraffin. Next, the spleen tissue was sliced into sections and stained with toluidine blue (TB) and GIMSA separately. The sections were observed using a Nikon Eclipse Ci microscope (Tokyo, Japan).

2.6. Detection of Allergy Indexes in Mice Serum

2.6.1. Detection of IgE Antibody in Mice Serum

The protein concentration in mice serum was detected by BCA protein assay kit. The serum (containing 40 µg protein) was added onto a nitrocellulose membrane (4 cm × 4 cm). After being dried, the membrane was incubated with 5% (*w/v*) BSA solution at 25 ± 2 °C for 1 h, and then incubated in 5% BSA solution containing 1:2000 diluted goat anti-mouse IgE antibody for 1 h. After washing the membrane with TBS-T solution, it was then incubated in 5% BSA solution containing 1:1500 diluted HRP-labeled donkey anti-goat IgG antibody for 1 h. After washing the membrane with TBS-T solution, the spots on the membrane were developed using a DAB color developing kit. Finally, the dried and developed membrane was imaged using an HP scanner, and the relatively quantitative analysis of dot intensity on the membrane was processed using Image J (Version 1.53).

2.6.2. Bioamine Detection via UPLC-QQQ-MS/MS

To detect the changes of bioamines in mice serum, 50 µL serum was mixed with 200 µL methanol, and then centrifuged at 13,000 g for 15 min. The supernatant was filtered with a nylon membrane (22 µm). Samples were separated using an Agilent 1290 Infinity II series UPLC system equipped with an Agilent Zobax Eclipse C18 Rapid Resolution HD column (2.1 mm × 100 mm, 1.8 µm). Mobile phases A and B were water (containing 2 mM ammonium formate and 0.1% *v/v* formic acid) and methanol, respectively. The gradient was set as follows: 1 min, 2% B; 4 min, 15% B; 4.5 min, 98% B; 6 min, 98% B; 6.1 min, 2% B; post time 3 min, 5% B. An Agilent 6470 ESI-QQQ system was adopted for MS acquisition. Mass spectrometry parameters were set as follows: 250 °C, gas temperature; 7 L/min, drying gas flow rate; 30 psi, nebulizer pressure; 325 °C, sheath gas temperature; 11 L/min sheath gas flow rate. The precursor ion, product ion, collision energy and fragmentor are listed in Table S1 in Supplementary Materials.

2.6.3. Metabolomics Analysis via UPLC-QTOF-MS/MS

To determine the changes of metabolites in mice serum, 50 µL serum was mixed with 200 µL methanol, and centrifuged at 13,000 g for 15 min. The supernatant was extracted and filtered with a nylon membrane (22 µm). Sample separation was accomplished using a UPLC system of Agilent 1290 Infinity II series. The UPLC system was equipped with an Agilent Eclipse Plus C18 Rapid Resolution HD column (2.1 mm × 100 mm, 1.8 m). The mobile phases and gradient were set as described [31]. For MS acquisition, an Agilent 6545 ESI-Q-TOF mass spectrometer was employed. The mass spectrum parameters were the same as our previous publication [31]. Reference ions 112.985587 and 1033.988109 were used for real-time calibration during acquisition in negative ionization mode. Metabolites were identified with the METLIN Database (DB). Metabolites with DB scores above 80 and mass error lower than 5 ppm (0.0005%) were screened as biomarkers for statistical analysis. Metabolic pathway analysis was conducted using MetaboAnalyst 4.0 and KEGG online platform.

2.7. Microbiota Analysis

The fecal DNA of mice was extracted using a commercial E.Z.N.A.® DNA kit from Omega Bio-Tek Inc. (Norcross, GA, USA). DNA concentration and purity were determined via NanoDrop2000 from Thermo Fisher Scientific Inc. (Pittsburgh, PA, USA). The 1% agarose gel electrophoresis was used for DNA quality assay. A thermocycler PCR system (ABI GeneAmp® 9700, Thermo Fisher Scientific, Waltham, MA, USA) was adopted for the amplification of 16 S rRNA gene in the hypervariable region of bacterial V3-V4, conducting with the universal primers 338F (5′-ACTCCTACGGGA GGCAGCAG-3′) and 806R (5′-GGACTACHVGGGTWTCTAAT-3′). An AxyPrep DNA gel extraction kit (Axygen Inc., Corning, NY, USA) was applied to purify the amplified products, and a Quantus™ fluorometer (Promega Inc., Madison, WI, USA) was used for the quantification of the purified products. An Illumina MiSeq System (Illumina Inc., San Diego, CA, USA) was adopted for the paired-end sequencing of amplicons. The raw Illumina data was quality-filtered and merged using FASTQ version 0.20.0 software and FLASH version 1.2.7 software, respectively. Sequences with more than 97% similarity were clustered into the same amplicon sequence variants (ASVs) by DADA2 plugin of Qiime2 version 2020.2 software. The ASVs taxonomic assignments were performed using the naïve Bayes consensus taxonomy classifier based on SILVA 16S rRNA database (v 138). The analysis of 16S rRNA microbiome sequencing data was conducted by bioinformatic tools on Majorbio i-Sanger cloud platform (http://en.majorbio.com). The alpha-diversity and beta-diversity analysis, Kruskal–Wallis H test, and linear discriminant analysis effect size (LEfSe) analysis were performed based on the Majorbio i-Sanger cloud platform.

2.8. Data Analysis

The variance t-test and ANOVA analysis were performed using SPSS version 21.0 software at a 95% confidence level. The Multi-experiment viewer (MEV) version 4.9 software was applied to heat-map analysis.

3. Results and Discussion

3.1. Enzyme-Treated Bee Pollen Alleviates Mice Scratching Behavior

The itch-associated response is a typical allergic reaction and can be used as an indicator for the evaluation of anaphylaxis levels [32,33]. To reflect the level of anaphylaxis, scratching frequency of mice was recorded. As shown in Figure 1B, the mice in OVA group exhibited the highest scratching frequency after injection. There was no significant difference in the scratching frequency of mice between OVA and BP groups. However, the mice in 2E-BP and 3E-BP groups exhibited significantly less scratching frequency than the mice in OVA and BP groups. This indicated that enzyme-treated bee pollen alleviates mice scratching behavior, and that enzymatic treatment can reduce the allergenicity of bee pollen.

3.2. Enzyme-Treated Bee Pollen Mitigates Histopathological Injury in Mice

Mast cells and granulocytes play important roles in food allergy. Specifically, FcεRI as a kind of IgE receptor existing in mast cells and basophils can be activated by crosslinking with allergen-specific IgE antibodies, leading to the release of allergic mediators responsible for the early- and late-phase of allergic reactions [16,34,35]. Herein, the mast cells and granulocytes in mice spleen were visualized by TB and GIEMSA staining, respectively. As shown in Figure 2, there are significantly more mast cells and granulocytes (marked with red arrows) in OVA and BP groups than in CK group, while there is no significant difference in the 2E-BP and 3E-BP groups compared with the CK group. This revealed that enzyme-treated bee pollen mitigates histopathological injury in mice.

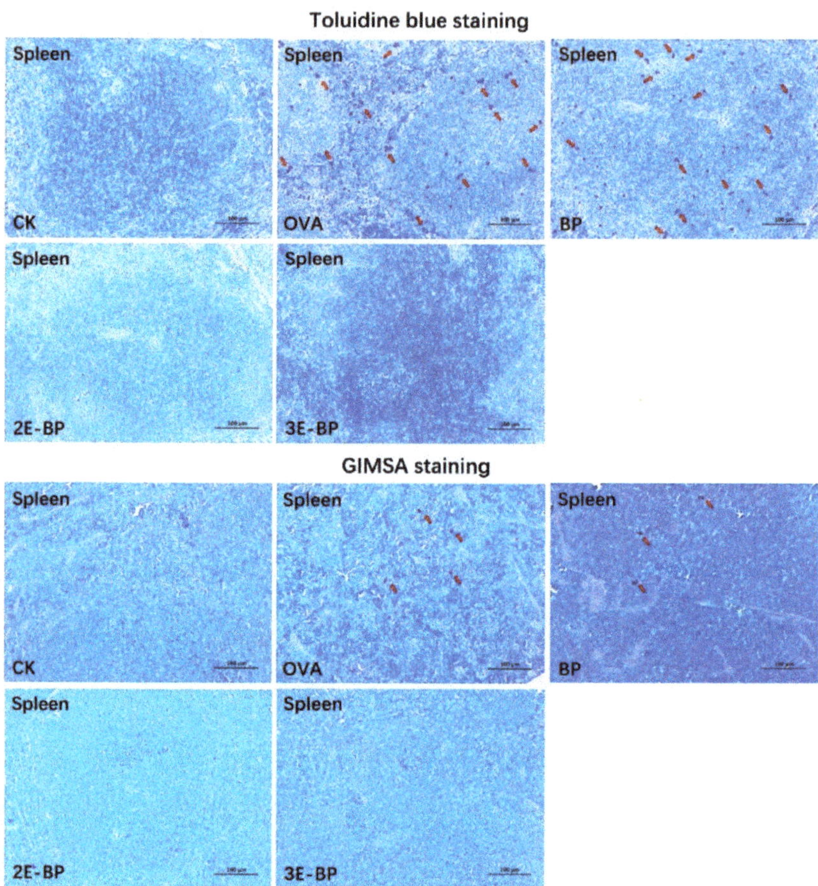

Figure 2. Histopathological changes in mouse spleen in different treatment groups. Mast cells and granulocytes stained separately with toluidine blue (TB) and GIMSA are marked with red arrows.

3.3. Enzyme-Treated Bee Pollen Decreases the Production of IgE Antibodies in Mice Serum

Generally, food allergy is an IgE-mediated type I hypersensitivity, which can induce immune cells to produce IgE antibodies [36–38]. Dot-blot assay was used to semi-quantitatively analyze the level of IgE antibodies in mice serum. As shown in Figure 3A, OVA and BP treatment induced a significant increase in mice serum IgE levels. However, mice serum IgE levels in the 2E-BP and 3E-BP groups exhibited a notable decrease compared with that in BP group. This suggested that enzyme-treated bee pollen could decrease the production of IgE antibodies in mice serum due to a reduction in allergic reactions.

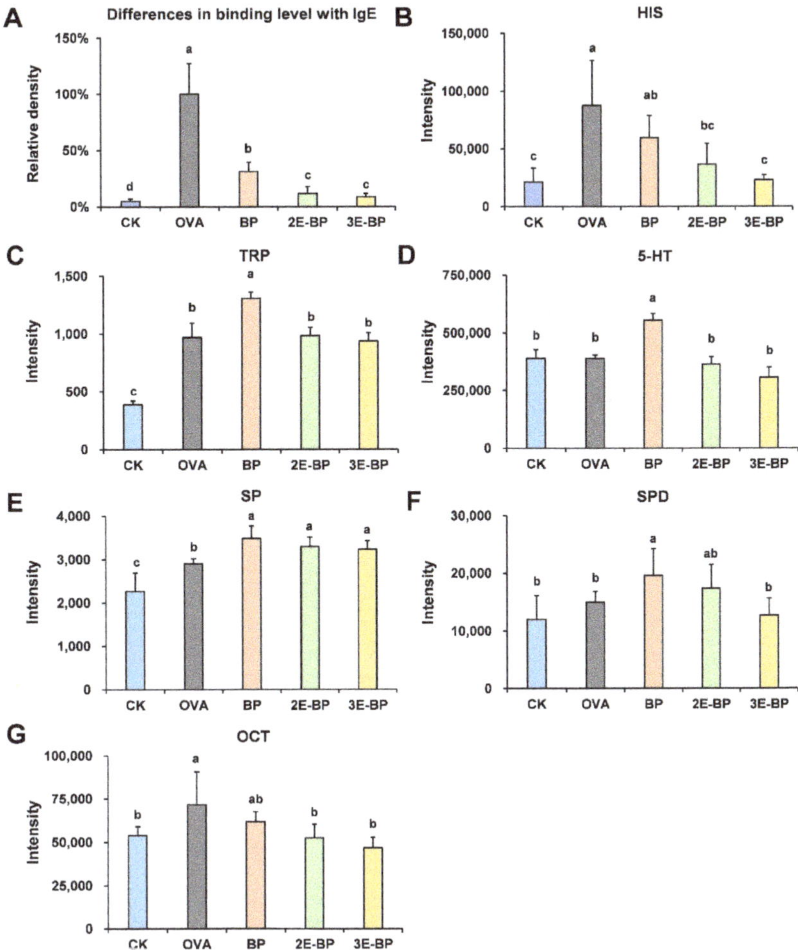

Figure 3. Changes in allergic mediators including IgE antibody (**A**), histamine (**B**), tryptamine (**C**), serotonin (**D**), spermine (**E**), spermidine (**F**) and octopamine (**G**) in mice serum of different treatment groups. Different letters indicate a significant difference among different groups ($p < 0.05$).

3.4. Enzyme-Treated Bee Pollen Modulates the Bioamine Level in Mice Serum

Serum bioamines are critical indicators that reveal the allergic status of the body. The UPLC-QQQ-MS/MS technique was conducted to discover the changes in levels of bioamines in mice serum. Histamine (HIS) is a key allergic mediator released by mast cells and basophils, which can regulate T helper (Th) lymphocytes to produce inflammatory cytokines [39,40]. As shown in Figure 3B, the HIS level in the OVA and BP groups was significantly higher than that in the CK group; while there was no significant difference among the CK, 2E-BP and 3E-BP groups. Regulating T lymphocytes (Treg) contributes to the acquisition of allergy tolerance [41]. Tryptamine (TRP) and 5-Hydroxymethyltryptamin (5-HT) can interact with immune cells, triggering the conversion of Tregs to Th17 cells [42]. As shown in Figure 3C,D, both the levels of TRP and 5-HT were significantly higher in the BP group than that in the CK, 2E-BP and 3E-BP groups, suggesting that 2E-BP and 3E-BP could alleviate allergic reactions in mice. Spermine (SP) and spermidine (SPD) have been reported to provide protective effects by inhibiting the development of allergic asthma [43]. As shown in Figure 3E,F, the level of SP in BP, 2E-BP and 3E-BP groups was significantly higher than that in the CK and OVA groups; and the level of SPD in BP and 2E-BP groups was significantly higher than that in the CK, OVA and 3E-BP groups. This was attributed to the fact that bee pollen contains a certain amount of SP and SPD which can increase their levels in serum after ingestion [44]. Octopamine (OCT) is also one of the allergic mediators with proinflammatory effects to the body [45]. As shown in Figure 3G, the OVA and BP groups exhibited a higher level of OCT than the CK, 2E-BP and 3E-BP groups, indicating the alleviatory effect of 2E-BP and 3E-BP on food allergy.

3.5. Enzyme-Treated Bee Pollen Regulates Metabolism in Mice Serum

Food allergy can induce metabolic disorders. Herein, the metabolomics analysis of mice serum was performed to explore the metabolism changes in mice serum. The metabolites with significant changes ($p < 0.05$; Fold Change > 2) among different groups were screened and enriched into corresponding pathways. As shown in Figure 4, the contents of (R)-3-hydroxybutanoic acid and abietic acid in CK, 2E-BP and 3E-BP groups were significantly higher than that in the OVA and BP groups; while the content of cholesterol sulfate in BP, 2E-BP and 3E-BP groups was notably lower than that in the CK and OVA groups. All of these significantly changed metabolites were enriched in two main metabolic pathways: (1) Steroid biosynthesis; (2) Butanoate metabolism.

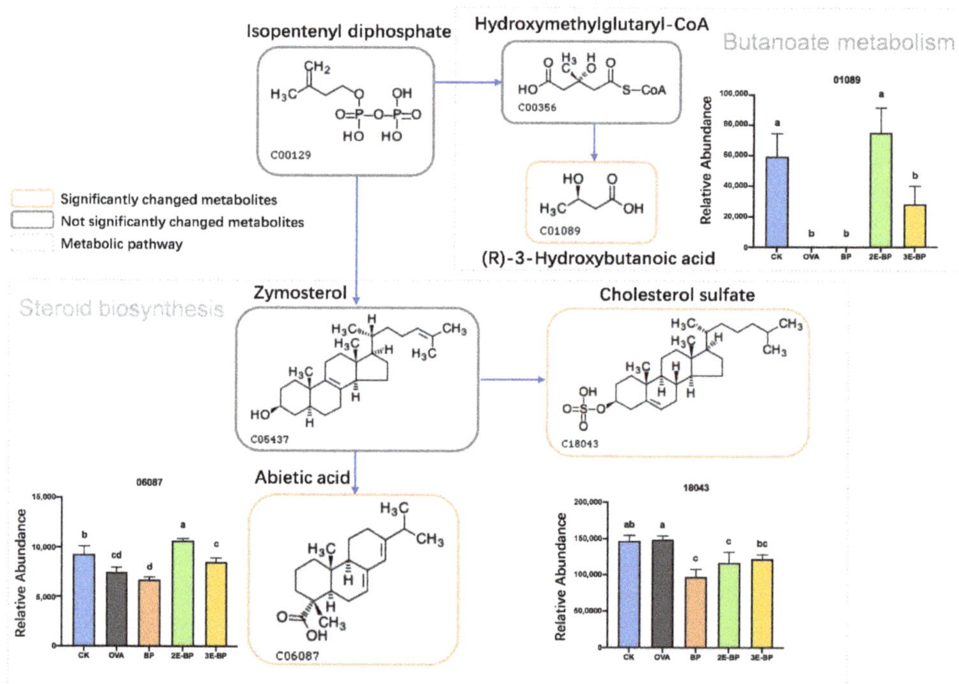

Figure 4. The changed metabolites and involved metabolic pathways in mice serum of different treatments groups. Different letters indicate a significant difference among different groups ($p < 0.05$).

As reported, steroid hormones exert various immunologic functions, for instance, steroid hormones can alleviate the clinical symptoms of allergic asthma [46,47]. Additionally, steroid hormones can contribute to the production of specific T cells, thereby exhibiting anti-inflammatory effects [48]. Butanoate metabolism was also closely associated with immune system function, for instance, butyrate helps enterocytes maintain their functionality and the integrity of intestinal mucosa, thereby preventing inflammation caused by pathogens [49]. Moreover, butyrate can reduce enterocyte inflammation by defending against oxidative stresses [50]. Moreover, the reduction of butyrate caused by the imbalance and dysfunction of gut microbiota leads to the aggravation of allergic reactions [51]. Therefore, enzyme-treated bee pollen might reduce the allergic reactions by regulating steroid biosynthesis and butanoate metabolism in mice serum.

3.6. Enzyme-Treated Bee Pollen Regulates the Composition of Gut Microbial Structures

Furthermore, food allergy can induce an imbalance of gut microbiota associated with the immune system. As shown in Figure 5A,C, the Ace, Chao and Shannon indices were higher in the BP, 2E-BP and 3E-BP groups than that in the OVA groups, indicating that bee pollen can increase α-diversity of gut microbiota in mice. As shown in Figure 5D,E, 2E-BP and 3E-BP groups were well-separated with OVA and BP groups, suggesting the significant changes in β-diversity of gut microbiota. Additionally, clustering analysis showed that the OVA and BP groups had similar microbial structures, while the CK, 2E-BP and 3E-BP groups exhibited analogous microbial structures (Figure 6A). In consideration of the above findings, a severe allergy was induced by OVA and BP but was alleviated by 2E-BP and 3E-BP, thereby leading to a similar microbial structure among CK, 2E-BP and 3E-BP groups. Kruskal–Wallis H test and LEfSe analysis showed that *Lachnospiraceae*, *Marinifilaceae* and *Helicobacteraceae* were significantly more abundant in the CK, 2E-BP and 3E-BP groups than in the OVA and BP groups; while, the abundance of *Bacillaceae* and *Akkermansiaceae* was significantly lower in the CK, 2E-BP and 3E-BP groups than in the OVA and BP groups (Figure 6B,C).

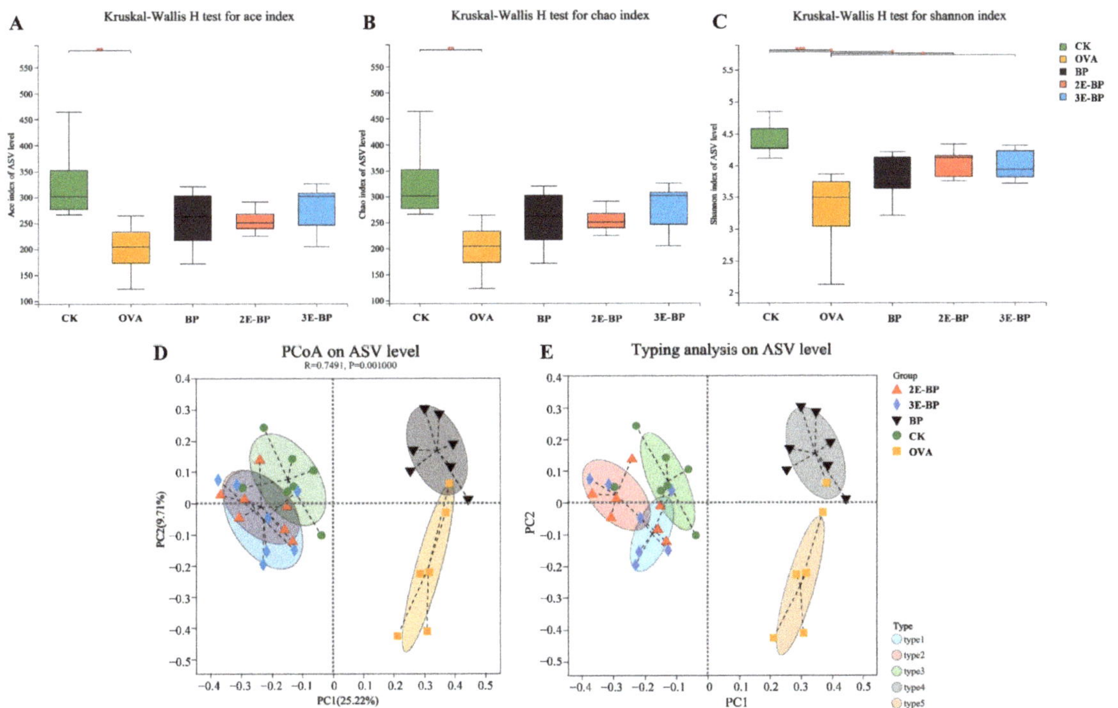

Figure 5. The alpha-diversity and beta-diversity analysis of different samples at ASV level, including (**A**) Ace index analysis; (**B**) Chao index analysis; (**C**) Shannon index analysis; (**D**) PCoA analysis; (**E**) Typing analysis. Asterisk * represents $p < 0.05$, ** represents $p < 0.01$, and *** depicts $p < 0.001$.

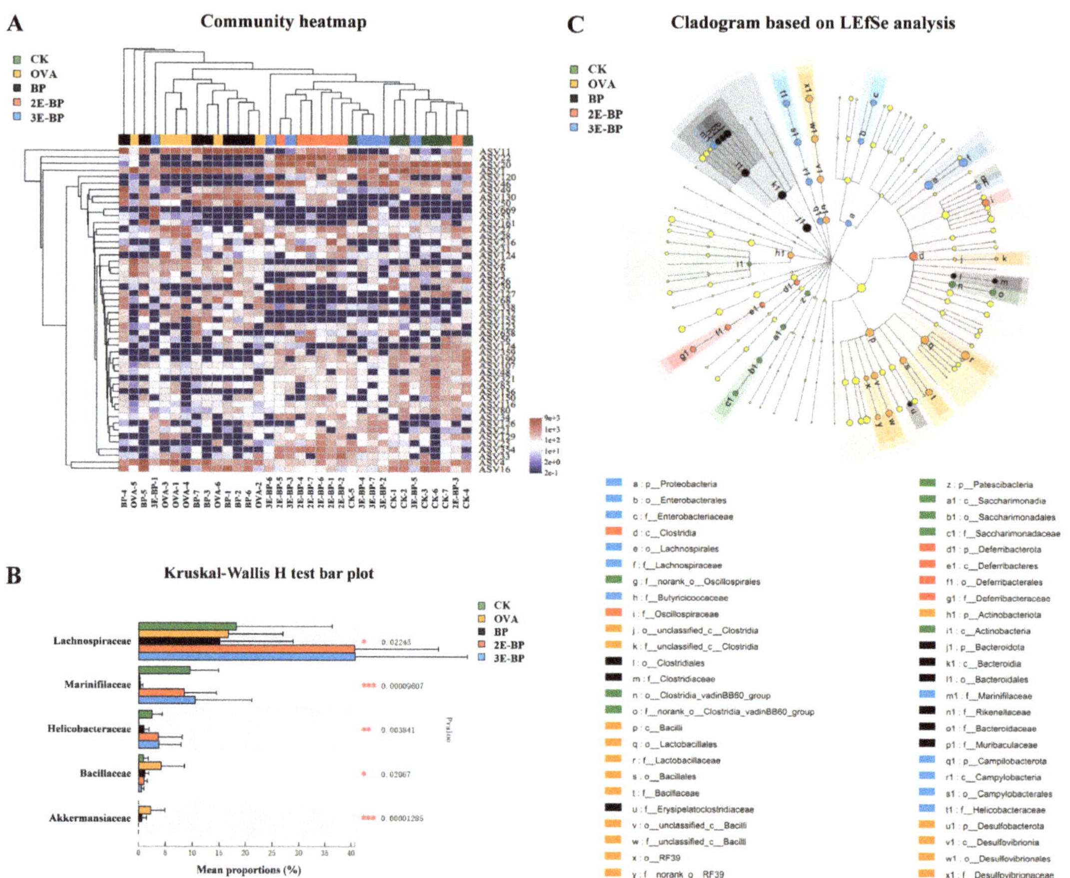

Figure 6. Heatmap analysis of gut microbiota community at ASV level (**A**). Kruskal–Wallis H test analysis of gut microbiota at family level (**B**). Cladogram based on LEfSe analysis of gut microbiota from phylum to class level (**C**). Asterisk * represents $p < 0.05$, ** represents $p < 0.01$, and *** depicts $p < 0.001$.

As reported, *Lachnospiracea* could be involved in food allergies [52]. Its abundance was significantly increased after the allergic mice received an allergen-specific Treg cell therapy compared with the no-treatment group [53]. *Lachnospiraceae* level was also significantly reduced in the gut microbiota of allergy sufferers compared with healthy population [54]. *Marinifilaceae* can be affected by intenstinal inflammation. Its abundance was reduced in the gut microbiota of colitis mice but recovered following anti-inflammatory therapy [55]. *Helicobacteraceae* shows beneficial effects against food allergy, for instance, the neutrophil-activating protein of *Helicobacter pylori* can inhibit peanut allergy by up-regulating the production of Tregs [56]. The level of *Bacillaceae* is related to gut inflammatory diseases [57]. It presented a higher level in the gut microbiota of Crohn's disease patients compared with healthy people [58]. Additionally, the level of *Akkermansiaceae* increased due to inflammatory gut injury and other gastorintestinal diseases [59]. Therefore, our findings correspond with those of previous reports. Enzyme-treated bee pollen can alleviate allergic reactions and regulate the composition of microbial gut structures.

4. Conclusions

Overall, in comparison with natural bee pollen, enzyme-treated bee pollen can reduce mice scratching frequency, spleen pathological injury, and serum IgE production. Moreover, it can additionally regulate bioamine serum levels, as well as modulate steroid biosynthesis and butanoate metabolism. Further, enzyme-treated bee pollen can regulate the composition of gut microbial structures by increasing the abundance of *Lachnospiraceae*, *Marinifilaceae* and *Helicobacteraceae*, while decreasing the abundance of *Bacillaceae* and *Akkermansiaceae*, which are involved in allergenicity mitigation. The findings suggest that enzymatic treatment has an alleviatory effect on the allergenicity of bee pollen, and provide a scientific basis for further development and utilization of hypoallergenic bee pollen products.

Supplementary Materials: The following supporting information can be downloaded at: https://www.mdpi.com/article/10.3390/foods11213454/s1, Table S1. The parameters for bioamines detection by UPLC-QQQ-MS/MS.

Author Contributions: Conceptualization, Q.L. and L.W.; Data curation, F.L. and L.M.; Formal analysis, Y.T. and E.Z.; Funding acquisition, Q.L. and L.W.; Investigation, Y.T. and E.Z.; Methodology, Q.L. and L.W.; Project administration, Q.L. and L.W.; Supervision, Q.L. and L.W.; Validation, F.L. and L.M.; Visualization, Y.T. and E.Z.; Writing—original draft, Y.T. and E.Z.; Writing—review and editing, Q.L. and L.W. All authors have read and agreed to the published version of the manuscript.

Funding: This work was supported by the National Natural Science Foundation of China (No. 32102605), and the Agricultural Science and Technology Innovation Program under Grant (CAAS-ASTIP-2020-IAR).

Institutional Review Board Statement: The animal study protocol was approved by the Animal Ethics Committee of Institute of Apicultural Research, Chinese Academy of Agricultural Sciences (Beijing, China) (protocol code: CAAS-IAR-ER0046, approval data: 25 August 2021).

Data Availability Statement: Data available on request due to privacy.

Conflicts of Interest: The authors declare no conflict of interest.

References

1. Komosinska-Vassev, K.; Olczyk, P.; Kaźmierczak, J.; Mencner, L.; Olczyk, K. Bee pollen: Chemical composition and therapeutic application. *Evid. Based Complement Alternat. Med.* **2015**, *2015*, 297425. [CrossRef] [PubMed]
2. Kostić, A.Ž.; Milinčić, D.D.; Barać, M.B.; Ali Shariati, M.; Tešić, Ž.L.; Pešić, M.B. The Application of Pollen as a Functional Food and Feed Ingredient-The Present and Perspectives. *Biomolecules* **2020**, *10*, 84. [CrossRef] [PubMed]
3. Végh, R.; Csóka, M.; Sörös, C.; Sipos, L. Food safety hazards of bee pollen—A review. *Trends Food Sci. Technol.* **2021**, *114*, 490–509. [CrossRef]
4. Choi, J.H.; Jang, Y.S.; Oh, J.W.; Kim, C.H.; Hyun, I.G. Bee Pollen-Induced Anaphylaxis: A Case Report and Literature Review. *Allergy Asthma Immunol. Res.* **2015**, *7*, 513–517. [CrossRef] [PubMed]
5. McNamara, K.B.; Pien, L. Exercise-induced anaphylaxis associated with the use of bee pollen. *Ann. Allergy Asthma Immunol.* **2019**, *122*, 118–119. [CrossRef]
6. Pitsios, C.; Chliva, C.; Mikos, N.; Kompoti, E.; Nowak-Wegrzyn, A.; Kontou-Fili, K. Bee pollen sensitivity in airborne pollen allergic individuals. *Ann. Allergy Asthma Immunol.* **2006**, *97*, 703–706. [CrossRef]
7. Nonotte-Varly, C. Allergenicity of Gramineae bee-collected pollen is proportional to its mass but is highly variable and depends on the members of the Gramineae family. *Allergol. Immunopathol.* **2016**, *44*, 232–240. [CrossRef]
8. Popescu, F.D. Cross-reactivity between aeroallergens and food allergens. *World J Methodol.* **2015**, *5*, 31–50. [CrossRef]
9. Abdullahi, N.; Atiku, M.K.; Umar, N.B. The roles of enzyme in food processing—An overview. *FUDMA J. Sci.* **2021**, *5*, 151–164. [CrossRef]
10. Yu, J.; Ahmedna, M.; Goktepe, I.; Cheng, H.; Maleki, S. Enzymatic treatment of peanut kernels to reduce allergen levels. *Food Chem.* **2011**, *127*, 1014–1022. [CrossRef]
11. Bu, G.; Huang, T.; Li, T. The separation and identification of the residual antigenic fragments in soy protein hydrolysates. *J. Food Biochem.* **2020**, *44*, e13144. [CrossRef] [PubMed]
12. Hu, J.; Liu, Y.; An, G.; Zhang, J.; Zhao, X.; Zhang, P.; Wang, Z. Characteristics of molecular composition and its anti-nutrition of β-conglycinin during flavorzyme proteolysis. *Food Biosci.* **2021**, *42*, 101039. [CrossRef]
13. Merz, M.; Kettner, L.; Langolf, E.; Appel, D.; Blank, I.; Stressler, T.; Fischer, L. Production of wheat gluten hydrolysates with reduced antigenicity employing enzymatic hydrolysis combined with downstream unit operations. *J. Sci. Food Agric.* **2016**, *96*, 3358–3364. [CrossRef] [PubMed]

14. Yang, Y.; Zhang, J.L.; Zhou, Q.; Wang, L.; Huang, W.; Wang, R.D. Effect of ultrasonic and ball-milling treatment on cell wall, nutrients, and antioxidant capacity of rose (Rosa rugosa) bee pollen, and identification of bioactive components. *J. Sci. Food Agric.* **2019**, *99*, 5350–5357. [CrossRef] [PubMed]
15. Tao, Y.; Yin, S.; Fu, L.; Wang, M.; Meng, L.; Li, F.; Xue, X.; Wu, L.; Li, Q. Identification of allergens and allergen hydrolysates by proteomics and metabolomics: A comparative study of natural and enzymolytic bee pollen. *Food Res. Int.* **2022**, *158*, 111572. [CrossRef]
16. Huang, J.; Liu, C.; Wang, Y.; Wang, C.; Xie, M.; Qian, Y.; Fu, L. Application of in vitro and in vivo models in the study of food allergy. *Food Sci. Hum. Wellness* **2018**, *7*, 235–243. [CrossRef]
17. Rupa, P.; Mine, Y. Oral immunotherapy with immunodominant T-cell epitope peptides alleviates allergic reactions in a Balb/c mouse model of egg allergy. *Allergy* **2012**, *67*, 74–82. [CrossRef]
18. Mine, Y.; Majumder, K.; Jin, Y.; Zeng, Y. Chinese sweet tea (Rubus suavissimus) polyphenols attenuate the allergic responses in a Balb/c mouse model of egg allergy. *J. Funct. Foods* **2020**, *67*, 103827. [CrossRef]
19. Adel-Patient, K.; Bernard, H.; Ah-Leung, S.; Créminon, C.; Wal, J.M. Peanut-and cow's milk-specific IgE, Th2 cells and local anaphylactic reaction are induced in Balb/c mice orally sensitized with cholera toxin. *Allergy* **2005**, *60*, 658–664. [CrossRef]
20. Untersmayr, E.; Schöll, I.; Swoboda, I.; Beil, W.J.; Förster-Waldl, E.; Walter, F.; Riemer, A.; Kraml, G.; Kinaciyan, T.; Spitzauer, S.; et al. Antacid medication inhibits digestion of dietary proteins and causes food allergy: A fish allergy model in BALB/c mice. *J. Allergy Clin. Immunol.* **2003**, *112*, 616–623. [CrossRef]
21. Chen, C.; Lianhua, L.; Nana, S.; Yongning, L.; Xudong, J. Development of a BALB/c mouse model for food allergy: Comparison of allergy-related responses to peanut agglutinin, β-lactoglobulin and potato acid phosphatase. *Toxicol. Res.* **2017**, *6*, 6251–6261. [CrossRef] [PubMed]
22. Yang, X.; Liu, Y.; Guo, X.; Bai, Q.; Zhu, X.; Ren, H.; Chen, Q.; Yue, T.; Long, F. Antiallergic activity of Lactobacillus plantarum against peanut allergy in a Balb/c mouse model. *Food Agric. Immunol.* **2019**, *30*, 762–773. [CrossRef]
23. Kato, Y.; Morikawa, T.; Kato, E.; Yoshida, K.; Imoto, Y.; Sakashita, M.; Osawa, Y.; Takabayashi, T.; Kubo, M.; Miura, K.; et al. Involvement of Activation of Mast Cells via IgE Signaling and Epithelial Cell–Derived Cytokines in the Pathogenesis of Pollen Food Allergy Syndrome in a Murine Model. *J. Immunol.* **2021**, *206*, 2791–2802. [CrossRef]
24. Lee-Sarwar, K.; Kelly, R.S.; Lasky-Su, J.; Moody, D.B.; Mola, A.R.; Cheng, T.Y.; Comstock, L.E.; Zeiger, R.S.; O'Connor, G.T.; Sandel, M.T.; et al. Intestinal microbial-derived sphingolipids are inversely associated with childhood food allergy. *J. Allergy Clin. Immunol.* **2018**, *142*, 335–338. [CrossRef] [PubMed]
25. Yadama, A.P.; Kelly, R.S.; Lee-Sarwar, K.; Mirzakhani, H.; Chu, S.H.; Kachroo, P.; Litonjua, A.A.; Lasky-Su, J.; Weiss, S.T. Allergic disease and low ASQ communication score in children. *Brain Behav. Immun.* **2020**, *83*, 293–297. [CrossRef] [PubMed]
26. Kong, J.; Chalcraft, K.; Mandur, T.S.; Jimenez-Saiz, R.; Walker, T.D.; Goncharova, S.; Gordon, M.E.; Naji, L.; Flader, K.; Larché, M.; et al. Comprehensive metabolomics identifies the alarmin uric acid as a critical signal for the induction of peanut allergy. *Allergy* **2015**, *70*, 495–505. [CrossRef]
27. Lee, K.H.; Song, Y.; Wu, W.; Yu, K.; Zhang, G. The gut microbiota, environmental factors, and links to the development of food allergy. *Clin. Mol. Allergy.* **2020**, *18*, 5. [CrossRef]
28. Rachid, R.; Chatila, T.A. The role of the gut microbiota in food allergy. *Curr. Opin. Pediatr.* **2016**, *28*, 748–753. [CrossRef]
29. Aitoro, R.; Paparo, L.; Amoroso, A.; Di Costanzo, M.; Cosenza, L.; Granata, V.; Di Scala, C.; Nocerino, R.; Trinchese, G.; Montella, M.; et al. Gut Microbiota as a Target for Preventive and Therapeutic Intervention against Food Allergy. *Nutrients* **2017**, *9*, 672. [CrossRef]
30. Hand, T.W.; Vujkovic-Cvijin, I.; Ridaura, V.K.; Belkaid, Y. Linking the Microbiota, Chronic Disease, and the Immune System. *Trends Endocrinol. Metab.* **2016**, *27*, 831–843. [CrossRef]
31. Yin, S.; Tao, Y.; Jiang, Y.; Meng, L.; Zhao, L.; Xue, X.; Li, Q.; Wu, L. A Combined Proteomic and Metabolomic Strategy for Allergens Characterization in Natural and Fermented Brassica napus Bee Pollen. *Front. Nutr.* **2022**, *9*, 822033. [CrossRef] [PubMed]
32. Nunomura, S.; Kitajima, I.; Nanri, Y.; Kitajima, M.; Ejiri, N.; Lai, I.S.; Okada, N.; Izuhara, K. The FADS mouse: A novel mouse model of atopic keratoconjunctivitis. *J. Allergy Clin. Immunol.* **2021**, *18*, 1596–1602. [CrossRef] [PubMed]
33. Lee, H.S.; Lee, B.C.; Ku, S.K. Effect of DHU001, a polyherbal formula, on dinitrofluorobenzene-induced contact dermatitis (type I allergy). *Toxicol. Res.* **2010**, *26*, 123–130. [CrossRef]
34. Rigoni, A.; Colombo, M.P.; Pucillo, C. Mast cells, basophils and eosinophils: From allergy to cancer. *Semin. Immunol.* **2018**, *35*, 29–34. [CrossRef] [PubMed]
35. Miyake, K.; Karasuyama, H. Emerging roles of basophils in allergic inflammation. *Allergol Int.* **2017**, *66*, 382–391. [CrossRef] [PubMed]
36. Gould, H.J.; Sutton, B.J.; Beavil, A.J.; Beavil, R.L.; McCloskey, N.; Coker, H.A.; Fear, D.; Smurthwaite, L. The biology of IGE and the basis of allergic disease. *Annu. Rev. Immunol.* **2003**, *21*, 579–628. [CrossRef]
37. Yamana, Y.; Yamana, S.; Uchio, E. Relationship among total tear IgE, specific serum IgE, and total serum IgE levels in patients with pollen-induced allergic conjunctivitis. *Graefes Arch. Clin. Exp. Ophthalmol.* **2022**, *260*, 281–287. [CrossRef]
38. Borish, L.; Chipps, B.; Deniz, Y.; Gujrathi, S.; Zheng, B.; Dolan, C.M.; TENOR Study Group. Total serum IgE levels in a large cohort of patients with severe or difficult-to-treat asthma. *Ann. Allergy Asthma Immunol.* **2005**, *95*, 247–253. [CrossRef]
39. Amin, K. The role of mast cells in allergic inflammation. *Respir. Med.* **2012**, *106*, 9–14. [CrossRef]

40. Thangam, E.B.; Jemima, E.A.; Singh, H.; Baig, M.S.; Khan, M.; Mathias, C.B.; Church, M.K.; Saluja, R. The Role of Histamine and Histamine Receptors in Mast Cell-Mediated Allergy and Inflammation: The Hunt for New Therapeutic Targets. *Front. Immunol.* **2018**, *9*, 1873. [CrossRef]

41. Yamashita, H.; Takahashi, K.; Tanaka, H.; Nagai, H.; Inagaki, N. Overcoming food allergy through acquired tolerance conferred by transfer of Tregs in a murine model. *Allergy* **2012**, *67*, 201–209. [CrossRef] [PubMed]

42. Yang, G.; Wu, G.; Yao, W.; Guan, L.; Geng, X.; Liu, J.; Liu, Z.; Yang, L.; Huang, Q.; Zeng, X.; et al. 5-HT is associated with the dysfunction of regulating T cells in patients with allergic rhinitis. *Clin. Immunol.* **2022**, *243*, 109101. [CrossRef] [PubMed]

43. Wawrzyniak, M.; Groeger, D.; Frei, R.; Ferstl, R.; Wawrzyniak, P.; Krawczyk, K.; Pugin, B.; Barcik, W.; Westermann, P.; Dreher, A.; et al. Spermidine and spermine exert protective effects within the lung. *Pharmacol. Res. Perspect.* **2021**, *9*, e00837. [CrossRef] [PubMed]

44. Gardana, C.; Del Bo', C.; Quicazán, M.C.; Correa, A.R.; Simonetti, P. Nutrients, phytochemicals and botanical origin of commercial bee pollen from different geographical areas. *J. Food Compos. Anal.* **2018**, *73*, 29–38. [CrossRef]

45. Ionescu, J.G. Personalized anti-inflammatory diets for allergic and skin disorders. *EPMA J.* **2014**, *5*, A160. [CrossRef]

46. McCleary, N.; Nwaru, B.I.; Nurmatov, U.B.; Critchley, H.; Sheikh, A. Endogenous and exogenous sex steroid hormones in asthma and allergy in females: A systematic review and meta-analysis. *J. Allergy Clin. Immunol.* **2018**, *141*, 1510–1513. [CrossRef] [PubMed]

47. Sympson, A.B.; Yousef, E. Association between milk allergy, steroid use, and rate of hospitalizations in children with asthma. *J. Allergy Clin. Immunol.* **2007**, *119*, S116. [CrossRef]

48. Dong, X.; Bachman, L.A.; Kumar, R.; Griffin, M.D. Generation of antigen-specific, interleukin-10-producing T-cells using dendritic cell stimulation and steroid hormone conditioning. *Transpl. Immunol.* **2003**, *11*, 323–333. [CrossRef]

49. Asselin, C.; Gendron, F.P. Shuttling of information between the mucosal and luminal environment drives intestinal homeostasis. *FEBS Lett.* **2014**, *588*, 4148–4157. [CrossRef]

50. Russo, I.; Luciani, A.; de Cicco, P.; Troncone, E.; Ciacci, C. Butyrate attenuates lipopolysaccharide-induced inflammation in intestinal cells and Crohn's mucosa through modulation of antioxidant defense machinery. *PLoS ONE* **2012**, *7*, e32841. [CrossRef]

51. Macia, L.; Mackay, C.R. Dysfunctional microbiota with reduced capacity to produce butyrate as a basis for allergic diseases. *J. Allergy Clin. Immunol.* **2019**, *144*, 1513–1515. [CrossRef] [PubMed]

52. Zimmermann, P.; Messina, N.; Mohn, W.W.; Finlay, B.B.; Curtis, N. Association between the intestinal microbiota and allergic sensitization, eczema, and asthma: A systematic review. *J. Allergy Clin. Immunol.* **2019**, *143*, 467–485. [CrossRef] [PubMed]

53. Noval Rivas, M.; Burton, O.T.; Wise, P.; Zhang, Y.Q.; Hobson, S.A.; Garcia Lloret, M.; Chehoud, C.; Kuczynski, J.; DeSantis, T.; Warrington, J.; et al. A microbiota signature associated with experimental food allergy promotes allergic sensitization and anaphylaxis. *J. Allergy Clin. Immunol.* **2013**, *131*, 201–212. [CrossRef] [PubMed]

54. Li, X.; Luo, J.; Nie, C.; Li, Q.; Sun, X.; Li, H.; Zhang, Y. Altered vaginal microbiome and relative co-abundance network in pregnant women with penicillin allergy. *Allergy Asthma Clin. Immunol.* **2020**, *16*, 79. [CrossRef] [PubMed]

55. Xu, Y.; Xie, L.; Zhang, Z.; Zhang, W.; Tang, J.; He, X.; Zhou, J.; Peng, W. Tremella fuciformis Polysaccharides Inhibited Colonic Inflammation in Dextran Sulfate Sodium-Treated Mice via Foxp3+ T Cells, Gut Microbiota, and Bacterial Metabolites. *Front. Immunol.* **2021**, *12*, 648162. [CrossRef] [PubMed]

56. Mai, J.; Liang, B.; Xiong, Z.; Ai, X.; Gao, F.; Long, Y.; Yao, S.; Liu, Y.; Gong, S.; Zhou, Z. Oral administration of recombinant *Bacillus subtilis* spores expressing *Helicobacter pylori* neutrophil-activating protein suppresses peanut allergy via up-regulation of Tregs. *Clin. Exp. Allergy* **2019**, *49*, 1605–1614. [CrossRef]

57. De Fazio, L.; Cavazza, E.; Spisni, E.; Strillacci, A.; Centanni, M.; Candela, M.; Praticò, C.; Campieri, M.; Ricci, C.; Valerii, M.C. Longitudinal analysis of inflammation and microbiota dynamics in a model of mild chronic dextran sulfate sodium-induced colitis in mice. *World J. Gastroenterol.* **2014**, *20*, 2051–2061. [CrossRef]

58. Marlow, G.; Ellett, S.; Ferguson, I.R.; Zhu, S.; Karunasinghe, N.; Jesuthasan, A.C.; Han, D.Y.; Fraser, A.G.; Ferguson, L.R. Transcriptomics to study the effect of a Mediterranean-inspired diet on inflammation in Crohn's disease patients. *Hum. Genom.* **2013**, *7*, 24. [CrossRef]

59. Bennett, C.J.; Henry, R.; Snipe, R.; Costa, R. Is the gut microbiota bacterial abundance and composition associated with intestinal epithelial injury, systemic inflammatory profile, and gastrointestinal symptoms in response to exertional-heat stress? *J. Sci. Med. Sport* **2020**, *23*, 1141–1153. [CrossRef]

 *foods*

Article

Botanical Origin of Galician Bee Pollen (Northwest Spain) for the Characterization of Phenolic Content and Antioxidant Activity

Sergio Rojo, Olga Escuredo *, María Shantal Rodríguez-Flores and María Carmen Seijo

Department of Vegetal Biology and Soil Sciences, Faculty of Sciences, University of Vigo, 32004 Ourense, Spain
* Correspondence: oescuredo@uvigo.es

Abstract: Bee pollen is considered a natural product, relevant for its nutritional and antioxidant properties. Its composition varies widely depending on its botanical and geographical origins. In this study, the botanical characteristics of 31 bee pollen samples from Galicia (Northwest Spain) were analyzed; samples have not been studied until now from this geographical area. The study focused on the evaluation of the influence of plant origin on total phenol and flavonoid contents and antioxidant activity measured by radical scavenging methods. The multivariate statistical treatment showed the contribution of certain pollen types in the extract of bee pollen as to phenols, flavonoids and antioxidant capacity. Specifically, the bee pollen samples with higher presence of *Castanea*, *Erica*, *Lythrum* and *Campanula* type indicated higher total phenol and flavonoid contents and antioxidant activities according to the principal component analysis. On the contrary, *Plantago* and *Taraxacum officinale* type contributed a lower content of these compounds and radical scavenging activity. The cluster analysis classified the bee pollen samples into three groups with significant differences ($p >$ 0.05) for the main pollen types, total phenol and flavonoid contents and antioxidant capacities. These results demonstrate the richness and botanical diversity in the pollen spectrum of bee pollen and enhance the possible beneficial nutraceutical properties of this beekeeping product.

Keywords: *Apis mellifera* L.; bee pollen; botanical source; phenols; flavonoids; radical scavenging activity

Citation: Rojo, S.; Escuredo, O.; Rodríguez-Flores, M.S.; Seijo, M.C. Botanical Origin of Galician Bee Pollen (Northwest Spain) for the Characterization of Phenolic Content and Antioxidant Activity. *Foods* **2023**, *12*, 294. https://doi.org/10.3390/foods12020294

Academic Editors: Liming Wu and Qiangqiang Li

Received: 7 December 2022
Revised: 28 December 2022
Accepted: 5 January 2023
Published: 8 January 2023

1. Introduction

Plant pollination is accomplished by transferring pollen grains from the flower stamen of one plant to the stigma of another plant with agents such as wind, water, and insects [1–3]. Bees are actively involved in pollination by collecting and dispersing pollen from flower to flower. In addition to this essential function of ecosystem maintenance, the bees, during the collection of pollen, mix the grains with their own secretions, agglutinating it as small pellets on the hind legs of the insect, which are then transported to the hive. In beekeeping, these pellets generated by collecting pollen from flower stamens by the European honeybee *Apis mellifera* L. are known as bee pollen.

The pollen grains are recognized in apiculture as a source important of proteins, minerals and fats and are used mainly as food for the larvae and younger bees in the early stage of development inside the hive [1]. Moreover, since ancient times, pollen loads have been daily consumed throughout the world. Currently, due to both a trend towards natural diet supplementation and medical applications taking advantage of its numerous anticancer, anti-obesity, antimicrobial, anti-inflammatory and antioxidant properties, it is gaining more attention [4–7].

Bee pollen as a mixture of floral pollens collected by bees widely varies in shape, color, size, weight and in chemical composition. In the group of basic chemical substances, there are proteins, amino acids, lipids and fatty acids, carbohydrates, phenolic compounds and enzymes, as well as vitamins and bioelements [1,8,9]. Considering this excellent nutrient profile, bee pollen provides a significant daily intake of nutrients and complements the human diet. At the same time, this bee product comprises many compounds, especially rich in biologically active substances that differ according to the origin of the plant species visited by the bees. Therefore, in addition to its importance as a functional food, bee pollen is gaining special attention due to its active natural metabolites, especially derivatives of essential amino acids, polyphenolic substances, vitamins and lipids [4,6,9]. Although several natural metabolites interfere in free radical scavenging activity, it appears that phenolic acids and flavonoids are responsible for most of the antioxidant properties [4,5,10,11]. However, the deviations in the antioxidant activity and polyphenolic content between pollens are remarkable, as a consequence of the particularities of the plant species source and different geographical areas [3,12].

On the other hand, consumers increasingly demand quality, safe and healthy food. This is the consequence of a proper monitoring of traceability and physicochemical characterization of food. Even though bee pollen is known as a potent natural food, its physicochemical characteristics and nutritional composition are ambiguous and vary greatly depending on bee species, and the botanical and geographical origins [3,4,6]. Palynology analysis is the most widely used method worldwide for identifying the botanical origins of bee products, as well bee pollen [4]. However, research studies that include a detailed pollen profile of this type of matrix and its relationship with the nutritional composition are infrequent [2,13–15].

In recent years, bee pollen as a commercial product has gained significant profitability in the beekeeping sector, taking advantage of its several functional properties as a health ingredient of multiples food [7]. Spain is the leading producer of bee pollen in the European Union. The unique qualities that contribute to beekeeping products are derived from the diversity in the flora, climate and soil of the Spanish territory. However, research on bee pollen from Spain is scarce from the point of view of its physicochemical and botanical characteristics, in comparison with other bee products such the honey. Polyphenolic and flavonoid compounds [16–18], in addition to profiling of amino acids, sugars, alkaloids and nucleic acids [18], and the amino acid content [19] of bee pollen collected in Spain were reported.

In order to guarantee well-characterized products, in recent years geographical indication has been included in the commercialization of bee products as an essential tool. Thus, the evaluation of the quality parameters and nutritional compounds of bee pollen from different geographical areas could provide differentiating information for producers in the sector and consumers [8]. Specifically, in the Northwest of Spain (Galicia) there are several unifloral honeys characterized and protected at European level by the quality scheme of Protected Geographical Indication (PGI) *Miel de Galicia* [20]. In this context, this study contributes to the knowledge of the bee pollen's characteristics, provides information for the inclusion as a quality product and consequently, favors the diversification of bee products.

Due to the variability in the chemical composition of bee products related to botanical origin, a set of bee pollen samples produced in different locations of Galicia was analyzed. The aim of this study was to evaluate the botanical diversity of bee pollen samples from Northwest Spain by means of palynological analysis and their influence on the total phenol content (TPC), total flavonoid content (TFC) and radical scavenging activity (DPPH and ABTS).

2. Materials and Methods

2.1. Bee Pollen Samples

In this study, 31 bee pollen samples from Galicia (Northwest Spain) were analyzed. The samples were collected using pollen traps from *Apis mellifera* hives in 19 different apiaries. The geographical origin of the pollen samples collected is shown in Figure 1. The apiaries were distributed in the four provinces of the Galician autonomous community (A Coruña, Lugo, Pontevedra and Ourense). Specifically, in A Coruña the samples were collected from 4 apiaries (9 samples), in Lugo from 3 apiaries (4 samples), in Pontevedra from 3 apiaries (4 samples) and in Ourense from 9 apiaries (14 samples). After collection, the bee pollen samples were cleaned of impurities (such as parts of dead bees or remains of wax), and until further analysis were stored at −20 °C.

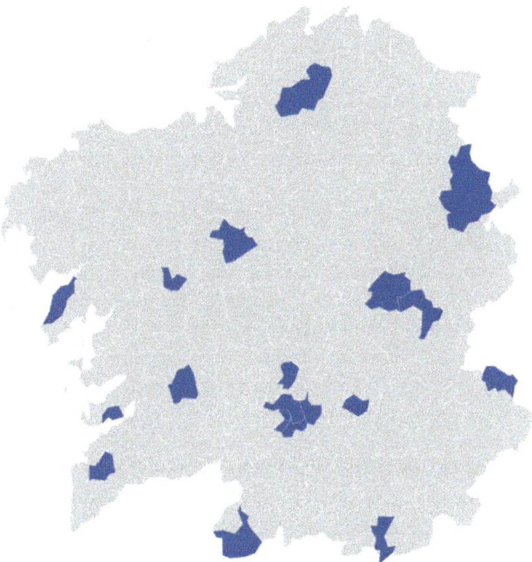

Figure 1. Distribution of geographical origin by municipality in blue of the bee pollen samples in Galicia (Northwest Spain). Created with Datawrapper.

2.2. Determination of Botanical Origin

Botanical origin was determined using a melissopalynological procedure. First, the original samples were conveniently homogenized, then 1 g was weighed and placed in separate vials. Subsequently, a colorimetric separation was carried out to obtain different subsamples, which were weighted and dissolved in distilled water (15 mL). The solutions were shaken for 10 min and at 4500 rpm for 5 min centrifuged. An aliquot of 100 µL was taken from the sediment to prepare the slides. An optical microscope (Nikon Optiphot II, UK Ltd., London, UK) was used to identify the botanical origin of the different subsamples. The pollen spectra of the samples were determined considering the weight of each subsample and its botanical origin. The results were expressed in percentages.

2.3. Preparation of Bee Pollen Extracts

The extracts of the bee pollen samples were prepared according to the method of Gabriele et al. [21], with minor modifications. 0.5 g of each pollen sample was dissolved with 80% ethanol to a concentration of 0.01 g/mL. These extracts were gently shaken in the dark for 5 h and subsequently macerated for 24 h. After this time, the extracts were centrifuged for 10 min at 4500 rpm, and subsequently properly stored in amber-colored glass containers to avoid direct incidence of light.

2.4. Assessment of Total Phenol Content

The determination of the total phenolic content (TPC) was carried out based on the method developed by Singleton and Rossi [22] adapted to bee pollen. 1 mL of the bee pollen extract solution (0.01 g/mL) was dissolved with 1 mL of Folin-Ciocalteu reagent and 10 mL of distilled water. After gently stirring and standing for 2 min, 4 mL of 7% Na_2CO_3 solution was added and made up to 25 mL with distilled water. The solutions were measured using a UV–Vis spectrophotometer (Jenway 6305, Fisher Scientific, Loughborough, UK) at an absorbance of 765 nm after being kept in the dark for 1 h. Gallic acid solutions were used to obtain the calibration curve. The results of TPC as gallic acid in mg/100 g were expressed.

2.5. Assessment of Total Flavonoid Content

The total flavonoid content (TFC) was measured using the method of Arvouet-Grand et al. [23] adapted for pollen. An aluminum chloride solution is used for its reaction with the flavonoid compounds present in the bee pollen solutions. Thus, 2 mL of the bee pollen extract solution (0.01 g/mL) was dissolved with 0.5 mL of 5% aluminum chloride and distilled water to final volume of 25 mL. After 30 min in the dark, the prepared solutions turned yellow, at which time the absorbance at 425 nm with UV–Vis spectrophotometer was measured (Jenway 6305, Fisher Scientific, Loughborough, UK). Quercetin solutions were used as reference pattern and the results of TFC were calculated as quercetin in mg/100 g.

2.6. Assessment of Antioxidant Activity by Radical Scavenging Assay: DPPH and ABTS

The radical scavenging capacity of the bee pollen extracts was determined based on the scavenging ability of the antioxidants towards the stable 2, 2-diphenyl-1-picrylhydrazyl radical known as DPPH method [24]. The scavenging activity on pollen extracts, mixed with 2.7 mL of a DPPH solution (6×10^{-5} M) was measured. The pollen sample solution and the blank-DPPH solution were incubated at room temperature for 30 min in the dark. The absorbance with a UV–vis spectrophotometer was measured at 517 nm.

The radical scavenging activity by ABTS assay was determined according to a method reported by Re et al. [25]. The ABTS solution was prepared by reacting ABTS 7 mM in water with 2.45 mM potassium persulfate. ABTS stock solution was left at room temperature for 12–16 h in the dark until it reached a stable oxidative state. ABTS stock solution was prepared by dilution with ethanol to give an absorbance of 0.70 ± 0.02 at 734 nm. Then, 980 µL of this solution was mixed with 20 µL of the ethanolic extract of bee pollen sample and finally, the absorbance was measured at 734 nm.

The antioxidant activity of bee pollen samples was expressed as percentage of DPPH and ABTS calculated using the following equation: Scavenging activity (%) = [(AbsB−AbsS)/AbsB] × 100, where AbsB is the absorbance of the blank solution according to radical used and AbsS is the absorbance of the pollen extract solution.

2.7. Data Analyses

The significant differences between the pollen types identified by palynological analysis, TPC, TFC and antioxidant variables of bee pollen samples set were determined using a Student's *t*-test. The level of statistical significance was taken into account given a *p*-value (*p*) less than 0.05. Principal component analysis (PCA) was applied with the objective of providing a reduced interpretation of the variance of the data of the studied variables (main pollen types, TPC, TFC and antioxidant activity) in the bee pollen samples. With the aim to provide a simplified interpretation of the variance of the data set of the main analyzed variables (main pollen types, TPC, TFC and antioxidant activity), a principal component analysis (PCA) was applied. The data matrix was reduced to a small number of principal components to analyze the significant relationships between the variables. At the same time, groups of pollen samples were established using multivariate cluster analysis. This statistical approach grouped samples based on a data set of variables from cases with similar characteristics. Differences between the groups were tested using the Bonferroni test through post hoc comparison (*p* < 0.05). STATGRAPHICS Centurion XVI software (Statpoint Technologies, Inc., Warrenton, VA, USA) was used for treatment of data.

3. Results

3.1. Botanical Preference for Bee Pollen Production

The results of palynological analysis of the studied bee pollen samples are summarized in Figure 2 and Table 1. In the set of samples analyzed were identified fifty-one pollen types belonging to 32 families.

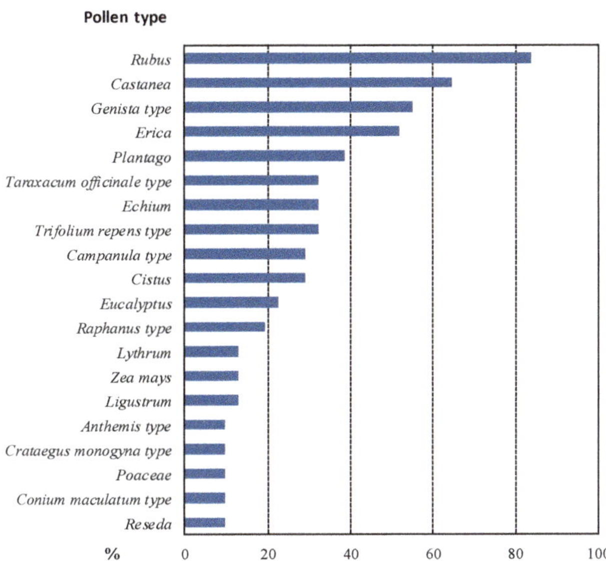

Figure 2. Frequency of the pollen types identified in the bee pollen samples.

Considering the distribution of the different pollen types in samples, pollen grains from *Rubus*, *Castanea*, *Genista* type and *Erica* were present in more than 50% of the samples (Figure 2). Other frequent pollen types were from *Taraxacum officinale* type, *Echium* and *Trifolium repens* type (found in 30% of the samples). With regards the abundance of pollen types in each sample, some of the well-distributed pollen types such as *Rubus*, *Castanea*, *Genista* type, *Taraxacum officinale* type and *Lythrum* were found, at least in one sample, as dominant pollen (>45%) (Table 1). However, *Rubus* and *Castanea* were the most representative pollen types (26 and 20 samples, respectively), and those with the highest percentage counted in the pollen spectrum, with mean values above 22% and maximum value of more than 90% (Table 1).

3.2. Concentration of Total Phenol, Flavonoid and Antioxidant Capacity of Bee Pollen

The mean content of TPC and TFC was 1612.6 mg/100 g and 256.8 mg/100 g, respectively (Table 2). The range for TPC was between 771.8 mg/100 g and 2638.9 mg/100 g and for TFC between 90.8 mg/100 g and 639.3 mg/100 g. The antioxidant activity expressed as DPPH and ABTS had a mean value of 65.7% and 57.4%, respectively. The maximum value of DPPH found in the samples was 88.2% and for ABTS, of 79.3%. Significant differences were found between the mean values for all the variables analyzed in the bee pollen samples ($p < 0.05$).

3.3. Contribution of the Botanical Origin to the Phenolic Content and Antioxidant Activity of Pollen

The relationships between the botanical origin of the bee pollens and the polyphenol and flavonoid contents and the antioxidant capacity have been evaluated using a PCA. This multivariate technique reduced the dataset and revealed a six-component model with 81.47% of the variance of the data (Table 3). The first three components explained 53.40% of the data variability. The variables with higher weight in the first component were *Taraxacum officinale* type, *Plantago*, DPPH, TPC and *Castanea* (with coefficients above 0.30). In the second component, the higher coefficients corresponded to *Erica*, TFC and *Campanula* type (above 0.35), while the third component was related with the higher coefficients of TFC, DPPH, *Echium*, *Trifolium repens* type and *Rubus* (coefficients above 0.26).

Table 1. Most-representative families and pollen types in the bee pollen samples produced in Galicia.

Family	Pollen Type	N	Mean	Standard Deviation	Maximum
Rosaceae	*Rubus*	26	29.3 *	27.4	94.5
	Crataegus monogyna type	3	0.5	2.1	11.7
Fagaceae	*Castanea*	20	22.6 *	27.7	91.1
	Quercus	2	1.8	7.9	42.1
Fabaceae	*Genista* type	17	9.8 *	19.0	61.2
	Trifolium repens type	10	2.6	7.7	39.5
Ericaceae	*Erica*	16	5.4 *	10.5	44.8
	Calluna vulgaris	2	0.6	2.3	11.2
Plantaginaceae	*Plantago*	12	2.9 *	7.3	27.0
Asteraceae	*Taraxacum officinale* type	10	5.9 *	15.7	60.3
	Anthemis type	3	1.0	4.7	26.3
Boraginaceae	*Echium*	10	2.6 *	5.7	22.2
Campanulaceae	*Campanula* type	9	3.1	8.8	37.5
Cistaceae	*Cistus*	9	1.8	5.3	28.2
	Cistus psilosepalus	2	0.2	0.7	3.6
Myrtaceae	*Eucalyptus*	7	0.7	1.8	7.8
Brassicaceae	*Raphanus* type	6	1.7	4.5	18.5
	Brassica	2	0.4	1.9	10.3
Lythraceae	*Lythrum*	4	2.2	9.8	54.5
Oleaceae	*Ligustrum*	4	0.5	1.6	7.3
Poaceae	*Zea mays*	4	0.3	0.9	3.3
	Poaceae	3	0.3	1.2	6.6
Apiaceae	*Conium maculatum* type	3	0.4	1.7	9.3
	Foeniculum vulgare type	2	0.5	2.9	16.0
Resedaceae	*Reseda*	3	0.1	0.3	1.5
	Sesamoides	2	0.5	1.9	10.2
Chenopodiaceae	*Chenopodiaceae*	2	0.2	1.0	5.4

N = number of pollen samples containing it. * Significant differences according to Student's *t*-test ($p < 0.05$).

Table 2. Descriptive analysis of TPC, TFC and antioxidant activity expressed as DPPH and ABTS.

	Mean	Standard Deviation	Minimum	Maximum
TPC (mg/100 g)	1612.6 *	531.0	771.8	2638.9
TFC (mg/100 g)	256.8 *	150.0	90.8	639.3
DPPH (%)	65.7 *	20.5	17.0	88.2
ABTS (%)	57.4 *	12.6	32.8	79.3

* Significant differences according to Student's *t*-test ($p < 0.05$).

The projection of the relationships among the palynological and chemical variables on the three first components is shown in Figure 3. The graphic representation shows the close relationship of the pollen variables *Erica*, *Castanea*, *Trifolium repens* type and *Echium* with TPC and DPPH, whereas *Campanula* type and *Lythrum* are closely related to ABTS and TFC. On the contrary, *Genista* type, *Plantago* and *Taraxacum officinale* type had an inverse relationship. Therefore, the bee pollen samples with higher presence of *Erica*, *Castanea*, *Echium* and *Trifolium repens* type had higher TPC and antioxidant activity. Bee pollen with high presence of *Campanula* pollen type and *Lythrum* were characterized by higher concentration of TFC and antioxidant content.

Table 3. Number of components extracted and component weights for each variable included in PCA.

Components	1	2	3	4	5	6
Eigenvalue	3.68	2.32	2.01	1.73	1.37	1.11
Variance (%)	24.55	15.45	13.40	11.52	9.13	7.41
Variance cumulative (%)	24.55	40.01	53.40	64.93	74.06	81.47
Component weights						
TPC	0.40	−0.19	−0.01	−0.38	−0.06	0.07
TFC	0.22	0.46	0.27	−0.03	−0.09	0.24
DPPH	0.41	−0.10	−0.27	0.11	0.05	0.08
ABTS	0.29	0.23	0.19	0.02	0.01	−0.43
Taraxacum officinale type	−0.45	0.03	0.08	−0.22	−0.23	−0.08
Echium	0.00	0.10	−0.48	0.00	−0.46	0.11
Campanula type	0.12	0.38	0.25	−0.04	0.12	−0.30
Cistus	−0.06	−0.12	0.26	0.40	−0.24	0.23
Erica	−0.01	0.46	−0.20	0.10	0.20	−0.16
Genista type	−0.13	−0.15	−0.24	0.04	0.69	0.13
Trifolium repens type	0.06	0.30	−0.50	0.20	−0.22	−0.03
Castanea	0.31	−0.26	−0.02	−0.44	−0.17	−0.21
Lythrum	0.09	0.28	0.16	−0.26	0.04	0.70
Plantago	−0.43	0.08	0.05	−0.29	−0.15	−0.10
Rubus	0.13	−0.23	0.27	0.49	−0.18	−0.04

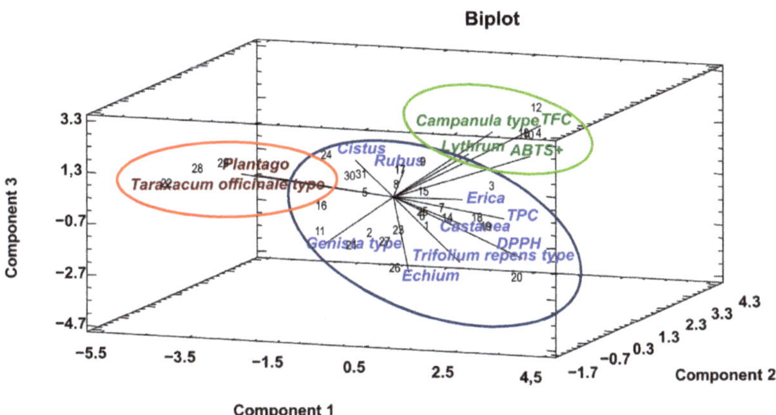

Figure 3. Plot of the first three principal components with the palynological and chemical variables obtained by PCA.

The cluster analysis was carried out with the palynological variables of greater representation in the pollen spectra, TPC, TFC and antioxidant activity (DPPH and ABTS).

The results of this multifactorial analysis classified the pollen samples into three groups (Figure 4).

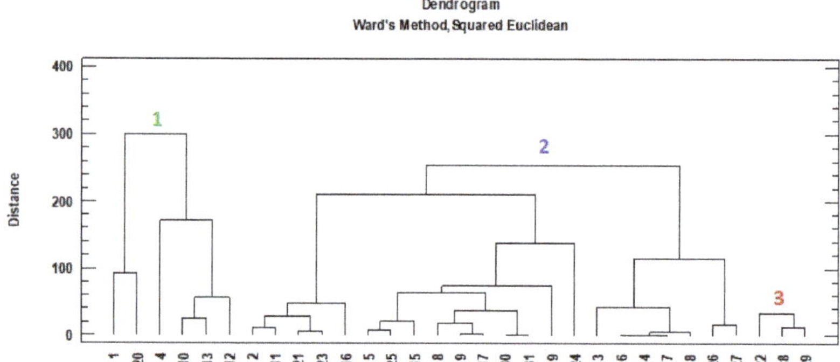

Figure 4. Dendrogram of cluster analysis with the distribution of bee pollen samples (1: group 1; 2: group 2; 3: group 3).

The first group (1) included six pollen samples with significantly higher percentages in *Erica* (20.1%), *Campanula* type (14.5%), *Lythrum* (11.5%), significantly higher in TFC (454.1 mg/100 g) in comparison to the other two groups (two and three), and DPPH (69.3%) respect to group three (21.0%) (Table 4). In group two, there is the largest number of pollen samples (22) and it was characterized by significantly higher mean percentage in the *Genista* type (12.1%), *Castanea* (28.4%) and *Rubus* (37.7%) pollen types, TPC (1741.4 mg/100 g), TFC (219.2 mg/100 g) and antioxidant activity (70.8%) compared to group three (21.0%). Finally, group three (with three bee pollen samples) included the samples with the significantly lower mean values in TPC, TFC and DPPH, but significantly higher proportions of the pollen types *Taraxacum officinale* type (50.2%) and *Plantago* (24.3%).

Table 4. Groups of bee pollen samples of each class obtained by the cluster analysis.

Groups	1	2	3
N (%)	6 (19.3)	22 (71.0)	3 (9.7)
TPC (mg/100 g)	1527.5	1741.4 a	838.3 a
TFC (mg/100 g)	454.1 a b	219.2 b	137.2 a
DPPH (%)	69.3 a	70.8 b	21.0 a b
ABTS (%)	65.0	56.9	46.2
Pollen types (%)			
Taraxacum officinale type	0.5 a	1.3 b	50.2 a b
Echium	4.0	2.2	3.1
Campanula type	14.5 a b	0.5 a	0.0 b
Cistus	0.0	2.5	0.0
Erica	20.1 a b	1.7 a	2.6 b
Genista type	2.2	12.1	8.6
Trifolium repens type	9.0	1.2	0.0
Castanea	12.5	28.4	0.0
Lythrum	11.5 a b	0.0 a	0.0 b
Plantago	1.5 a	0.4 b	24.3 a b
Rubus	13.3	37.7 a	0.0 a

N: number of samples. Same letter shows the significant differences between groups by Bonferroni test ($p < 0.05$).

4. Discussion

The trend in the human diet is to consume foods with a high nutritional value, replacing the more conventional foods, as well as using them as supplements to provide one's diet the energy and essential nutrients required for proper mental and physical development [26]. In terms of human nutrition, bee pollen is considered a natural substance that constitutes a potential source of compounds with diverse nutritional and antioxidant relevance. Since pollen comes from different plant species, the evaluation of quality, safety and its characterization depend on the botanical and geographical origins. Hence the importance of their characterization and differentiation [1,15,27,28]. According to Campos et al. [29] the correct control of the processing procedures declares a consistent composition and could be considered as an indicator of outstanding quality and properties of food products. Some countries are establishing internal regulations with the intention of favoring the quality control of this product. In addition, local growth and the potential for the sale of bee pollen through exports has motivated the creation of international regulations to standardize the analytical methods for the physical-chemical and nutritional analysis of bee pollen [1].

The beneficial functions of phenolic compounds for human health have been demonstrated by reducing oxidative stress and inhibiting macromolecular oxidation; they positively collaborate in reducing the risk of degenerative diseases [30]. Within the phenolic compounds, the flavonoids in the bee pollen matrix have been recognized as quality factors in terms of antioxidant capacity [3,29]. However, phenolic composition of bee products is conditioned by their botanical origin, hence the need to evaluate the particular botanical characteristics of bee pollen based on geographical origin. In the present study, the bee pollen samples with higher presence of pollen types from *Erica*, *Castanea*, *Trifolium repens* type and *Echium* had the highest TPC and RSA expressed in DPPH. The bee pollen with the highest TFC had the highest antioxidant capacity expressed as ABTS, coinciding with the pollen types with the highest presence of *Campanula* type and *Lythrum*.

The Galician territory is characterized by a transition zone composed of different types of climate, resulting an environmental diversity that favors the abundance of plant resources for bees. The plant species of greatest beekeeping interest in Northwest Spain are from the families Fagaceae, Rosaceae, Fabaceae and Ericaceae. *Castanea* and *Rubus* taxa (Fagaceae and Rosaceae, respectively) produce a high quantity of nectar and pollen during the flowering stage (between May and July in the lowest lands and the mountains, respectively), with important productions of unifloral honeys of this botanical origin [31,32]. Bramble and blackberry plants are the most common of the *Rubus* genus (Rosaceae), with almost 300 species growing in Central European. *Rubus ulmifolius*, *R. caesius*, *R. sampaianus*, *R. praecox* or *R. henriquesii*, among others, are the most widespread in Galicia [33,34]. *Castanea* trees produce a great amount of nectar and pollen, and are considered one of the best beekeeping resources in this geographical area, given the important production of honeys of this botanical origin [32]. Species of the family Ericaceae, and other taxa of Fabaceae, mainly *Trifolium repens* and *Genista* type, constitute the typical scrub of the area, with important beekeeping interest [35]. Undoubtedly, the botanical richness that characterizes the Galician territory confers the chemical particularities of bee products.

The TFC quantified in bee pollen from Northwest Spain was similar to that found in samples from locations in southern Spain [17] and Portugal [7,30,36]. Pascoal et al. [15] reported a higher TPC in bee pollen produced in Portugal (Northeast area), with a dominant abundance of pollen types *Erica*, *Echium* and *Castanea*. The abundance of these pollen types probably increased the phenol content in the set of bee pollen samples, coinciding with the significant relationship found in the PCA of this research. *Erica*, *Castanea* and *Echium* turned out to be the plant variables with weight in PCA, coinciding with the variables close to TPC (Figure 3). Mârghitas et al. [37] reported similar TPC in honeybee-collected pollen pellets from Romania separated by colorimetry (*Crataegus monogyna*, *Centaurea cyanus*, *Salix*), but TFC was higher than the bee pollen analyzed in this study.

On the other hand, bee pollen produced in the Sonoran Desert (North of Tucson, USA) with a pollen spectra characterized by the presence of *Prosopis, Yucca, Washingtonia, Larrea, Mimosa* and Chenopodiaceae had higher TPC [2] compared to the pollen samples of the present study. Other researchers also reported higher TPC in bee pollen from Brazil, with a dominant pollen representation of *Cecropia, Eucalyptus, Mimosa pudica, Elaeis, Eupatorium* and *Scoparia* [10], and with dominant pollen *Cocos nucifera, Miconia, Spondias* and *Eucalyptus* [13]. Kostic et al. [28] characterized sunflower bee pollen (*Helianthus annuus*) with *Taraxacum officinale* as accompanying pollen from Serbia by palynological analysis, with lower TPC and TFC than bee pollen from Galicia. *Taraxacum officinale* was also found in the pollen spectrum of studied bee pollen in our study, coinciding with the samples of lower TPC and TFC (Figure 3).

There are several antioxidant compounds involved in the oxidation of the bee products, and the total antioxidant activity is the most accurate measurement [38]. The objective of the determination of the antioxidant activity of bee pollen is to analyze the generation of free radicals due to the disappearance of antioxidants. The DPPH and ABTS methods are the most used and most stable to evaluate the antioxidant capacity of hydrogen-donating antioxidants (aqueous radical scavengers) and chain-breaking antioxidants (lipid peroxyl radical scavengers), although they also show differences [38]. The positive relationship between TFC and antioxidant activity of the bee pollen samples was referenced [10]. The abundance of specific species identified in the extract of pollen samples with high radical scavenging activity and TPC was related, such as *Sinapis alba, Robinia pseudoacacia* [12], *Salix alba* [37] or *Mimosa* [2]. In our study, the abundance of *Castanea, Erica, Rubus, Campanula* type and *Genista* type differentiated the bee pollen samples with higher radical scavenging activity, highly related to TPC and TFC (Figure 4 and Table 4). Bee pollen of *Taraxacum officinale* from Romania [37] and Poland [12] despite high TPC showed low antioxidant capacity by DPPH assay. Several researchers have concluded that the antioxidant capacities are not clearly associated with its total phenolic content [12,30,37]. It is possible that the presence of particular phenolic compounds determines the increase in antioxidant activity [13]. The polyphenolic profile is variable in bee pollen, and the antioxidant activity of polyphenols depends on the number and location of the hydroxyl groups it contains in its chemical structure [38]. Its chemical structure is conducive to scavenging free radicals, because the hydrogen atom from the aromatic hydroxyl group readily donates to the radical species and the stability of the quinone structure it turned out to support an unpaired electron [39]. This strong association is mainly attributed to flavonoids and cinnamic acid derivatives [3,12]. Hence, the importance of relating the botanical origin with the individual phenolic compounds, because they can contribute to the discrimination of the antioxidant capacity of some pollen samples based on their floral origin.

It has also been documented that differences in environmental conditions, soil or plant physiology may interfere in the free radical reactions and the ability to remove reactive oxygen species in this bee product [3]. Some authors supported the close relationship between the antioxidant capacity and the collection period of bee pollen, highlighting a higher antioxidant activity in bee pollen produced in a period of more UV-intense, specifically from the beginning to the end of summer [2]. Therefore, in addition to correct management practices, the time of collection of the bee product will influence the chemical and functional characteristics (closely linked to the flowering period) and must be taken into account by the beekeeper.

5. Conclusions

The botanical characterization of bee pollen is essential for the particular identification of its chemical composition. The results of multivariate statistical treatment applied to the bee pollen sample set revealed the influence of botanical origin on TPC, TFC and antioxidant capacity. *Castanea, Erica, Lythrum* and *Campanula* type have been characterized as the pollen types with the greatest influence on TPC, TFC and antioxidant activity (as indicated by the first two main components of PCA) of bee pollen produced in the Northwest of Spain.

On the contrary, *Plantago* and *Taraxacum officinale* type contributed a lower content of these compounds in this geographical territory. These results provide some evidence for the healthy potential in which the free radicals of bee pollen are involved, promoting the consumption of this traditional food. Expanding these results with a larger number of samples of this botanical origin will help to confirm these conclusions.

Author Contributions: S.R., O.E. and M.C.S. conceived and designed the methodology. S.R. and M.S.R.-F. carried out the experiment. O.E., M.S.R.-F. and M.C.S. analyzed and interpreted data. O.E. and M.C.S., writing original draft preparation. O.E., M.S.R.-F. and M.C.S., writing review and editing. M.C.S., project administration. All authors have read and agreed to the published version of the manuscript.

Funding: This study was supported by Xunta de Galicia (Rural Development Programme 2014/2020, FEADER 2020/048A, "Innovation on the productive process for a sustainable apiculture").

Data Availability Statement: The datasets that were generated during and/or analyzed during the current study are available from the corresponding author on reasonable request.

Acknowledgments: The authors would like to thank the beekeepers and *Agrupación Apícola de Galicia* for their collaboration in this study.

Conflicts of Interest: The authors declare no conflict of interest.

References

1. Campos, M.G.; Bogdanov, S.; de Almeida-Muradian, L.B.; Szczesna, T.; Mancebo, Y.; Frigerio, C.; Ferreira, F. Pollen composition and standardisation of analytical methods. *J. Apic. Res.* **2008**, *47*, 154–161. [CrossRef]
2. LeBlanc, B.W.; Davis, O.K.; Boue, S.; DeLucca, A.; Deeby, T. Antioxidant activity of Sonoran Desert bee pollen. *Food Chem.* **2009**, *115*, 1299–1305. [CrossRef]
3. Aličić, D.; Šubarić, D.; Jašić, M.; Pašalić, H.; Ačkar, Đ. Antioxidant properties of pollen. *Hrana Zdr. Boles. Znan. Stručni Časopis Nutr. Dijetetiku* **2014**, *3*, 6–12.
4. Li, Q.-Q.; Wang, K.; Marcucci, M.C.; Sawaya, A.C.H.F.; Hu, L.; Xue, X.-F.; Wu, L.-M.; Hu, F.-L. Nutrient-rich bee pollen: A treasure trove of active natural metabolites. *J. Funct. Foods* **2018**, *49*, 472–484. [CrossRef]
5. Kaškonienė, V.; Adaškevičiūtė, V.; Kaškonas, P.; Mickienė, R.; Maruška, A. Antimicrobial and antioxidant activities of natural and fermented bee pollen. *Food Biosci.* **2020**, *34*, 100532. [CrossRef]
6. Thakur, M.; Nanda, V. Composition and functionality of bee pollen: A review. *Trends Food Sci. Technol.* **2020**, *98*, 82–106. [CrossRef]
7. Aylanc, V.; Ertosun, S.; Russo-Almeida, P.; Falcão, S.I.; Vilas-Boas, M. Performance of green and conventional techniques for the optimal extraction of bioactive compounds in bee pollen. *Int. J. Food Sci. Technol.* **2022**, *57*, 3490–3502. [CrossRef]
8. Komosinska-Vassev, K.; Olczyk, P.; Kaźmierczak, J.; Mencner, L.; Olczyk, K. Bee Pollen: Chemical Composition and Therapeutic Application. *Evid. Based Complement. Altern. Med.* **2015**, *2015*, 297425. [CrossRef]
9. Ares, A.M.; Valverde, S.; Bernal, J.L.; Nozal, M.J. Extraction and determination of bioactive compounds from bee pollen. *J. Pharm. Biomed. Anal.* **2018**, *147*, 110–124. [CrossRef]
10. Freire, K.R.L.; Lins, A.C.S.; Dórea, M.C.; Santos, F.A.R.; Camara, C.A.; Silva, T.M.S. Palynological Origin, Phenolic Content, and Antioxidant Properties of Honeybee-Collected Pollen from Bahia, Brazil. *Molecules* **2012**, *17*, 1652–1664. [CrossRef]
11. Borycka, K.; Grabek-Lejko, D.; Kasprzyk, I. Antioxidant and antibacterial properties of commercial bee pollen products. *J. Apic. Res.* **2015**, *54*, 491–502. [CrossRef]
12. Leja, M.; Mareczek, A.; Wyżgolik, G.; Klepacz-Baniak, J.; Czekońska, K. Antioxidative properties of bee pollen in selected plant species. *Food Chem.* **2007**, *100*, 237–240. [CrossRef]
13. Araújo, J.S.; Chambó, E.D.; Costa, M.A.P.D.C.; Da Silva, S.M.P.C.; De Carvalho, C.A.L.; Estevinho, L.M. Chemical Composition and Biological Activities of Mono- and Heterofloral Bee Pollen of Different Geographical Origins. *Int. J. Mol. Sci.* **2017**, *18*, 921. [CrossRef]
14. Zou, Y.; Hu, J.; Huang, W.; Zhu, L.; Shao, M.; Dordoe, C.; Ahn, Y.J.; Wang, D.; Zhao, Y.; Xiong, Y.; et al. The botanical origin and antioxidant, anti-BACE1 and antiproliferative properties of bee pollen from different regions of South Korea. *BMC Complement. Med. Ther.* **2020**, *20*, 1–14. [CrossRef]
15. Pascoal, A.; Chambó, D.; Estevinho, L.M. Botanical origin, physicochemical characterization, and antioxidant activity of bee pollen samples from the northeast of Portugal. *J. Apic. Res.* **2022**, 1–11. [CrossRef]
16. Tomás-Lorente, F.; Garcia-Grau, M.M.; Nieto, J.L.; Tomás-Barberán, F.A. Flavonoids from Cistus ladanifer bee pollen. *Phytochemistry* **1992**, *31*, 2027–2029. [CrossRef]
17. Bonvehí, J.S.; Torrentó, M.S.; Lorente, E.C. Evaluation of Polyphenolic and Flavonoid Compounds in Honeybee-Collected Pollen Produced in Spain. *J. Agric. Food Chem.* **2001**, *49*, 1848–1853. [CrossRef] [PubMed]

18. Lu, P.; Takiguchi, S.; Honda, Y.; Lu, Y.; Mitsui, T.; Kato, S.; Kodera, R.; Furihata, K.; Zhang, M.; Okamoto, K.; et al. NMR and HPLC profiling of bee pollen products from different countries. *Food Chem. Mol. Sci.* **2022**, *5*, 100119. [CrossRef] [PubMed]
19. Ares, A.M.; Toribio, L.; Tapia, J.A.; González-Porto, A.V.; Higes, M.; Martín-Hernández, R.; Bernal, J. Differentiation of bee pollen samples according to the apiary of origin and harvesting period based on their amino acid content. *Food Biosci.* **2022**, *50*, 102092. [CrossRef]
20. Commission Regulation (EC) No 868/2007 of 23 July 2007 entering a designation in the Register of protected designations of origin and protected geographical indications (Miel de Galicia or Mel de Galicia PGI). *O.J.E.U.* **2007**, L 192/11–L 192/18.
21. Gabriele, M.; Parri, E.; Felicioli, A.; Sagona, S.; Pozzo, L.; Biondi, C.; Domenici, V.; Pucci, L. Phytochemical composition and antioxidant activity of Tuscan bee pollen of different botanic origins. *Ital. J. Food Sci.* **2015**, *27*, 248.
22. Singleton, V.L.; Rossi, J.A. Colorimetry of total phenolics with phosphomolybdic-phosphotungstic acid reagents. *Am. J. Enol. Vitic.* **1965**, *16*, 144–158.
23. Arvouet-Grand, A.; Vennat, B.; Pourrat, A.; Legret, P. Standardization of propolis extract and identification of principal constituents. *J. Pharm. Belg.* **1994**, *49*, 462–468. [PubMed]
24. Brand-Williams, W.; Cuvelier, M.E.; Berset, C. Use of a free radical method to evaluate antioxidant activity. *LWT Food Sci. Technol.* **1995**, *28*, 25–30. [CrossRef]
25. Re, R.; Pellegrini, N.; Proteggente, A.; Pannala, A.; Yang, M.; Rice-Evans, C. Antioxidant activity applying an improved ABTS radical cation decolorization assay. *Free Radic. Biol. Med.* **1999**, *26*, 1231–1237. [CrossRef]
26. Kieliszek, M.; Piwowarek, K.; Kot, A.M.; Błażejak, S.; Chlebowska-Śmigiel, A.; Wolska, I. Pollen and bee bread as new health-oriented products: A review. *Trends Food Sci. Technol.* **2018**, *71*, 170–180. [CrossRef]
27. Šarić, A.; Balog, T.; Sobočanec, S.; Kušić, B.; Šverko, V.; Rusak, G.; Likić, S.; Bubalo, D.; Pinto, B.; Reali, D.; et al. Antioxidant effects of flavonoid from Croatian *Cystus incanus* L. rich bee pollen. *Food Chem. Toxicol.* **2009**, *47*, 547–554. [CrossRef]
28. Kostić, A.; Milinčić, D.D.; Gašić, U.M.; Nedić, N.; Stanojević, S.P.; Tešić, L.; Pešić, M.B. Polyphenolic profile and antioxidant properties of bee-collected pollen from sunflower (*Helianthus annuus* L.) plant. *LWT Food Sci. Technol.* **2019**, *112*, 108244. [CrossRef]
29. Campos, M.; Markham, K.R.; Mitchell, K.A.; da Cunha, A.P. An approach to the characterization of bee pollens via their flavonoid/phenolic profiles. *Phytochem. Anal.* **1997**, *8*, 181–185. [CrossRef]
30. Morais, M.; Moreira, L.; Feás, X.; Estevinho, L.M. Honeybee-collected pollen from five Portuguese Natural Parks: Palynological origin, phenolic content, antioxidant properties and antimicrobial activity. *Food Chem. Toxicol.* **2011**, *49*, 1096–1101. [CrossRef]
31. Seijo, M.C.; Escuredo, O.; Fernández-González, M. Fungal diversity in honeys from northwest Spain and their relationship to the ecological origin of the product. *Grana* **2011**, *50*, 55–62. [CrossRef]
32. Rodríguez-Flores, S.; Escuredo, O.; Seijo, M.C. Characterization and antioxidant capacity of sweet chestnut honey produced in North-West Spain. *J. Apic. Sci.* **2016**, *60*, 19–30. [CrossRef]
33. Monasterio-Huelin, E. Avances del estudio del género *Rubus* L. (Rosaceae) en la Península Ibérica. *Anales Jard. Bot. Madrid* **1991**, *48*, 274–281.
34. Escuredo, O.; Silva, L.R.; Valentão, P.; Seijo, M.C.; Andrade, P.B. Assessing *Rubus* honey value: Pollen and phenolic compounds content and antibacterial capacity. *Food Chem.* **2012**, *130*, 671–678. [CrossRef]
35. Rodríguez-Flores, M.S.; Escuredo, O.; Seijo-Rodríguez, A.; Seijo, M.C. Characterization of the honey produced in heather communities (NW Spain). *J. Apic. Res.* **2018**, *58*, 84–91. [CrossRef]
36. Feás, X.; Vázquez-Tato, M.P.; Estevinho, L.; Seijas, J.A.; Iglesias, A. Organic Bee Pollen: Botanical Origin, Nutritional Value, Bioactive Compounds, Antioxidant Activity and Microbiological Quality. *Molecules* **2012**, *17*, 8359–8377. [CrossRef]
37. Mărghitaş, L.A.; Stanciu, O.G.; Dezmirean, D.S.; Bobiş, O.; Popescu, O.; Bogdanov, S.; Campos, M.G. In vitro antioxidant capacity of honeybee-collected pollen of selected floral origin harvested from Romania. *Food Chem.* **2009**, *115*, 878–883. [CrossRef]
38. Kuskoski, E.M.; Asuero, A.G.; Troncoso, A.M.; Mancini-Filho, J.; Fett, R. Aplicación de diversos métodos químicos para determinar actividad antioxidante en pulpa de frutos. *Food Sci. Technol.* **2005**, *25*, 726–732. [CrossRef]
39. Rice-Evans, C.A.; Miller, N.J.; Paganga, G. Structure-antioxidant activity relationships of flavonoids and phenolic acids. *Free Radic. Biol. Med.* **1996**, *20*, 933–956. [CrossRef]

 *foods*

Communication

Effects of Bee Pollen Derived from *Acer mono* Maxim. or *Phellodendron amurense* Rupr. on the Lipid Composition of Royal Jelly Secreted by Honeybees

Enning Zhou [1,2,†], Qi Wang [1,†], Xiangxin Li [2], Dan Zhu [3], Qingsheng Niu [1,*], Qiangqiang Li [2,*] and Liming Wu [2]

1   Apiculture Science Institute of Jilin Province, Jilin 132011, China
2   Institute of Apicultural Research, Chinese Academy of Agricultural Sciences, Beijing 100093, China
3   Department of Food Science, University of Otago, Dunedin 9016, New Zealand
*   Correspondence: apis1969@163.com (Q.N.); liqiangqiang@caas.cn (Q.L.); Tel.: +86-13943233663 (Q.N.); +86-13269495300 (Q.L.)
†   These authors contributed equally to this work.

Abstract: Royal jelly is a specific product secreted by honeybees, and has been sought after to maintain health because of its valuable bioactive substances, e.g., lipids and vitamins. The lipids in royal jelly come from the bee pollen consumed by honeybees, and different plant source of bee pollen affects the lipid composition of royal jelly. However, the effect of bee pollen consumption on the lipid composition of royal jelly remains unclear. Herein, we examined the influence of two factors on the lipid composition of royal jelly: first, two plant sources of bee pollen, i.e., *Acer mono* Maxim. (BP-Am) and *Phellodendron amurense* Rupr. (BP-Pa); secondly, different feeding times. Lipidomic analyses were conducted on the royal jelly produced by honeybees fed BP-Am or BP-Pa using ultra-high performance liquid chromatography (UPLC)-Q-Exactive Orbitrap mass spectrometry. The results showed that the phospholipid and fatty acid contents differed in royal jelly produced by honeybees fed BP-Am compared to those fed BP-Pa. There were also differences between timepoints, with many lipid compounds decreasing in abundance soon after single-pollen feeding began, slowly increasing over time, then decreasing again after 30 days of single-pollen feeding. The single bee pollen diet destroyed the nutritional balance of bee colonies and affected the development of hypopharyngeal and maxillary glands, resulting in differences in royal jelly quality. This study provides guidance for optimal selection of honeybee feed for the production of high-quality royal jelly.

Keywords: royal jelly; bee pollen; lipidomics; UPLC-Q-Exactive Orbitrap mass spectrometry

Citation: Zhou, E.; Wang, Q.; Li, X.; Zhu, D.; Niu, Q.; Li, Q.; Wu, L. Effects of Bee Pollen Derived from *Acer mono* Maxim. or *Phellodendron amurense* Rupr. on the Lipid Composition of Royal Jelly Secreted by Honeybees. *Foods* **2023**, *12*, 625. https://doi.org/10.3390/foods12030625

Academic Editor: Vladimir Tomović

Received: 6 December 2022
Revised: 13 January 2023
Accepted: 20 January 2023
Published: 2 February 2023

1. Introduction

In the dietary structure of honeybees, honey and bee pollen are the main food sources. Honey is the primary source of carbohydrates, whereas bee pollen can provide a variety of nutrients for growth and development, including proteins, carbohydrates, lipids, vitamins, and minerals [1]. These substances provide bee pollen with nutritional and medical properties, such as antibacterial, antioxidant, and immunoenhancement activities [2]. Bee pollen is also an important source of proteins and lipids for production of royal jelly [3], a viscous, milky substance that is secreted by honeybee hypopharyngeal and maxillary glands. It is nutrient-rich, containing substances such as proteins, amino acids, organic acids, lipids, and minerals [4]. Royal jelly not only promotes the growth and development of queen bee larvae, prolongs the honeybee lifespan [5], benefits animal reproduction and breeding [6], but it can also serve as a human functional food and dietary supplement, and is therefore of great significance in human healthcare.

In recent years, many studies have been conducted on the nutritional components and functional activities of royal jelly. The active substances in royal jelly (such as proteins, peptides, and lipids) can endow it with various functional activities, including anti-microbial, anti-oxidative, anti-aging, anti-inflammatory, and hypoglycemic activities [7,8]. Lipids are small molecules that play important roles in maintaining the physiological dynamic balance. The LIPID MAPS database classifies lipids into eight categories, such as glycerolipids (GLs), glycerophospholipids (GPs), and sphingolipids (SPs) [9]. Royal jelly is rich in lipids, which account for 4–8% of fresh royal jelly and up to 15–30% of freeze-dried royal jelly products [3]. Free fatty acids (FAs) comprise ~90% of royal jelly lipids, which are of great significance to honeybee colonies. They not only provide nutrition for the queen and larvae during the developmental period, but also have antibacterial and mite repelling functions [10,11]. The free FAs in royal jelly also have various health-promoting activities and can be used as preventive and supplemental drugs for human health. Studies have shown that the lipid components of royal jelly can inhibit tumor cells, regulate the human immune system, prevent skin aging, and induce neurogenesis [12]. It is necessary to study the lipid composition of royal jelly to promote its development and utilization in human health applications.

The lipids in royal jelly are derived from the bee pollen consumed by honeybees. The lipid content of bee pollen is up to 13% of the dry weight, and the types and levels of lipids in bee pollen vary based on the plant source [13]. Importantly, it has been shown that the type of plant from which bee pollen is produced affects the lipid composition of royal jelly [5]. However, there has been a limited number of studies related to the effects of bee pollen consumption on the lipid composition of royal jelly. The traditional Chinese herbs *Acer mono* Maxim. and *Phellodendron amurense* Rupr. have a wide range of biological activities, including anti-oxidant, anti-tumor, anti-microbial, anti-inflammatory, anti-diabetic, and hepatoprotective effects [14,15]. Bee pollen derived from *A. mono* Maxim. (BP-Am) and from *P. amurense* Rupr. (BP-Pa) also retain these nutritional and pharmacological properties. However, the effects of BP-Am and BP-Pa consumption on royal jelly lipid composition have not been clarified.

We speculated that different plant source of bee pollen and different feeding time would significantly affect the lipid composition in royal jelly. Lipidomics technology has been widely used to identify the lipid composition in various foods and biological samples [16]. In this study, we applied a lipidomics approach to investigate the effects of two types of bee pollen (BP-Am and BP-Pa) on the lipid composition of royal jelly. Royal jelly secreted by honeybees with a natural mixed bee pollen diet was used as the control. We conducted analyses of royal jelly collected at several time points to determine changes in lipid composition during different stages of bee pollen consumption. An ultra-high performance liquid chromatography (UPLC)-Q-Exactive Orbitrap mass spectrometer was used to identify and quantify lipids in royal jelly samples. Our findings would provide guidance for the selection of the optimal feed for honeybees for the production of high-quality royal jelly.

2. Materials and Methods

2.1. Reagents

Acetonitrile (ACN), isopropanol (IPA), chloroform ($CHCl_3$), ethanol, methanol (MeOH), and ammonium acetate were obtained from Fisher Scientific, Inc. (Pittsburgh, PA, USA). All solvents used as UPLC mobile phases or for lipid extraction were chromatographic grade. Other chemicals were of analytical grade and purchased from Sangon Biotechnology Co., Ltd. (Shanghai, China). Ultrapure water from a milli-Q water purification system (Millipore, Bedford, MA, USA) was used for lipidomics.

2.2. Bee Pollen Collection

BP-Am was collected in April 2021 (during the *A. mono* Maxim. flower season) from an apiary located at the *A. mono* Maxim. planting base in Jilin Province, China. BP-Am samples were confirmed through palynological analysis [17] to have about 82% *A. mono* Maxim. pollen grains. BP-Pa was collected in May 2021 (during the *P. amurense* Rupr. flowering season) from an apiary located at the *P. amurense* Rupr. planting base in Jilin Province,

China. BP-Pa samples were confirmed to have about 86% *P. amurense* Rupr. pollen grains. Our samples met the requirements for classification as monofloral bee pollen (>80%) [18]. Bee pollen samples were ground and lyophilized into powder, sterilized by irradiation at 7 kGy, then stored at $-80\ ^\circ$C prior to further study.

2.3. Royal Jelly Collection

One month before the experiment, six honeybee colonies were prepared and populations were adjusted to maintain a constant number of worker bees in each colony. Empty combs were added to each colony and stimulative feeding was performed to maintain consistent worker bee ages and larvae numbers. Royal jelly was collected from the six colonies before the honeybees were fed single bee pollen; these samples were used as the controls. Traps were then set at hive entrances to prevent honeybees from carrying pollen in. The six honeybee colonies were randomly divided into two groups of three colonies each. Each colony was regularly and separately fed with equal amounts of BP-Am or BP-Pa to ensure normal royal jelly production. An artificial royal jelly collection frame was placed into the hive. After honeybees became familiar with it, larvae were transferred into the frame to induce nurse bees to secrete royal jelly into it. Royal jelly was collected from the frame at 72 h after larvae were transferred. Royal jelly samples were stored at $-80\ ^\circ$C prior to further study.

2.4. Lipid Extraction

Lipids were extracted from bee pollen and royal jelly as previously described [13]. Briefly, each 300-mg sample (BP-Am, BP-Pa, *A. mono* Maxim. royal jelly [RJ-Am], or *P. amurense* Rupr. royal jelly [RJ-Pa]) was mixed with 1.2 mL of 1:2 MeOH:CHCl$_3$ (v/v). After vortexing for 5 min, 400 µL of ultrapure water was added, then samples were vortexed for 1 min. After centrifugation at 3000× g for 15 min, the lower organic phase was transferred to a clean glass tube and dried under a stream of nitrogen. Dried extracts were then redissolved in 200 µL of 1:2 MeOH:CHCl$_3$ (v/v) for UPLC-Q-Exactive Orbitrap/MS analysis.

2.5. UPLC-MS Analyses

A UPLC system combined with a Q-Exactive Orbitrap mass spectrometer (Thermo Fisher Scientific, Waltham, MA, USA) equipped with a heated electrospray ionization (HESI) probe was used for lipidomic profiling as previously reported [13]. An XSelect CSH C18 column (100 × 2.1 mm, 2.5 µm) (Waters, Milford, MA, USA) was used for lipid separation. Mobile phase A was 3:2 ACN:H$_2$O (v/v) and mobile phase B was 9:1 IPA:ACN (v/v), both of which contained 10 mM ammonium acetate. Previously published elution procedures and mass spectrometer parameters were used [16]. Data acquisition was performed in negative ionization mode with an m/z range of 200–2000. The resolutions of the full scan and fragment spectra were 70,000 and 17,500, respectively.

2.6. Data Analysis

LipidSearch v4.0 (Thermo Fisher Scientific, Massachusetts, MA, USA) was used for peak recognition, peak extraction, and lipid identification. Lipids were identified by matching to a database based on the retention time, characteristic parent ions (with an m/z tolerance of 5 ppm) and product ions (with an m/z tolerance of 8 ppm). Lipid filtration was performed using peak area and m-score thresholds of <1e^5 and <10, respectively.

2.7. Statistics

One-way analysis of variance (ANOVA) was performed to test the significance of differences in lipid levels between samples using SPSS v21.0 (Statistical Product and Service Solutions, Illinois, IL, USA). Differences were considered significant at $p < 0.05$.

3. Results

3.1. Lipidomic Profiles in Different Types of Royal Jelly and Bee Pollen

Phospholipids and FAs are functional lipids that can regulate human immunity and metabolism. The composition and concentrations of functional lipids are important indexes by which food nutritional quality can be evaluated [19]. The negative ionization mode of mass spectrometry is suitable for phospholipid and FA measurements [13]. We here used negative ionization mode to analyze phospholipids and FAs in royal jelly and bee pollen samples. There were 20 kinds of ceramides (Cers), 14 kinds of phosphatidylcholines (PCs), 25 kinds of phosphatidylethanolamines (PEs), 3 kinds of sphingomyelins (SMs), and 51 kinds of FAs detected in royal jelly and bee pollen (Table S1). There were significant differences in the abundance of Cers, PCs, PEs, SMs, and monounsaturated FAs (MUFAs) between BP-Am and BP-Pa (Table 1). However, there were no significant differences in the abundance of saturated FAs (SFAs), and polyunsaturated FAs (PUFAs). We hypothesized that differences in bee pollen lipid composition may contribute to differences in the corresponding royal jelly lipid composition. Therefore, differences were analyzed in the lipid composition of two types of royal jelly (RJ-Am and RJ-Pa) obtained from honeybees fed BP-Am or BP-Pa, respectively.

Table 1. Relative abundance of different kinds of lipids in single bee pollen derived from *Acer mono* Maxim. and *Phellodendron amurense* Rupr.

Lipids	BP-Am	BP-Pa	*p* Value
Cer	$2.89 \times 10^9 \pm 3.42 \times 10^8$	$3.29 \times 10^9 \pm 3.09 \times 10^8$	0.038
PC	$8.71 \times 10^9 \pm 2.36 \times 10^8$	$5.97 \times 10^9 \pm 4.18 \times 10^8$	0.030
PE	$1.11 \times 10^{10} \pm 1.94 \times 10^9$	$2.59 \times 10^9 \pm 5.74 \times 10^8$	0.012
SM	$1.72 \times 10^6 \pm 1.03 \times 10^5$	$5.60 \times 10^6 \pm 5.52 \times 10^4$	0.005
SFA	$6.40 \times 10^{10} \pm 4.60 \times 10^9$	$3.93 \times 10^{10} \pm 1.72 \times 10^9$	0.052
MUFA	$2.94 \times 10^{10} \pm 5.08 \times 10^9$	$1.32 \times 10^{10} \pm 1.11 \times 10^9$	0.027
PUFA	$8.07 \times 10^{10} \pm 8.85 \times 10^9$	$8.64 \times 10^{10} \pm 3.09 \times 10^9$	0.624
FA(18:0)	$8.91 \times 10^9 \pm 1.24 \times 10^9$	$2.17 \times 10^9 \pm 1.66 \times 10^8$	0.012
FA(18:1)	$2.55 \times 10^{10} \pm 3.94 \times 10^9$	$9.87 \times 10^9 \pm 8.81 \times 10^8$	0.024
FA(18:2)	$3.22 \times 10^{10} \pm 4.83 \times 10^9$	$2.15 \times 10^{10} \pm 1.04 \times 10^9$	0.082
FA(18:3)	$5.79 \times 10^{10} \pm 7.58 \times 10^9$	$6.66 \times 10^{10} \pm 4.64 \times 10^9$	0.060

Note: Cer, ceramide; PC, phosphatidylcholine; PE, phosphatidylethanolamine; SM, sphingomyelin; SFA, saturated fatty acid; MUFA, monounsaturated fatty acid; PUFA, polyunsaturated fatty acid; FA(18:0), FA(18:1), FA(18:2), and FA(18:3) are classified as 18-carbon fatty acids; BP-Am, bee pollen derived from *Acer mono* Maxim.; BP-Pa, bee pollen derived from *Phellodendron amurense* Rupr.

3.2. Differences in Royal Jelly Phospholipid and Sphingolipid Composition

Phospholipids and sphingolipids are important components of cell membranes and have a variety of physiological functions that are beneficial to human health [20]. To further understand the effects of the BP-Pa and BP-Am diets on the composition of phospholipids and sphingolipids in royal jelly, we quantified SMs, Cers, PCs, and PEs in the two types of royal jelly (RJ-Am and RJ-Pa). There were clear trends in the relative content of SMs, Cers, PCs, and PEs in RJ-Am and RJ-Pa (Figure 1). In their natural state, bee colonies gather various kinds of pollen for food. After honeybees were fed with a single pollen source instead of natural pollen for six days, levels of SMs, Cers, PCs, and PEs were significantly lower in the single-pollen royal jelly compared to the control. Lipid levels in the single-pollen royal jelly began to increase after 12 days of feeding; the highest lipid levels occurred on day 18 or 24, but were not as high as lipid levels in the control royal jelly. On day 30, the lipid content of royal jelly began to decrease significantly. Notably, the average PC content was significantly higher in RJ-Am than in RJ-Pa, which might be attributed to the significantly higher PC content in BP-Am than in BP-Pa.

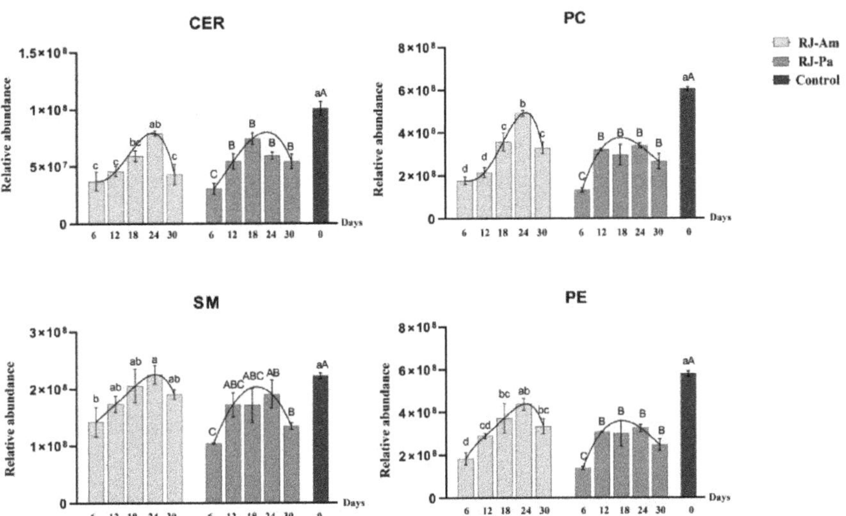

Figure 1. Changes in levels of phosphatidylcholines (PCs), phosphatidylethanolamines (PEs), ceramides (Cers), and sphingomyelins (SMs) in royal jelly secreted by honeybees fed bee pollen from a single plant source, *Acer mono* Maxim. (BP-Am) or *Phellodendron amurense* Rupr. (BP-Pa). Samples were collected on days 6, 12, 18, 24, and 30 after honeybees began feeding on single bee pollen. RJ-Am, royal jelly secreted by honeybees fed BP-Am; RJ-Pa, royal jelly secreted by honeybees fed BP-Pa. Significant differences in lipid content of RJ-Am or RJ-Pa secreted by honeybees fed BP-Am or BP-Pa at different timepoints are indicated with lowercase letters (a, b, c, d) and uppercase letters (A, B, C), respectively.

3.3. Differences in Royal Jelly Fatty Acid Composition

In the present study, a relative quantitative analysis of SFAs, MUFAs, and PUFAs was carried out in royal jelly. Trends in levels of SFAs, MUFAs, and PUFAs were similar to those of SMs, Cers, PCs, and PEs (Figure 2). Specifically, SFA, MUFA, and PUFA contents were significantly decreased in royal jelly after initial feeding with single bee pollen. However, as honeybees adapted to the single-pollen diet, levels of SFAs, MUFAs, and PUFAs gradually increased. The highest levels were on day 18 or 24, but levels decreased on day 30.

In addition, trends in 10-hydroxydec-2-enoic acid (10-HDA) levels were consistent with those of other lipids in royal jelly (Figure 2): 10-HDA content was significantly decreased after initial feeding with single bee pollen; levels gradually increased as honeybees adapted to the single-pollen diet; the highest level of 10-HDA occurred on day 18 or 24; and levels were decreased on day 30. The average 10-HDA content was significantly higher in RJ-Am than in RJ-Pa. These results indicated that continuous feeding of different types of single bee pollen had significant effects on 10-HDA content in royal jelly.

Furthermore, we compared the relative abundance of 18-carbon fatty acids in bee pollen and found that the content of FA(18:0) and FA(18:1) in BP-Am was significantly higher than that in BP-Pa (Table 1). However, there's no significant difference in the relative abundance of 18-carbon fatty acids between RJ-Am and RJ-Pa (Figure 3), which might be attributed to the participation of 18-carbon fatty acids in lipid metabolism and the synthesis of 10-HDA. We also analyzed the ratio of free and bound 18-carbon saturated and unsaturated FAs in RJ-Am and RJ-Pa on day 18. We found that there were no significant differences in the ratios of FA(18:0), FA(18:1), FA(18:2), or FA(18:3) between the two royal jelly samples (Figure 4). This suggested that different types of bee pollen did not affect levels of 18-carbon saturated and unsaturated FAs in royal jelly, which might be related to lipid metabolism and regulation of maxillary glands during royal jelly secretion.

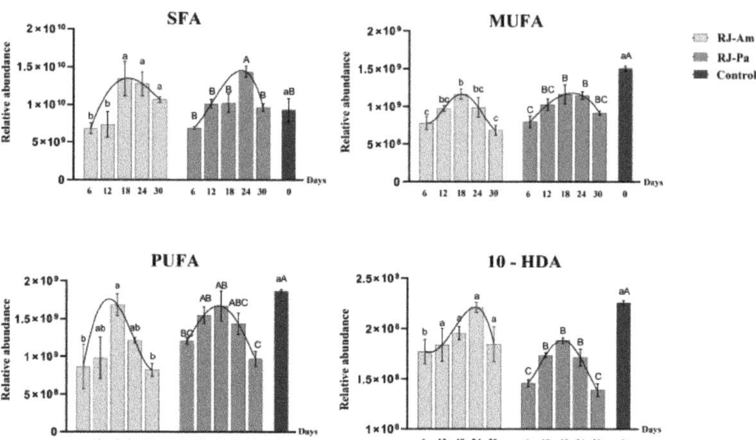

Figure 2. Changes in levels of saturated fatty acids (SFAs), monounsaturated fatty acids (MUFAs), polyunsaturated fatty acids (PUFAs), and 10-hydroxydec-2-enoic acid (10-HDA) in royal jelly samples. Samples were collected on days 6, 12, 18, 24, and 30 after honeybees began feeding on single bee pollen. BP-Am, single bee pollen derived from *A. mono* Maxim.; BP-Pa, single bee pollen derived from *P. amurense* Rupr.; RJ-Am, royal jelly secreted by honeybees fed BP-Am; RJ-Pa, royal jelly secreted by honeybees fed BP-Pa. Significant differences in lipid content of RJ-Am or RJ-Pa secreted by honeybees fed BP-Am or BP-Pa at different timepoints are indicated with lowercase letters (a, b, c) and uppercase letters (A, B, C), respectively.

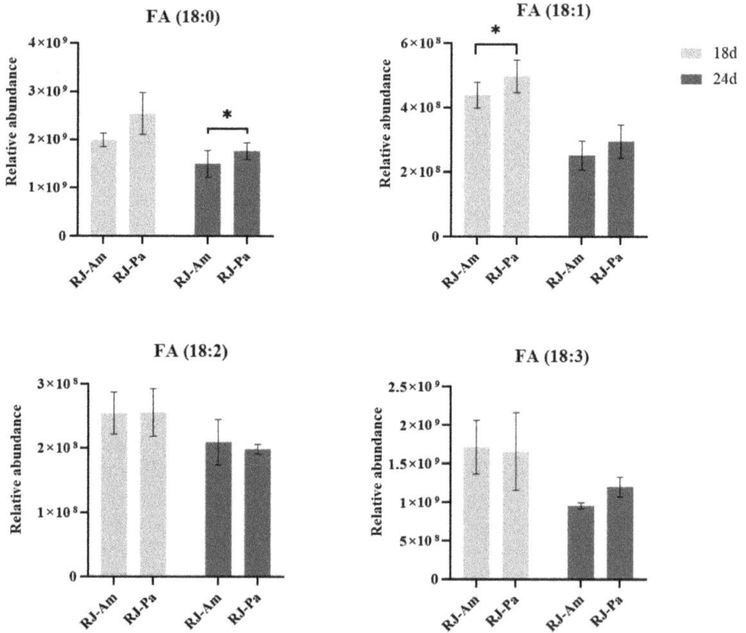

Figure 3. Changes in levels of 18-carbon saturated and unsaturated fatty acids in royal jelly secreted by honeybees fed single bee pollen derived from *Acer mono* Maxim. (BP-Am) or *Phellodendron amurense* Rupr. (BP-Pa). Samples were collected on days 18 and 24 after honeybees began feeding on single bee pollen. RJ-Am, royal jelly secreted by honeybees fed BP-Am; RJ-Pa, royal jelly secreted by honeybees fed BP-Pa. *, *p*-value < 0.05.

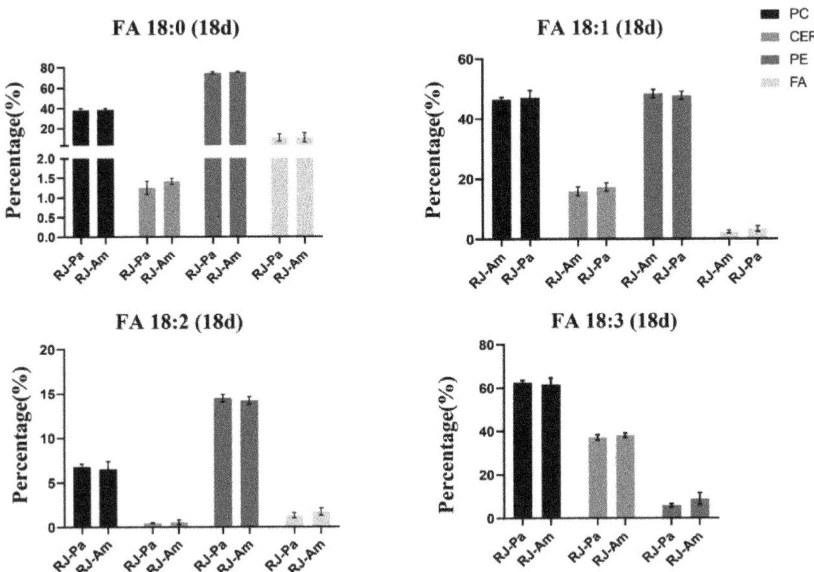

Figure 4. The ratio of free and bound 18-carbon saturated and unsaturated fatty acids in royal jelly secreted by honeybees fed single bee pollen derived from *Acer mono* Maxim. (BP-Am) or *Phellodendron amurense* Rupr. (BP-Pa). RJ-Am, royal jelly secreted by honeybees fed BP-Am; RJ-Pa, royal jelly secreted by honeybees fed BP-Pa.

4. Discussion

In this study, UPLC-Q-Exactive Orbitrap/MS analysis was conducted to explore the effects of different types of bee pollen consumption by honeybees on the lipid composition of royal jelly. PC and PE belong to the phospholipids, Cer and SM belong to the sphingolipids. PC is known as the "third nutrient" and has a variety of biological activities. It can regulate lipid metabolism, prevent vascular diseases, counteract inflammation and oxidation, improve brain and nerve function, and prevent senile dementia [21,22]. PEs maintain the structure and function of cells, repair nerve cell membranes, and promote normal metabolism in brain neurons. The PC/PE ratio is an important factor affecting metabolic dysfunction and insulin sensitivity [23]. SMs are phospholipids containing sphingosine or dihydrosphingosine; these are primarily metabolized in the small intestine and colon and have various physiological functions. For example, SM can reduce intestinal absorption of cholesterol, down-regulate proteins related to cholesterol absorption, and reduce blood cholesterol levels [24]. SMs can also promote colon cancer cell apoptosis, inhibit colon cancer cell proliferation, improve skin barrier function, and maintain physiological functions of skin [25,26]. SMs are important initial substrates of the sphingomyelin signaling pathway and can be hydrolyzed by sphingomyelinase to produce Cer. In the reverse reaction, sphingomyelinase synthesizes SMs from Cers [27]. Cer is an important signaling molecule in many basic cellular physiological and biochemical processes, such as inflammation, immune cell transport, stress responses, apoptosis, and autophagy [27]. Notably, PCs can be hydrolyzed into linoleic acid (FA 18:2) and α-linolenic acid (FA 18:3) which can participate in linoleic acid and α-linoleic acid metabolism. Stearic acid (FA 18:0) and oleic acid (FA 18:1) are also associated with glycerophospholipid metabolism through the FA biosynthesis pathway (Figure 5). Therefore, the nutritional and functional properties of royal jelly may be affected by changes in the lipid composition. Our findings suggested that the optimum time point for collecting royal jelly from artificial feeding bee pollen is 18~24 days, because the content of functional lipids in royal jelly is the most abundant during this period.

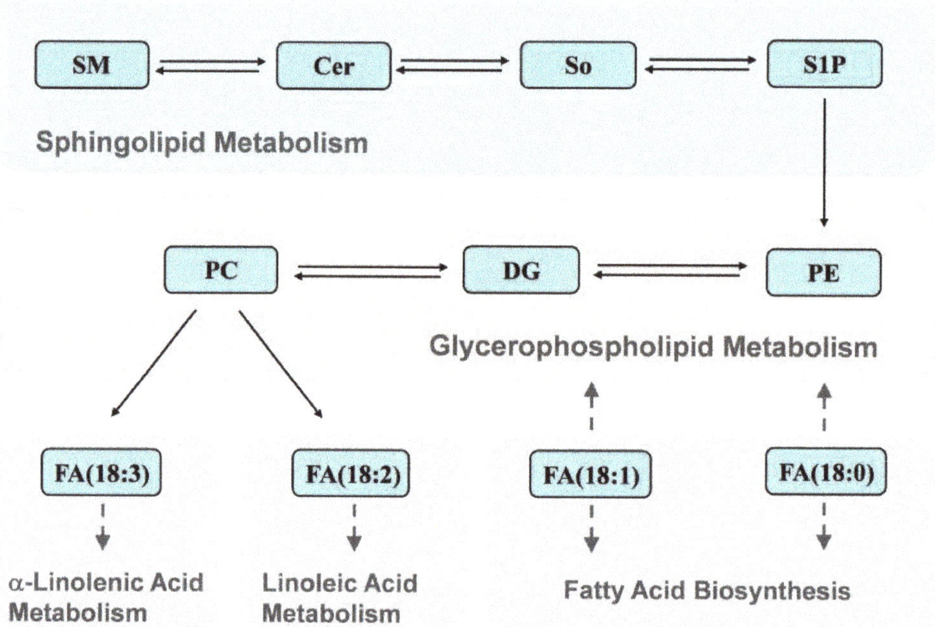

Figure 5. Kyoto Encyclopedia of Genes and Genomes (KEGG) biological pathway annotations for the detected lipids. SM, sphingomyelin; Cer, ceramide; So, sphingosine; S1P, sphingosine-1-phosphate; PE, phosphatidylethanolamine; DG, 1,2-diacyl-sn-glycerol; PC, phosphatidylcholine.

Studies have shown that honeybee foraging preferences may be related to the protein/lipid ratio in pollen. Honeybees can selectively balance their intake of amino acids and FAs during natural foraging [28]. Moreover, royal jelly is secreted by the hypopharyngeal and maxillary glands, development of which is positively correlated with the type and quantity of pollen ingested. A previous study found that consumption of mixed bee pollen is better for hypopharyngeal gland development than single pollen [29]. In addition, the increase in lipids before day 24 may therefore be attributed to a gradual adaptation to the single-pollen diet. However, the forced change from a natural mixed bee pollen diet to the single bee pollen diet may destroy the nutritional balance and therefore affect the development of honeybee hypopharyngeal glands over time, resulting in the observed decrease in lipids after 24 days of single bee pollen consumption.

Moreover, FAs can be used as fuel, bioactive lipid media precursors, and cell membrane components (in the form of phospholipids and glycolipids). SFAs do not contain unsaturated double bonds, which are generally considered the main trigger of high blood cholesterol, obesity, and coronary heart disease in humans [30,31]. Unsaturated FAs are classified as MUFAs or PUFAs, with classification depending on the number and positions of double bonds. Dietary unsaturated FAs have various physiological functions, such as inhibiting inflammatory mediators and cytokines, regulating lipid metabolism through transcription factors, and preventing cardiovascular diseases [32,33]. Ahad unique MUFA with antibacterial, antioxidative and anti-inflammatory activities, 10-HDA, accounts for over 50% of free FAs in royal jelly [12]. Honeybees that consume bee pollen from different plant sources secrete royal jelly with different levels of 10-HDA compared to those that consume single pollen [34]. In honeybee maxillary glands, 10-HDA is converted from an 18-carbon FA (stearic acid) through hydroxylation and β-oxidation to shorten the carbon chain [35]. Consumption of oleic acid can also affect 10-HDA levels in royal jelly [36].

Therefore, our results indicated that feeding bees with BP-Am was beneficial to increase the content of 10-HDA in royal jelly than feeding with BP-Pa, due to the significantly higher content of stearic acid and oleic acid in BP-Am than in BP-Pa.

5. Conclusions

In summary, the results indicated that levels of phospholipids and FAs were decreased in royal jelly secreted by honeybees fed with single bee pollen. However, as honeybees adapted to the single-pollen diet, levels of phospholipids and FAs gradually increased again. The highest levels appeared on days 18 or 24, but were decreased on day 30. Thus, it suggested that the optimum time point for collecting royal jelly from artificial feeding bee pollen is 18~24 days. Moreover, the average contents of PC and 10-HDA were significantly higher in RJ-Am than in RJ-Pa, indicating that continuous feeding of BP-Am was beneficial to increase the contents of PC and 10-HDA in royal jelly than BP-Pa. Additionally, our findings showed that a forced change from the natural mixed bee pollen diet to a single-pollen diet may negatively influence the quality of royal jelly. This study would provide a new perspective on bee pollen diets and scientific guidance for royal jelly production.

Supplementary Materials: The following supporting information can be downloaded at: https://www.mdpi.com/article/10.3390/foods12030625/s1, Table S1: Lipidomic profiles.

Author Contributions: Conceptualization, Q.L. and L.W.; Methodology, Q.L.; Software, E.Z.; Validation, X.L. and D.Z.; Formal analysis, E.Z., Q.W. and X.L.; Investigation, E.Z. and Q.W.; Resources, Q.W. and Q.N.; Data cu-ration, X.L. and D.Z.; Writing—original draft, E.Z. and Q.W.; Writing—review and editing, Q.L.; Visualization, E.Z.; Supervision, Q.N., Q.L. and L.W.; Project administration, Q.N. and L.W.; Funding acquisition, Q.N., Q.L. and L.W. All authors have read and agreed to the published version of the manuscript.

Funding: This work was supported by the National Natural Science Foundation of China (No. 31372386), the Fundamental Research Fund of Chinese Academy of Agricultural Sciences (No. 125161029000160013), and the Agricultural Science and Technology Innovation Program under Grant (CAAS-ASTIP-2021-IAR).

Institutional Review Board Statement: The animal study protocol was approved by the Animal Ethics Committee of Institute of Apicultural Research, Chinese Academy of Agricultural Sciences (Beijing, China) (protocol code: CAAS-IAR-ER0045, approval data: April 2021).

Informed Consent Statement: Not applicable.

Data Availability Statement: Data available on request due to privacy.

Conflicts of Interest: The authors declared no conflict of interest.

References

1. Tsuruda, J.M.; Chakrabarti, P.; Sagili, R.R. Honey Bee Nutrition. *Vet. Clin. N. Am. Food Anim. Pract.* **2021**, *37*, 505–519. [CrossRef] [PubMed]
2. Abdelnour, S.A.; Abd El-Hack, M.E.; Alagawany, M.; Farag, M.R.; Elnesr, S.S. Beneficial impacts of bee pollen in animal production, reproduction and health. *J. Anim. Physiol. Anim. Nutr.* **2019**, *103*, 477–484. [CrossRef] [PubMed]
3. Yan, S.; Wang, X.; Sun, M.; Wang, W.; Wu, L.; Xue, X. Investigation of the lipidomic profile of royal jelly from different botanical origins using UHPLC-IM-Q-TOF-MS and GC-MS. *LWT-Food Sci. Technol.* **2022**, *169*, 113894. [CrossRef]
4. Ramadan, M.F.; Al-Ghamdi, A. Bioactive compounds and health-promoting properties of royal jelly: A review. *J. Funct. Foods* **2012**, *4*, 39–52. [CrossRef]
5. Shi, J.L.; Liao, C.H.; Wang, Z.L.; Wu, X.B. Effect of royal jelly on longevity and memory-related traits of Apis mellifera workers. *J. Asia-Pac. Entomol.* **2018**, *21*, 1430–1433. [CrossRef]
6. Abdelnour, S.A.; Abd El-Hack, M.E.; Alagawany, M.; Taha, A.E.; Elnesr, S.S.; Abd Elmonem, O.M.; Swelum, A.A. Useful impacts of royal jelly on reproductive sides, fertility rate and sperm traits of animals. *J. Anim. Physiol. Anim. Nutr.* **2020**, *104*, 1798–1808. [CrossRef]
7. Mureşan, C.I.; Dezmirean, D.S.; Marc, B.D.; Suharoschi, R.; Pop, O.L.; Buttstedt, A. Biological properties and activities of major royal jelly proteins and their derived peptides. *J. Funct. Foods* **2022**, *98*, 105286. [CrossRef]
8. Guo, J.; Wang, Z.; Chen, Y.; Cao, J.; Tian, W.; Ma, B.; Dong, Y. Active components and biological functions of royal jelly. *J. Funct. Foods* **2021**, *82*, 104514. [CrossRef]

9. Fahy, E.; Subramaniam, S.; Murphy, R.C.; Nishijima, M.; Raetz, C.R.; Shimizu, T.; Spener, F.; van Meer, G.; Wakelam, M.J.; Dennis, E.A. Update of the LIPID MAPS comprehensive classification system for lipids. *J. Lipid Res.* **2009**, *50*, S9–S14. [CrossRef]
10. Spannhoff, A.; Kim, Y.K.; Raynal, N.J.; Gharibyan, V.; Su, M.B.; Zhou, Y.Y.; Li, J.; Castellano, S.; Sbardella, G.; Issa, J.P.; et al. Histone deacetylase inhibitor activity in royal jelly might facilitate caste switching in bees. *EMBO Rep.* **2011**, *12*, 238–243. [CrossRef]
11. Šedivá, M.; Laho, M.; Kohútová, L.; Mojžišová, A.; Majtán, J.; Klaudiny, J. 10-HDA, A major fatty acid of royal jelly, exhibits pH dependent growth-inhibitory activity against different strains of paenibacillus larvae. *Molecules* **2018**, *23*, 3236. [CrossRef] [PubMed]
12. Li, X.; Huang, C.; Xue, Y. Contribution of lipids in honeybee (*Apis mellifera*) royal jelly to health. *J. Med. Food* **2013**, *16*, 96–102. [CrossRef] [PubMed]
13. Li, Q.; Liang, X.; Zhao, L.; Zhang, Z.; Xue, X.; Wang, K.; Wu, L. UPLC-Q-Exactive Orbitrap/MS-Based lipidomics approach to Characterize lipid extracts from bee pollen and their in vitro anti-inflammatory properties. *J. Agric. Food Chem.* **2017**, *65*, 6848–6860. [CrossRef] [PubMed]
14. Wang, X.; Zhang, H.; Wang, Z.; Bai, H. Optimization of ultrasonic-assisted alkaline extraction of polysaccharides from Phellodendron amurense Rupr. pollen using response surface methodology and its structure features. *RSC Adv.* **2015**, *5*, 106800–106808. [CrossRef]
15. Bi, W.; Gao, Y.; Shen, J.; He, C.; Liu, H.; Peng, Y.; Zhang, C.; Xiao, P. Traditional uses, phytochemistry, and pharmacology of the genus Acer (maple): A review. *J. Ethnopharmacol.* **2016**, *189*, 31–60. [CrossRef]
16. Hyötyläinen, T.; Bondia-Pons, I.; Orešič, M. Lipidomics in nutrition and food research. *Mol. Nutr. Food Res.* **2013**, *57*, 1306–1318. [CrossRef]
17. Ren, C.; Wang, K.; Luo, T.; Xue, X.; Wang, M.; Wu, L.; Zhao, L. Kaempferol-3-O-galactoside as a marker for authenticating *Lespedeza bicolor* Turcz. monofloral honey. *Food Res. Int.* **2022**, *160*, 111667. [CrossRef]
18. Campos, M.G.R.; Bogdanov, S.; de Almeida-Muradian, L.B.; Szczesna, T.; Mancebo, Y.; Frigerio, C.; Ferreira, F. Pollen composition and standardisation of analytical methods. *J. Apic. Res.* **2008**, *47*, 154–161. [CrossRef]
19. Alabdulkarim, B.; Bakeet, Z.A.N.; Arzoo, S. Role of some functional lipids in preventing diseases and promoting health. *J. King Saud Univ.—Sci.* **2012**, *24*, 319–329. [CrossRef]
20. Küllenberg, D.; Taylor, L.A.; Schneider, M.; Massing, U. Health effects of dietary phospholipids. *Lipids Health Dis.* **2012**, *11*, 3. [CrossRef]
21. Tandy, S.; Chung, R.W.; Kamili, A.; Wat, E.; Weir, J.M.; Meikle, P.J.; Cohn, J.S. Hydrogenated phosphatidylcholine supplementation reduces hepatic lipid levels in mice fed a high-fat diet. *Atherosclerosis* **2010**, *213*, 142–147. [CrossRef] [PubMed]
22. Sun, N.; Chen, J.; Wang, D.; Lin, S. Advance in food-derived phospholipids: Sources, molecular species and structure as well as their biological activities. *Trends Food Sci. Technol.* **2018**, *80*, 199–211. [CrossRef]
23. Grapentine, S.; Bakovic, M. Significance of bilayer-forming phospholipids for skeletal muscle insulin sensitivity and mitochondrial function. *J. Biomed. Res.* **2019**, *34*, 1–13. [CrossRef]
24. Yang, F.; Chen, G.; Ma, M.; Qiu, N.; Zhu, L.; Li, J. Egg-Yolk Sphingomyelin and Phosphatidylcholine Attenuate Cholesterol Absorption in Caco-2 Cells. *Lipids* **2018**, *53*, 217–233. [CrossRef] [PubMed]
25. Moschetta, A.; Portincasa, P.; van Erpecum, K.J.; Debellis, L.; Vanberge-Henegouwen, G.P.; Palasciano, G. Sphingomyelin protects against apoptosis and hyperproliferation induced by deoxycholate: Potential implications for colon cancer. *Dig. Dis. Sci.* **2003**, *48*, 1094–1101. [CrossRef]
26. Haruta-Ono, Y.; Ueno, H.; Ueda, N.; Kato, K.; Yoshioka, T. Investigation into the dosage of dietary sphingomyelin concentrate in relation to the improvement of epidermal function in hairless mice. *Anim. Sci. J.* **2012**, *83*, 178–183. [CrossRef]
27. Maceyka, M.; Spiegel, S. Sphingolipid metabolites in inflammatory disease. *Nature* **2014**, *510*, 58–67. [CrossRef]
28. Vaudo, A.D.; Tooker, J.F.; Patch, H.M.; Biddinger, D.J.; Coccia, M.; Crone, M.K.; Fiely, M.; Francis, J.S.; Hines, H.M.; Hodges, M.; et al. Pollen protein: Lipid macronutrient ratios may guide broad patterns of bee species floral preferences. *Insects* **2020**, *11*, 132. [CrossRef]
29. Omar, E.; Abd-Ella, A.A.; Khodairy, M.M.; Moosbeckhofer, R.; Crailsheim, K.; Brodschneider, R. Influence of different pollen diets on the development of hypopharyngeal glands and size of acid gland sacs in caged honey bees (*Apis mellifera*). *Apidologie* **2017**, *48*, 425–436. [CrossRef]
30. Gu, Y.; Yin, J. Saturated fatty acids promote cholesterol biosynthesis: Effects and mechanisms. *Obes. Med.* **2020**, *18*, 100201. [CrossRef]
31. Li, B.; Leung, J.C.K.; Chan, L.Y.Y.; Yiu, W.H.; Tang, S.C.W. A global perspective on the crosstalk between saturated fatty acids and Toll-like receptor 4 in the etiology of inflammation and insulin resistance. *Prog. Lipid Res.* **2020**, *77*, 101020. [CrossRef] [PubMed]
32. Kotlyarov, S.; Kotlyarova, A. Anti-inflammatory function of fatty acids and involvement of their metabolites in the resolution of inflammation in chronic obstructive pulmonary disease. *Int. J. Mol. Sci.* **2021**, *22*, 12803. [CrossRef] [PubMed]
33. Abdelhamid, A.S.; Martin, N.; Bridges, C.; Brainard, J.S.; Wang, X.; Brown, T.J.; Hanson, S.; Jimoh, O.F.; Ajabnoor, S.M.; Deane, K.H.; et al. Polyunsaturated fatty acids for the primary and secondary prevention of cardiovascular disease. *Cochrane Database Syst. Rev.* **2018**, *7*, Cd012345. [CrossRef] [PubMed]
34. Pattamayutanon, P.; Peng, C.C.; Sinpoo, C.; Chantawannakul, P. Effects of pollen feeding on quality of royal jelly. *J. Econ. Entomol.* **2018**, *111*, 2974–2978. [CrossRef]

35. Plettner, E.; Slessor, K.N.; Winston, M.L. Biosynthesis of Mandibular Acids in Honey Bees (*Apis mellifera*): De novo Synthesis, Route of Fatty Acid Hydroxylation and Caste Selective β-Oxidation. *Insect Biochem. Mol. Biol.* **1998**, *28*, 31–42. [CrossRef]
36. Hu, X.Y. Effects of Acid Supplementation on 10-HDA Synthesis in *Apis mellifera ligustica* and Its Hypoglycemic Mechanism (In Chinese with English abstract). Doctoral Thesis, Shandong Agricultural University, Taian, China, 2021. [CrossRef]

 *foods*

Article

Pesticide Residues and Metabolites in Greek Honey and Pollen: Bees and Human Health Risk Assessment

Konstantinos M. Kasiotis *, Effrosyni Zafeiraki, Electra Manea-Karga, Pelagia Anastasiadou and Kyriaki Machera

Laboratory of Pesticides' Toxicology, Department of Pesticides Control and Phytopharmacy, Benaki Phytopathological Institute, 145 61 Kifissia, Greece
* Correspondence: k.kasiotis@bpi.gr; Tel.: +30-2108180357

Abstract: Background: Bees encounter a plethora of environmental contaminants during nectar and pollen collection from plants. Consequently, after their entrance into the beehives, the transfer of numerous pollutants to apicultural products is inevitable. Methods: In this context, during the period of 2015–2020, 109 samples of honey, pollen, and beebread were sampled and analyzed for the determination of pesticides and their metabolites. More than 130 analytes were investigated in each sample by applying two validated multiresidue methods (HPLC-ESI-MS/MS and GC-MS/MS). Results: Until the end of 2020, 40 determinations were reported in honey, resulting in a 26% positive to at least one active substance. The concentrations of pesticides ranged from 1.3 ng/g to 785 ng/g honey. For seven active substances in honey and pollen, maximum residue limits (MRLs) exceedances were observed. Coumaphos, imidacloprid, acetamiprid, amitraz metabolites (DMF and DMPF), and tau-fluvalinate were the predominant compounds detected in honey, while several pyrethroids such as λ-cyhalothrin, cypermethrin, and cyfluthrin were also found. Pollen and beebread, as expected, accumulated a higher number of active substances and metabolites (32 in total), exhibiting almost double the number of detections. Conclusions: Although the above findings verify the occurrence of numerous pesticide and metabolite residues in both honey and pollen, the human risk assessment in the majority of the cases does not raise any concerns, and the same applies to bee risk assessment.

Keywords: pesticides; metabolites; LC-MS/MS; GC-MS/MS; honey; pollen; risk assessment

Citation: Kasiotis, K.M.; Zafeiraki, E.; Manea-Karga, E.; Anastasiadou, P.; Machera, K. Pesticide Residues and Metabolites in Greek Honey and Pollen: Bees and Human Health Risk Assessment. *Foods* **2023**, *12*, 706. https://doi.org/10.3390/foods 12040706

Academic Editors: Liming Wu and Qiangqiang Li

Received: 29 December 2022
Revised: 23 January 2023
Accepted: 31 January 2023
Published: 6 February 2023

1. Introduction

Honey and apicultural products such as pollen, beebread, propolis, and royal jelly have been consumed since antiquity [1] and used in pharmaceutical products and supplements worldwide with broad approval of their beneficial effect on human health [2–5]. Therefore, they should be free from contaminants linked to undesired health effects that also reduce their quality and commercial value. Recent reports designate that the EU is only 60% honey self-sufficient, with an apparent negative equilibrium of imports/exports [6]. Therefore, it relies on imports to a large extent. Hence, the need to increase honey and related apiculture commodities production in the EU is apparent. Such an increase should be accompanied by the quality management of apiculture activity. A major category of contaminants is pesticides, considering that plant protection products (PPPs) are indispensable and broadly applied for the protection of cultivated crops. Evidently, pesticides are strongly related to bees and apicultural commodities since plants and flowers, as the major pollen and nectar resources, can contain such substances. Exposure of bees to contaminants not only affects honey and other apicultural products' quality [7] but also can have negative effects on their health, even at concentrations found in environmental compartments [8,9]. In order to ensure the quality and safety of apicultural products, such as food, with the main focus on honey, well-organized monitoring programs and sustainable management are necessary, combined with powerful and fully validated analytical methods making use of the most recent technologies in the area. Hence, analytical methods have been developed, validated, and applied by a plethora of research groups for the detection of pesticides and metabolites [10–12], other organic pollutants, pyrrolizidine alkaloids [13], and heavy metals [14,15]. The presented work, as a continuation of previous work of our research group [11], aims to provide an overview of the pesticides and metabolites occurrence in honey, pollen, and beebread samples during the period of 2015–2020 in Greece and add further data on the contribution of pesticide residues to the overall chemical burden related to food consumption, human and bee health.

2. Materials and Methods

2.1. Samples Collection

Honey, pollen, and beebread samples of the present study were collected from different areas in Greece during 2015–2020 (Figure 1). Individual beekeepers and public authorities proceeded to sample in the context of monitoring pesticide residues in these commodities related, in some cases, to bee intoxication incidents. After field sampling, the samples were sent to the lab to investigate the occurrence of pesticide residues and metabolites. In some cases, beebread was carefully isolated from beeswax. Due to the limited number of beebread samples, the latter was considered the same matrix with pollen (in total, 63 honey and 46 pollen and beebread were collected). Honey (multifloral honey) used as a control sample was obtained from beehives of organic apiculture origin. Similarly, for pollen, a sample previously checked and devoid of pesticide residues (of the presented scope) was selected as a blank sample.

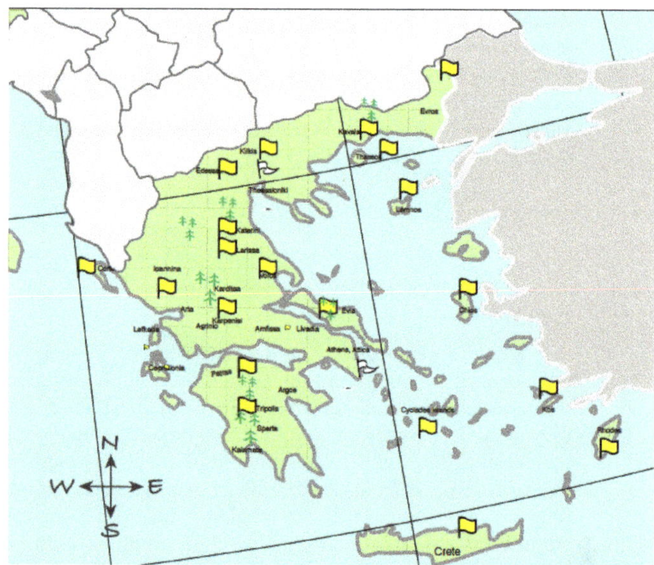

Figure 1. Sample collection points., created by Ortelius software Version 2.0.7.

2.2. Chemicals

Several pesticides and metabolites (more than 130 analytes) were monitored in the current study by applying liquid and gas chromatography–tandem mass spectrometry (LC-MS/MS, GC-MS/MS). Analytical standards and materials used are described in previous works of our group [11,15–17]. Carbendazim-d3, imidacloprid-d4, dimethoate-d6, and chlorpyrifos-d10 were used as mass-labeled internal standards for the quantification of the compounds measured by LC-MS/MS, and they were purchased from Sigma-Aldrich

(Seelze, Germany). Triphenyl phosphate (TPP), deltamethrin-d6, and dichlorvos-d6 were used in the GC-MS/MS analysis and acquired from Sigma-Aldrich (Seelze, Germany), respectively.

2.3. Sample Preparation

The sample preparation for the quantitative method applied for the analysis of all the honey, pollen, and beebread samples of the current study was previously described by our group [11,16,17]. Briefly, 1 g of each matrix was spiked with the six internal standards, carbendazim-d3, imidachloprid-d4, dimethoate-d6, chlorpyriphos-d10, deltamethrin-d6, and triphenyl phosphate (TPP). The extraction step was performed by the application of a modified QuEChERS method using C18, PSA, and Z-Sep. The extract was then centrifuged (4500 rpm, 10 min), and the supernatant was collected, and the final extract was divided into two parts. The two aliquots were evaporated till dryness, and then they were reconstituted with 1 mL of a 75:25 (v/v) MeOH:H$_2$O and pure acetone, respectively. The former aliquot was measured in LC-MS/MS, while the second in GC-MS/MS system.

2.4. Instrumental Analysis

All the samples were analyzed both in LC-MS/MS and GC-MS/MS. Regarding LC-MS/MS, an Agilent triple quadrupole (6410 QQQ) system was used, and an injection volume of 20 μL was applied. The GC-MS/MS analysis of the samples was performed on a Chromtech Evolution 3 MS/MS triple quadrupole mass spectrometer built on an Agilent 5975 B inert XL EI/CI MSD system using an injection volume of 2 μL. Chromatographic and mass spectrometric conditions are described in the Supplementary Data and also in previous works of our analytical group [11,16,17].

2.5. Quantification and Quality Assurance

The two methods (LC-MS/MS and GC-MS/MS) were validated for specificity, selectivity, reproducibility, repeatability, recovery, and sensitivity according to SANTE guidance documents (applicable at the time of the study) on analytical quality control and method validation procedures for pesticide residues analysis in food and feed [18,19]. For analyte confirmation, retention time (RT) and ion ratio were used. Recoveries were estimated using internal standards and were found to vary between 65 and 120% for all the analytes. The repeatability and reproducibility of both methods were tested by the analysis of multiple spiked with the mixture of analytical standards samples at three different concentrations (at LOQ, 10LOQ, and 100LOQ). The calculated limits of quantification LOQs and other validation metrics for each individual compound are presented in the Supplementary Data (Table S1) and in previous works of our group [11,16,17]. Honey and pollen used as control samples were also analyzed to monitor background contamination. Control samples were analyzed in every sequence of samples, and no pesticide residues and their metabolites were detected in any of them.

2.6. Human Health Noncarcinogenic Risk Assessment

The risk to human health posed by the pesticides detected in honey and pollen samples was assessed using the Hazard Quotient and Hazard Index approach [20]. The hazard quotient (*HQ*, unitless) was evaluated for each pesticide in honey and pollen, considering the dietary exposure via honey and pollen consumption. The *HQs* have been computed following Equations (1) and (2), as depicted below:

$$HQ = \frac{ADD}{ADI} \tag{1}$$

$$ADD = C \times \frac{IR}{BW} \tag{2}$$

where: *ADD* is the average daily pesticide intake (μg·kg^{-1}·d^{-1}), *ADI* is the acceptable daily intake (or daily reference dose, μg·kg^{-1}·d^{-1}) set by EFSA (peer review of pesticides risk assessment), *C* is the mean of pesticide concentration in honey and pollen (μg·kg^{-1}), *IR* is the daily honey pollen consumption rate (kg·person^{-1}·d^{-1}), (honey: 0.005 [21]; pollen: 0.02 for children and 0.04 for adults [22,23]), and *BW* is mean body weight (70 kg for adults, and 15 kg for children).

The *ADI* of an active substance (related to hazard identification and characterization) is based on the assessment of accessible toxicological data and is defined after the establishment of the no-observed-adverse-effect level (NOAEL) and use of the appropriate assessment factor. If the *HQ* is ≤1, it indicates that no adverse effect is likely to occur (health-protective). If *HQ* is >1, then a high level of concern is indicated for chronic effect occurrence. The higher the *HQ*, the higher the concern for chronic toxic effects, highlighting the need for immediate risk management actions.

For the estimation of the total risk from the simultaneous exposure to the mixture of chemicals that might be present in the commodity, the hazard index approach (*HI*, unitless) was applied to approximate the overall risk of multiple pesticides. In the specific approach, the hypothesis of dose additivity was assumed and calculated as the summation of individual *HQ* values (Equation (3)):

$$HI = \Sigma HQs = HQ1 + HQ2 + HQ3 + \ldots + HQn \tag{3}$$

2.7. Bee Health Risk Assessment

The risk assessment for bees was based on EFSA's bee guidance document [24] and the published risk assessment procedure by Sanchez-Bayo and Goka [25]. LD50 values for pesticides were retrieved from EFSA's publications on active substances and the Pesticide Properties Database of the University of Hertfordshire [26].

To address the subsequent risk to bees due to the consumption of contaminated pollen, the standard risk approach was followed, taking into account EFSA's bee guidance document [24] and a pertinent published work [25]. To determine the risk, the following equation was followed:

$$Risk = F \,(\%) \times \frac{dose}{LD50} \tag{4}$$

where *F*: the % detection frequency of the active substance in samples; *dose*: is the daily dose in µg per bee, considering pollen's daily consumption by bees (maximum consumption as a worst-case scenario) and concentration (average and maximum) obtained in this work; *LD50*: median lethal dose per bee (oral and contact, in µg per bee). The frequency of detection is essential since it depicts, as mentioned by previous research group [25], the probability of exposure of bees to the determined pollutants.

3. Results

3.1. Pesticide and Metabolites Residues

The analytical results showed a 26% positive to at least one active substance honey samples. In these samples, 19 active substances were detected in total, while the most common combination comprised coumaphos (an organophosphorus insecticide and acaricide approved as a veterinary medicinal product [27]), imidacloprid, and DMF (a metabolite of the acaricidal active substance amitraz approved only for veterinary use [27]) (Table 1 and Figure 2). In pollen and beebread, a higher number of active substances were identified (32), accompanied by a superior number of determinations (including a higher number of fungicides detected compared to honey) and an advanced proportion of positive samples (65%) (Table 2 and Figure 3). The latter designates that pollen constitutes a better environmental marker compared to honey, which is reasonable since pollen and nectar are the primary nutrition sources for bees, unsheltered from a plethora of organic and inorganic pollutants [15,16,28].

Table 1. Active substances, determinations, and concentration ranges in honey samples.

Active Substance	Determination	Concentration Range (ng/g)	MRL (ng/g)	Authorization at Sampling Period
Coumaphos	9	4.3–**88.7** [a]	100	Yes (approved as veterinary medicinal product)
Acetamiprid	5	3.1–20.5	50	Yes
Imidacloprid	4	25.1–**784.7** [a]	50	Yes
Coumaphos oxon	2	2.8–**12.8** [a]	100	Coumaphos metabolite
Cypermethrin	2	5.1–8.2	50	Yes
DMF *	2	4.9–11.2	200 *	Yes, amitraz metabolite (amitraz approved as veterinary medicinal product)
DMPF *	1	6.9	200 *	Yes, amitraz metabolite
λ-Cyhalothrin	2	4–7.2	50	Yes
Cyprodinil	1	31	50	Yes
Penconazole	1	15.9	50	Yes
Pirimiphos-methyl	1	**53.7** [a]	50	Yes
Malathion	1	26.5	50	Yes
Cadusafos	1	1.8	10	No
Ethoprofos	1	1.3	NA	No
Tricyclazole	1	1.4	50	No
Cyfluthrin	1	3.7	50	No **
Etofenprox	1	35	50	Yes
Tau-fluvalinate	1	10.3	50	Yes
Imidacloprid olefin	1	34.5		Imidacloprid metabolite

* Amitraz metabolite, MRL, applies only for its use as veterinary substance; ** β-cyfluthrin was approved, [a]: in bold, MRL exceedance, for coumaphos it applies after summation with the highest concentration of its metabolite coumaphos oxon.

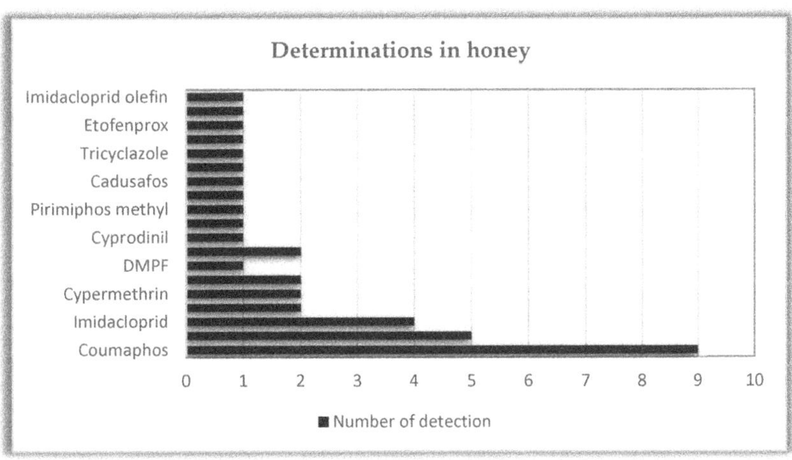

Figure 2. Graph showing determinations of active substances in honey (imidacloprid olefin is a metabolite of imidacloprid).

Table 2. Active substances, determinations, and concentration ranges in pollen beebread samples#.

Active Substance	Determination in Pollen Beebread	Concentration Range (ng/g)	MRL (ng/g)	Frequency of Detection (%)
Clothianidin [a]	7	8.9–**136.4** *	50	8.5
Carbendazim [a]	7	3.4–18	1000	8.5
Coumaphos	6	14.5–**511.3** *	100	7.3
Chlorpyrifos ethyl [a]	6	5.8–35.9	50	7.3
Tau-fluvalinate	5	<LOQ-180	NA	6.1
DMF **	4	4.9–14	200	4.9
Dimethoate [a]	4	7.9–**210** *	10	4.9
Cypermethrin	3	5.9–7.9	50	3.7
Cyfluthrin	3	3.7–11.8	50	3.7
Coumaphos oxon	2	2.2–10.7	100	2.4
Methomyl [b]	2	**85.6–154.6** *	10	2.4
Boscalid [a]	2	5.3–10.4	50	2.4
Propiconazole	2	31.6–31.9	50	2.4
Hexaconazole [b]	2	1.3–17.9	NA	2.4
Terbuthylazine [a]	2	45–53.2	50	2.4
Tebuconazole [a]	2	7.1–**150** *	50	2.4
Imidacloprid [a]	2	4.5–11.8	50	2.4
Omethoate [c]	2	**13–30** *	10	2.4
Acetamiprid	2	0.9–1.8	50	2.4
Trifloxystrobin [a]	2	1,6–18	50	2.4
Azoxystrobin [a]	2	2.1–3.1	50	2.4
Pyraclostrobin [a]	2	6.6–13	50	2.4
Permethrin [b]	2	11–30	NA	2.4
λ-Cyhalothrin [a]	1	7.2	50	1.2
Thiacloprid [a]	1	172	200	1.2
DMPF **	1	6.9	200	1.2
Chlorpyrifos oxon [d]	1	9.7	50	1.2
Dimethomorph [a]	1	15.2	50	1.2
Pendimethalin [a]	1	10.9	50	1.2
Pirimiphos-methyl	1	**170** *	50	1.2
Fenpropathrin [b]	1	<LOQ	NA	1.2
Acrinathrin [a]	1	9.9	50	1.2

\# For authorization status of common active substances, see Table 1; * in bold, MRL exceedance; ** amitraz metabolite; NA, not available; [a] authorized/approved at time of sampling; [b] not approved; [c] dimethoate metabolite; [d] chlorpyrifos ethyl metabolite.

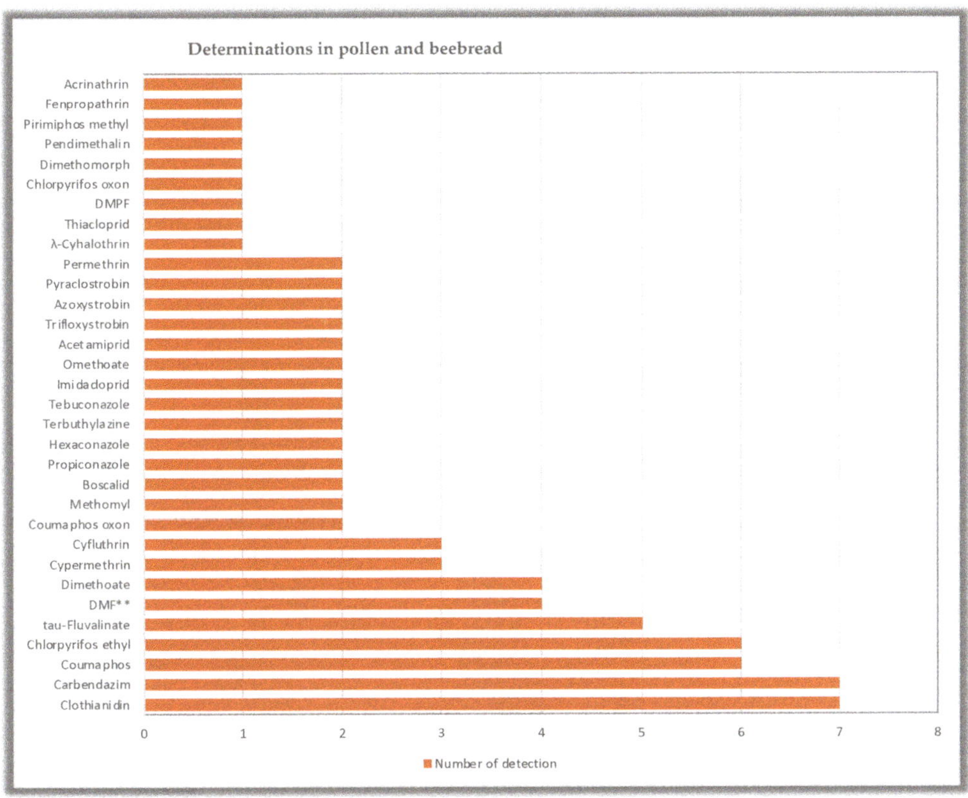

Figure 3. Graph showing determinations of active substances in pollen beebread.

In honey, MRLs were surpassed for three active substances. More specifically, in one case, coumaphos (as the sum of coumaphos and its metabolite coumaphos oxon) was quantified slightly above its MRL, while the other exceedances were registered for imidacloprid (two **cases** at 286.8 and 784.7 ng/g) and pirimiphos-methyl (Table 2). In all cases in which exceptionally high concentrations of pesticides encountered were associated with honey samples originating from honeycombs, honeybee death incidents were observed. The latter might postulate misapplications of PPPs, drift phenomena carrying substances away from application fields, applications during bees flying, or even deliberate application of these PPPs to harm the honeybee colonies. It is noteworthy that banned active substances were also detected at quantifiable concentrations. Among them were two organophosphates, cadusafos and ethoprofos, the triazolobenzothiazole active substance tricyclazole, and the pyrethroid cyfluthrin. Nevertheless, for some of the detections in the presented research, we cannot exclude the previously contaminated honeycomb as a potential contributing factor since incoming information from the beekeepers reported long-term use of the same honeycombs. Similarly, MRL exceedances were observed in pollen for clothianidin, coumaphos (the highest concentration observed in one beebread sample), dimethoate, omethoate, tebuconazole, methomyl and pirimiphos-methyl (Table 2).

Banned pesticides were also detected in pollen and beebread samples. These were fenpropathrin, methomyl, and permethrin. The above results confirm the higher pesticide load of pollen in comparison to the respective levels in honey. Hence, pollen, though far less consumed than honey, deserves noticeable attention, as it is more prone to environmental contaminants. Another valuable conclusion is that pollen is a better marker of environmental contamination of the sampled areas, particularly of pesticides, but also inorganic contaminants, as reported in recent works of our group [15,28].

3.2. Human Health Risk Assessment

Prior to commenting on the health risk assessment results, a basic admission on the consumption of pollen was contemplated. Daily consumption of honey is established in the European Union at 5 g per day. On the contrary, due to the rarity of pollen consumption data, published works mostly on therapeutic uses of pollen were considered [22,29] and also embraced in previous work of our group [15]. Yet, an overestimation of the risk due to elevated pollen consumption is expected.

Regarding the health risk assessment, the average daily intake of each pesticide, ADI, and HQ values are presented in Table 3. The only active substance for which the health risk assessment showed (in two samples) alarming levels was coumaphos. More specifically, when mean coumaphos concentration in pollen was incorporated in respective endpoint calculations, it led to an HQ value for children above the threshold value. Similar conclusions were derived when HI values were subtracted.

Considering honey, an exemplary "worst-case" sample contained coumaphos (and coumaphos oxon as the sum, at 101.5 ng/g) accompanied by imidacloprid (286.8 ng/g) and acetamiprid (20.5 ng/g). Nevertheless, after individual calculations of HQs, their summation led to an HI value of 0.04 and 0.13 for adults and children, correspondingly. Consequently, no risk was posed for humans after the potential consumption of this honey.

3.3. Bee health Risk Assessment

The risk calculations (Table 4) (see Table S2 for bumblebees and Table S3 for solitary bees) demonstrated negligible risk for bees considering the available consumption data for pollen, including honey bees, bumblebees, and solitary bees [24]. With regard to honeybee consumption, the results refer only to the nurse bee (for foragers, zero consumption is considered).

Table 3. Human health risk assessment for honey and pollen consumption.

	Honey						Pollen				
	Average Daily Intake (µg/kg/day)		ADI (µg/kg/day)	HQ			Average Daily Intake (µg/kg/day)		ADI (µg/kg/day)	HQ	
Active Substance	Adults	Children		Adults	Children	Active Substance	Adults	Children		Adults	Children
Acetamiprid	8.43×10^{-4}	0.004	25	3.37×10^{-5}	1.57×10^{-4}	Boscalid	0.005	0.010	40	1.03×10^{-4}	5.01×10^{-4}
Cadusafos	1.29×10^{-4}	6.00×10^{-4}	0.4	3.21×10^{-4}	0.001	Carbendazim	0.006	0.012	20	0.0003	0.0012
Coumaphos *	3.78×10^{-3}	0.018	0.25	0.015	0.070	Chlorpyrifos *	0.017	0.034	1	0.017	0.068
Cyfluthrin	2.64×10^{-4}	0.001	10	2.64×10^{-5}	1.23×10^{-4}	Clothianidin	0.048	0.097	97	4.28×10^{-4}	0.001
λ-Cyhalothrin	5.14×10^{-4}	0.002	2.5	2.06×10^{-4}	9.60×10^{-4}	Coumaphos *	0.176	0.352	0.25	0.603	1.408
Cypermethrin	4.93×10^{-4}	0.002	5	9.86×10^{-5}	4.60×10^{-4}	Cyfluthrin	0.005	0.010	10	4.43×10^{-4}	0.001
Cyprodinil	0.002	0.010	30	7.38×10^{-5}	3.44×10^{-4}	λ-Cyhalothrin	0.005	0.010	2.5	0.002	0.004
DMF	5.75×10^{-4}	0.003	3	1.92×10^{-4}	8.94×10^{-4}	Cypermethrin	0.005	0.009	5	7.89×10^{-4}	0.002
DMPF	4.93×10^{-4}	0.002	3	1.64×10^{-4}	7.67×10^{-4}	Dimethomorph	0.010	0.020	50	1.74×10^{-5}	4.05×10^{-4}
Ethoprofos	9.29×10^{-5}	4.00×10^{-4}	0.4	2.32×10^{-4}	0.001	DMF	0.006	0.013	3	0.002	0.004
Tau-fluvalinate	7.36×10^{-4}	0.003	5	1.47×10^{-4}	6.87×10^{-4}	DMPF	0.005	0.009	3	0.001	0.003
Imidacloprid	0.029	0.135	60	4.82×10^{-4}	2.25×10^{-3}	Hexaconazole	0.006	0.013	5	0.001	0.003
Malathion	0.002	0.009	30	6.31×10^{-5}	2.94×10^{-4}	Methomyl	0.080	0.160	2.5	0.032	0.064
Penconazole	0.001	0.005	30	3.79×10^{-5}	1.77×10^{-4}	Propiconazole	0.021	0.042	40	4.54×10^{-4}	0.001
Pirimiphos-methyl	0.004	0.018	4	9.59×10^{-4}	0.004	Tebuconazole	0.005	0.009	30	1.35×10^{-4}	3.16×10^{-4}
Tricyclazole	1.00×10^{-4}	4.67×10^{-4}	30	3.33×10^{-6}	1.56×10^{-5}	Terbuthylazine	0.033	0.065	4	0.008	0.016
Etofenprox	5.57×10^{-4}	0.003	30	1.86×10^{-5}	8.67×10^{-5}	Thiacloprid	0.115	0.229	10	0.010	0.023
						Imidacloprid	0.015	0.030	60	2.11×10^{-4}	4.92×10^{-4}
						Pendimethalin	0.007	0.015	125	4.98×10^{-5}	1.16×10^{-4}
						Dimethoate	0.012	0.024	1	0.012	0.024

* Coumaphos oxon and chlorpyrifos oxon concentrations are added to the parent compounds' concentration (for concomitant detections with the parent). DMF and DMPF are presented for the cases in which separate detections were observed. Their summation has negligible effect on the HQ (in cases in which both were detected).

Table 4. Risk calculations for honeybees based on the active substances (mean and maximum concentrations) quantified in pollen samples#.

	LD50 (Oral, µg per Bee)	LD50 (Contact, µg per Bee)	Oral		Contact	
			RISK (Mean)	RISK (Maximum)	RISK (Mean)	RISK (Maximum)
Clothianidin	0.004	0.044	0.019	0.035	0.002	0.003
Carbendazim	100	50	1.09×10^{-7}	1.84×10^{-7}	2.19×10^{-7}	3.69×10^{-7}
Coumaphos	na	100	-	-	2.31×10^{-6}	4.49×10^{-6}
Chlorpyrifos ethyl	0.15	0.068	1.22×10^{-4}	2.10×10^{-4}	2.69×10^{-4}	4.64×10^{-4}
Tau-fluvalinate	12.6	12	5.26×10^{-6}	1.04×10^{-5}	5.51×10^{-6}	1.10×10^{-5}
Amitraz (sum of DMF + DMPF)	na	50	-	-	1.51×10^{-7}	2.45×10^{-7}
Dimethoate	0.1	0.1	6.38×10^{-4}	1.23×10^{-4}	6.38×10^{-4}	6.38×10^{-4}
Cypermethrin	0.172	0.023	1.76×10^{-5}	2.02×10^{-5}	1.32×10^{-4}	1.51×10^{-4}
Cyfluthrin	0.05	0.001	6.80×10^{-5}	1.04×10^{-4}	0.003	0.005
Methomyl	0.28	0.16	1.26×10^{-4}	1.62×10^{-4}	2.20×10^{-4}	2.83×10^{-4}
Boscalid	160	200	1.44×10^{-8}	1.90×10^{-8}	1.15×10^{-8}	1.52×10^{-8}
Propiconazole	100	100	9.29×10^{-8}	9.34×10^{-8}	9.29×10^{-8}	9.34×10^{-8}
Hexaconazole	100	na	2.81×10^{-8}	5.24×10^{-8}	-	-
Terbuthylazine	22.6	32	6.36×10^{-7}	6.89×10^{-7}	4.49×10^{-7}	4.87×10^{-7}
Tebuconazole	83.05	200	2.77×10^{-7}	5.29×10^{-7}	1.15×10^{-7}	2.19×10^{-7}
Imidacloprid	0.0037	0.081	6.45×10^{-4}	9.33×10^{-4}	2.95×10^{-5}	4.26×10^{-5}
Acetamiprid	14.53	8.09	2.72×10^{-8}	3.63×10^{-8}	4.88×10^{-8}	6.51×10^{-8}
Trifloxystrobin	200	200	1.43×10^{-8}	2.63×10^{-8}	1.43×10^{-8}	2.63×10^{-8}
Azoxystrobin	25	200	3.04×10^{-8}	3.63×10^{-8}	3.80×10^{-9}	4.54×10^{-9}
Pyraclostrobin	110	100	2.61×10^{-8}	3.46×10^{-8}	2.87×10^{-8}	3.80×10^{-8}
Permethrin	0.13	0.024	4.61×10^{-5}	6.75×10^{-5}	2.50×10^{-4}	3.66×10^{-4}
λ-Cyhalothrin	0.91	0.038	1.16×10^{-6}	1.16×10^{-6}	2.77×10^{-5}	2.77×10^{-5}
Thiacloprid	17.32	38.82	1.45×10^{-6}	1.45×10^{-6}	6.48×10^{-7}	6.48×10^{-7}
Dimethomorph	32.4	102	6.86×10^{-8}	6.86×10^{-8}	2.18×10^{-8}	2.18×10^{-8}
Pendimethalin	na	100	-	-	1.95×10^{-8}	1.59×10^{-8}
Pirimiphos-methyl	0.22	na	1.13×10^{-4}	1.13×10^{-4}	-	-
Fenpropathrin	na	0.05	-	-	2.93×10^{-6}	2.93×10^{-6}
Acrinathrin	0.077	0.084	1.88×10^{-5}	1.88×10^{-5}	1.72×10^{-5}	1.72×10^{-5}

For the combinations DMF + DMPF (DMA was not detected), chlorpyrifos + chlorpyrifos oxon, and coumaphos + coumaphos oxon, concomitant determinations were considered the sum for risk assessment; na, not available in the open literature.

4. Discussion

Multiresidue methods (LC, GC-MS/MS) constitute the mainstay of analytical laboratories involved in pesticide residue analysis in a multitude of matrices. The latter was verified in this work through the detection and quantitation of numerous active substances and metabolites in both pollen beebread and honey. It is important to point out that the presented results are the outcome of the chemical analysis of randomly collected and dispatched honey and pollen samples. Hence, it cannot be viewed under an organized sampling/monitoring scheme in which PPP applications could be straightforwardly connected to specific crops via concomitant monitoring of residues in apicultural commodities, crops, and soil. Nevertheless, based on the incoming information from individual authorities that sent the samples, it was endeavored to proceed to potential associations of active substances detected with the predominant crops in some of the regions and the timing of sampling. In the same context, honey and pollen (and beebread) samples associated with bees' intoxication incidents (in which deliberate application cannot be excluded) were not included in these interpretations since it would possibly lead to misleading assumptions for PPP applications in the specified areas. A characteristic example stems from one of the few lowland areas of the island region of Chios Island (northern Aegean Sea), in which the cultivation of vegetables and watermelons is documented. From this region, two honey samples were found positive for pesticides. The detection of coumaphos in one sample can be attributed to acaricidal treatments. However, detecting the fungicide active substance penconazole in both honey samples designates the potential uptake of this chemical by bees in the nearby cultivations and subsequent transfer to the beehive, or potential transfer due to pesticides' drift favored by environmental conditions. Penconazole formulations were approved to control the fungi *Sphaerotheca Fuliginea*, *Erysiphe cichoracearum* in watermelon, and other flowering plant species of the Cucurbitaceae family. Plants of this family, such as watermelon (abundant in the region), are pollinated by bees; therefore, this finding can be a logical hypothesis. In the same samples, the organophosphate fumigant insecticide pirimiphos-methyl was also detected. Its presence seems disconnected from the predominant (active) cultivations of the area, yet, this active substance is usually applied to control pests, such as *Sitophilus granarius* and *Oryzaephilus surinamensis*, of stored seeds. Nevertheless, in the past, the specific region had significant production of wheat, which can interplay in the specific finding through the potential application of pirimiphos-methyl in stored wheat areas. Acetamiprid and etofenprox (both insecticides) detection can be attributed to field applications to combat aphids such as *Macrosiphum euphorbiae* in vegetables, or Lepidoptera such as *Pieris brasiccae* in cabbage and European grapevine moth *Lobesia botrana*. In two additional characteristic samples originating from Northern Greece (Chalkidiki region), 11 active substances and metabolites (chlorpyrifos ethyl, carbendazim, dimethoate, omethoate, tebuconazole, trifloxystrobin, acetamiprid, pyraclostrobin, azoxystrobin, pirimiphos-methyl, and fenpropathrin) in total were identified in two samples of pollen (9 and 10 synchronous detections in the two samples correspondingly). The detection of these compounds can be attributed to PPP applications related to crops of the respective area, such as olives, grapes, and cereals. From the above-mentioned substances, dimethoate (and its metabolite omethoate), along with chlorpyrifos, were authorized at the time of the study and widely used for the control of various pests, such as the olive fruit fly (*Dacus oleae*). The strobilurin fungicides detected are also used in the olive crop and grape for the control of mildew or black rot in grapes. Acetamiprid is a neonicotinoid active substance with a wide range of applications, such as the control of *Lobesia botrana* (major threat to grapes) and *Scaphoideus titanus* in grapes. In addition, the documented substances are applied in crops that bees visit for pollen and nectar, such as stone fruits that are cultivated in the specific region (i.e., peach tree). Consequently, it is possible to rationalize the presented results in the context of land use of the specific regions and PPP applications. For coumaphos and amitraz (via its metabolites detection), their prevalence is expected due to their use as veterinary medicinal products to control the parasitic mite *Varroa destructor*. Last decade's literature verifies that these compounds are still detected in hive products [30–34], demonstrating in some cases exceedances of the ascribed MRLs (see indicatively [31]). As regards the other detected substances, the findings are largely in line with the recent bibliography. More specifically, in the recent literature concerning honey, pollen, and beebread [35–38], substances such as boscalid, carbendazim, pyraclostrobin, chlorpyrifos, tau-fluvalinate, cypermethrin, fenpropathrin, and λ-cyhalothrin that were detected in the presented study are also reported. Nevertheless, compounds such as fluopyram and chlorothalonil that are reported in the bibliography were not included in the scope of the presented analytical method.

Regarding the human health risk assessment, the current findings should not be neglected due to the less frequent consumption of pollen/beebread. The last 20-year trend in the consumption of raw, unprocessed food [39], especially of organic origin with proven health benefits, involves apicultural products and is an attitude that steadily increases. The inclination to these products is further strengthened by the scientific reviews on the beneficial effects of the consumption of apicultural commodities on human health [40]. Therefore, it is anticipated that more adults and children are expected to consume such products in all their variations (e.g., pure pollen, beebread, honey, or wax-containing products). Another viewpoint is the consumption of beebread or honeycomb containing it, especially in agricultural communities where the proximity to "raw, unprocessed" food is easier. Honeycombs usually are not devoid of Varroa mites, leading to inevitable applications of coumaphos and other acaricides belonging to the class of organic chemicals. Last but not least, bees' susceptibility to mixtures of chemicals via pollen consumption is also apparent and can affect their longevity and survival. Consequently, any work or report on residue finding is of utmost importance and should be disseminated to increase awareness for the protection of such a pivotal insect.

5. Conclusions

Extensive monitoring of pesticides and their metabolites (due to the broad applications of plant protection products) in apiculture commodities is pivotal due to their consumption by both adults and children and the consumption of pollen by bees. In the same context, monitoring these matrices can unveil, to a certain degree, the contamination that occurs in the agroenvironment. The presented study, after implementing two multiresidue methods, depicted the residual prevalence of pesticides and metabolites in Greek honey and pollen samples between 2015–2020, corroborating the detection of 40 active substances in total (including metabolites) in an overall concentration range of <LOQ to 785 ng/g. Consequently, a risk assessment was conducted to evaluate the potential health effects on humans and bees. The health risk assessment demonstrated, in the majority of cases, negligible risk for bees and humans. Nevertheless, the more than 30 active substances detected in pollen, though anticipated, confirm the occurrence of a plethora of contaminants in an important bee nutrition matrix that deserves attention both from health risk assessment and environmental perspectives.

Supplementary Materials: The following supporting information can be downloaded at: https://www.mdpi.com/article/10.3390/foods12040706/s1, Text: chromatographic and mass spectrometric conditions, Table S1: active substances and LOQs (pollen and honey), Table S2: risk calculations for bumble bees based on the active substances (mean and maximum concentrations) quantified in pollen samples, Table S3: risk calculations for solitary bees based on the active substances (mean and maximum concentrations).

Author Contributions: Conceptualization, K.M. and K.M.K.; methodology, K.M.K. and E.Z.; software, K.M.K., E.Z. and E.M.-K.; validation, K.M.K. and E.Z.; formal analysis, K.M.K.; investigation, K.M.K., E.Z., E.M.-K. and P.A.; resources, K.M.; data curation, K.M.K. and E.Z.; writing—original draft preparation, K.M.K.; writing—review and editing, K.M.K., E.Z. and K.M.; visualization, K.M.K. and E.Z.; supervision, K.M. and K.M.K.; project administration, K.M.; funding acquisition, K.M. All authors have read and agreed to the published version of the manuscript.

Funding: This research received no external funding.

Institutional Review Board Statement: Not applicable.

Informed Consent Statement: Not applicable.

Data Availability Statement: All data available are presented in this manuscript.

Acknowledgments: The authors thank all beekeepers and authorities that sent samples for chemical analysis.

Conflicts of Interest: The authors declare no conflict of interest.

References

1. Adamson, M.W. *Regional Cuisines of Medieval Europe: A Book of Essays*; Routledge: New York, NY, USA, 2002.
2. Zhao, H.; Cheng, N.; Wang, Q.; Zhou, W.; Liu, C.; Liu, X.; Chen, S.; Fan, D.; Cao, W. Effects of honey-extracted polyphenols on serum antioxidant capacity and metabolic phenotype in rats. *Food Funct.* **2019**, *10*, 2347–2358. [CrossRef] [PubMed]
3. Cianciosi, D.; Forbes-Hernández, T.Y.; Afrin, S.; Gasparrini, M.; Quiles, J.L.; Gil, E.; Bompadre, S.; Simal-Gandara, J.; Battino, M.; Giampieri, F. The Influence of in Vitro Gastrointestinal Digestion on the Anticancer Activity of Manuka Honey. *Antioxid. Basel* **2020**, *9*, 64. [CrossRef] [PubMed]
4. Afrin, S.; Haneefa, S.M.; Fernandez-Cabezudo, M.J.; Giampieri, F.; Al-Ramadi, B.K.; Battino, M. Therapeutic and preventive properties of honey and its bioactive compounds in cancer: An evidence-based review. *Nutr. Res. Rev.* **2020**, *33*, 50–76. [CrossRef]
5. Šedík, P.; Pocol, C.B.; Horská, E.; Fiore, M. Honey: Food or medicine? A comparative study between Slovakia and Romania. *Br. Food J.* **2019**, *121*, 1281–1297. [CrossRef]
6. EU-Honey. Report from the Commission to the European Parliament and the Council on the Implementation of Apiculture Programmes, Brussels, 17.12.2019 COM(2019) 635 Final. Available online: https://ec.europa.eu/info/sites/default/files/food-farming-fisheries/animals_and_animal_products/documents/report-implementation-measures-apiculuture-sector_2019-12-17_en.pdf (accessed on 10 November 2021).
7. Böhme, F.; Bischoff, G.; Zebitz, C.P.W.; Rosenkranz, P.; Wallner, K. From field to food—Will pesticide-contaminated pollen diet lead to a contamination of royal jelly? *Apidologie* **2018**, *49*, 112–119. [CrossRef]
8. Dolezal, A.G.; Carrillo-Tripp, J.; Miller, W.A.; Bonning, B.C.; Toth, A.L. Pollen Contaminated with Field-Relevant Levels of Cyhalothrin Affects Honey Bee Survival, Nutritional Physiology, and Pollen Consumption Behavior. *J. Econ. Entomol.* **2016**, *109*, 41–48. [CrossRef]
9. Leska, A.; Nowak, A.; Nowak, I.; Górczyńska, A. Effects of Insecticides and Microbiological Contaminants on Apis mellifera Health. *Molecules* **2021**, *26*, 5080. [CrossRef]
10. Gbylik-Sikorska, M.; Sniegocki, T.; Posyniak, A. Determination of neonicotinoid insecticides and their metabolites in honey bee and honey by liquid chromatography tandem mass spectrometry. *J. Chromatography. B Anal. Technol. Biomed. Life Sci.* **2015**, *990*, 132–140. [CrossRef]
11. Kasiotis, K.M.; Anagnostopoulos, C.; Anastasiadou, P.; Machera, K. Pesticide residues in honeybees, honey and bee pollen by LC–MS/MS screening: Reported death incidents in honeybees. *Sci. Total Environ.* **2014**, *485–486*, 633–642. [CrossRef]
12. Chienthavorn, O.; Dararuang, K.; Sasook, A.; Ramnut, N. Purge and trap with monolithic sorbent for gas chromatographic analysis of pesticides in honey. *Anal. Bioanal. Chem.* **2012**, *402*, 955–964. [CrossRef]
13. Sixto, A.; Niell, S.; Heinzen, H. Straightforward Determination of Pyrrolizidine Alkaloids in Honey through Simplified Methanol Extraction (QuPPE) and LC-MS/MS Modes. *ACS Omega* **2019**, *4*, 22632–22637. [CrossRef] [PubMed]
14. Malhat, F.; Kasiotis, K.M.; Hassanin, A.; Shokr, S. An MIP-AES study of heavy metals in Egyptian honey: Toxicity assessment and potential health hazards to consumers. *J. Elementol.* **2019**, *24*, 473–488. [CrossRef]
15. Zafeiraki, E.; Kasiotis, K.M.; Nisianakis, P.; Manea-Karga, E.; Machera, K. Occurrence and human health risk assessment of mineral elements and pesticides residues in bee pollen. *Food Chem. Toxicol.* **2022**, *161*, 112826. [CrossRef]
16. Kasiotis, K.M.; Zafeiraki, E.; Kapaxidi, E.; Manea-Karga, E.; Antonatos, S.; Anastasiadou, P.; Milonas, P.; Machera, K. Pesticides residues and metabolites in honeybees: A Greek overview exploring Varroa and Nosema potential synergies. *Sci. Total Environ.* **2021**, *769*, 145213. [CrossRef]
17. Kasiotis, K.M.; Tzouganaki, Z.D.; Machera, K. Chromatographic determination of monoterpenes and other acaricides in honeybees: Prevalence and possible synergies. *Sci. Total Environ.* **2018**, *625*, 96–105. [CrossRef]
18. *SANTE/11945/2015*; Guidance Document on Analytical Quality Control and Method Validation Procedures for Pesticides Residues Analysis in Food and Feed. DG SANTE, European Commission: Brussels, Belgium, 2015.
19. *SANTE/11813/2017*; Guidance Document on Analytical Quality Control and Method Validation Procedures for Pesticides Residues Analysis in Food And Feed. European Commission: Brussels, Belgium, 2017.
20. US-EPA. *Risk Based Concentration Table*; United States Environmental Protection Agency: Philadelphia, PA, USA, 2000.
21. *SANTE/11956/2016*; Technical Guidelines for Determining the Magnitude of Pesticide Residues in Honey and Setting Maximum Residue Levels in Honey. European Commission: Brussels, Belgium, 2018.
22. Khalifa, S.; Elashal, M.; Yosri, N.; Du, M.; Musharraf, S.; Nahar, L.; Sarker, S.; Guo, Z.; Cao, W.; Zou, X.; et al. Bee Pollen: Current Status and Therapeutic Potential. *Nutrients* **2021**, *13*, 1876. [CrossRef] [PubMed]
23. Kostić, A.; Milinčić, D.D.; Barać, M.B.; Shariati, M.A.; Tešić, L.; Pešić, M.B. The Application of Pollen as a Functional Food and Feed Ingredient—The Present and Perspectives. *Biomolecules* **2020**, *10*, 84. [CrossRef]
24. EFSA-Guidance. EFSA Guidance Document on the Risk Assessment of Plant Protection Products on Bees (Apis mellifera, Bombus spp. and solitary bees). *EFSA J.* **2013**, *11*, 3295.
25. Sanchez-Bayo, F.; Goka, K. Pesticide Residues and Bees—A Risk Assessment. *PLoS ONE* **2014**, *9*, e94482. [CrossRef]
26. *PPBD: Pesticide Properties Database*; The University of Hertfordshire: Hatfield, UK, 2022; Available online: http://sitem.herts.ac.uk/aeru/ppdb/en (accessed on 5 September 2022).
27. EFSA. Reasoned opinion on the setting of maximum residue levels for amitraz, coumaphos, flumequine, oxytetracycline, permethrin and streptomycin in certain products of animal origin. *EFSA J.* **2016**, *14*, 39.

28. Zafeiraki, E.; Sabo, R.; Kasiotis, K.M.; Machera, K.; Sabová, L.; Majchrák, T. Adult Honeybees and Beeswax as Indicators of Trace Elements Pollution in a Vulnerable Environment: Distribution among Different Apicultural Compartments. *Molecules* **2022**, *27*, 6629. [CrossRef] [PubMed]
29. Komosinska-Vassev, K.; Olczyk, P.; Kaźmierczak, J.; Mencner, L.; Olczyk, K. Bee Pollen: Chemical Composition and Therapeutic Application. *Evid. Based Complement. Altern. Med.* **2015**, *2015*, 297425. [CrossRef] [PubMed]
30. Valdovinos-Flores, C.; Gaspar-Ramírez, O.; Heras-Ramírez, M.E.; Lara-Álvarez, C.; Dorantes-Ugalde, J.A.; Saldaña-Loza, L.M. Boron and Coumaphos Residues in Hive Materials Following Treatments for the Control of Aethina tumida Murray. *PLoS ONE* **2016**, *11*, e0153551. [CrossRef] [PubMed]
31. Pohorecka, K.; Kiljanek, T.; Antczak, M.; Skubida, P.; Semkiw, P.; Posyniak, A. Amitraz marker residues in honey from honeybee colonies treated with Apiwarol. *J. Vet. Res.* **2018**, *62*, 297–301. [CrossRef] [PubMed]
32. Chaimanee, V.; Johnson, J.; Pettis, J.S. Determination of amitraz and its metabolites residue in honey and beeswax after Apivar® treatment in honey bee (Apis mellifera) colonies. *J. Apic. Res.* **2022**, *61*, 213–218. [CrossRef]
33. Christodoulou, D.L.; Kanari, P.; Kourouzidou, O.; Constantinou, M.; Hadjiloizou, P.; Kika, K.; Constantinou, P. Pesticide residues analysis in honey using ethyl acetate extraction method: Validation and pilot survey in real samples. *Int. J. Environ. Anal. Chem.* **2015**, *95*, 894–910. [CrossRef]
34. Lozano, A.; Hernando, M.; Uclés, S.; Hakme, E.; Fernández-Alba, A. Identification and measurement of veterinary drug residues in beehive products. *Food Chem.* **2019**, *274*, 61–70. [CrossRef]
35. Kiljanek, T.; Niewiadowska, A.; Małysiak, M.; Posyniak, A. Miniaturized multiresidue method for determination of 267 pesticides, their metabolites and polychlorinated biphenyls in low mass beebread samples by liquid and gas chromatography coupled with tandem mass spectrometry. *Talanta* **2021**, *235*, 122721. [CrossRef]
36. Xiao, J.; He, Q.; Liu, Q.; Wang, Z.; Yin, F.; Chai, Y.; Yang, Q.; Jiang, X.; Liao, M.; Yu, L.; et al. Analysis of honey bee exposure to multiple pesticide residues in the hive environment. *Sci. Total Environ.* **2022**, *805*, 150292. [CrossRef]
37. Hakme, E.; Lozano, A.; Gómez-Ramos, M.; Hernando, M.; Fernández-Alba, A. Non-target evaluation of contaminants in honey bees and pollen samples by gas chromatography time-of-flight mass spectrometry. *Chemosphere* **2017**, *184*, 1310–1319. [CrossRef]
38. Murcia-Morales, M.; Heinzen, H.; Parrilla-Vázquez, P.; Gómez-Ramos, M.D.M.; Fernández-Alba, A.R. Presence and distribution of pesticides in apicultural products: A critical appraisal. *TrAC Trends Anal. Chem.* **2022**, *146*, 116506. [CrossRef]
39. Koebnick, C.; Garcia, A.L.; Dagnelie, P.C.; Strassner, C.; Lindemans, J.; Katz, N.; Leitzmann, C.; Hoffmann, I. Long-Term Consumption of a Raw Food Diet Is Associated with Favorable Serum LDL Cholesterol and Triglycerides But Also with Elevated Plasma Homocysteine and Low Serum HDL Cholesterol in Humans. *J. Nutr.* **2005**, *135*, 2372–2378. [CrossRef] [PubMed]
40. Mărgăoan, R.; Strant, M.; Varadi, A.; Topal, E.; Yücel, B.; Cornea-Cipcigan, M.; Campos, M.G.; Vodnar, D.C. Bee Collected Pollen and Bee Bread: Bioactive Constituents and Health Benefits. *Antioxidants* **2019**, *8*, 568. [CrossRef] [PubMed]

Article

Quality of Honey Imported into the United Arab Emirates

Tareq M. Osaili [1,2,*], Wael A. M. Bani Odeh [3], Maryam S. Al Sallagi [3], Ahmed A. S. A. Al Ali [4], Reyad S. Obaid [1], Vaidehi Garimella [5], Fatema Saeed Bin Bakhit [5], Hayder Hasan [1], Richard Holley [6] and Nada El Darra [7]

[1] Department of Clinical Nutrition and Dietetics, College of Health Sciences, University of Sharjah, Sharjah P.O. Box 27272, United Arab Emirates

[2] Department of Nutrition and Food Technology, Faculty of Agriculture, Jordan University of Science and Technology, P.O. Box 3030, Irbid 22110, Jordan

[3] Studies and Risk Assessment Unit, Dubai Municipality, Dubai P.O. Box 67, United Arab Emirates

[4] Food Studies and Policies Section, Dubai Municipality, Dubai P.O. Box 67, United Arab Emirates

[5] Dubai Central Laboratories Department, Dubai Municipality, Dubai P.O. Box 67, United Arab Emirates

[6] Department of Food Science and Human Nutrition, University of Manitoba, Winnipeg, MB R3T 2N2, Canada

[7] Department of Nutrition and Dietetics, Faculty of Health Sciences, Beirut Arab University, Tarik El Jedidah—Beirut, P.O. Box 115020, Riad El Solh, Beirut 1107 2809, Lebanon

* Correspondence: tosaili@sharjah.ac.ae or tosaili@just.edu.jo

Abstract: This study was performed to assess the physicochemical quality characteristics of honey imported by the United Arab Emirates (UAE) via Dubai ports between 2017 and 2021. There were 1330 samples analyzed for sugar components, moisture, hydroxymethylfurfural (HMF) content, free acidity, and diastase number. Of the honey tested, 1054 samples complied with the Emirates honey standard, but 276 (20.8%) did not; this was due to non-compliance with one or more quality parameters, thus suggesting some level of adulteration, improper storage or inappropriate heat treatment. For the non-compliant samples, the average values of sucrose content ranged from 5.1 to 33.4%; the sum of glucose and fructose ranged from 19.6 to 88.1%; the moisture content varied from 17.2 to 24.6%; the HMF occurred in a range from 83.2 to 663.0 mg/kg, and the acidity varied from 52 to 85 meq/kg. The non-compliant honey samples were grouped according to their country of origin. India was shown to be the country having the highest percentage of non-compliant samples at 32.5% and Germany had the lowest at 4.5%. This study emphasized that the inspection of honey samples traded internationally should involve physicochemical analysis. A comprehensive inspection of honey at the Dubai ports should reduce incidents of adulterated products being imported.

Keywords: honey; adulteration; non-compliant quality; hydroxymethylfurfural; sugar; imported

Citation: Osaili, T.M.; Odeh, W.A.M.B.; Al Sallagi, M.S.; Al Ali, A.A.S.A.; Obaid, R.S.; Garimella, V.; Bakhit, F.S.B.; Hasan, H.; Holley, R.; El Darra, N. Quality of Honey Imported into the United Arab Emirates. *Foods* **2023**, *12*, 729. https://doi.org/10.3390/foods12040729

Academic Editors: Liming Wu and Qiangqiang Li

Received: 6 December 2022
Revised: 24 January 2023
Accepted: 30 January 2023
Published: 7 February 2023

1. Introduction

Honey is an ancient natural product valued for its health benefits, medical characteristics and biological properties [1,2]. According to the definition of Codex Alimentarius [3] and a European Community Directive [4], honey is a natural substance produced by honeybees from the nectar of flowers without any added ingredients. Usually, the quality of honey is assessed by several parameters, including its moisture, sugar content, pH, total acidity, hydroxymethylfurfural (HMF) content and other factors [3,5,6]. The sweet taste of honey is primarily due to its high fructose content [4,6]. Consumers generally perceive it as a natural and healthier sweetening alternative to table sugar [7].

Due to its high price, honey is frequently adulterated [8]. It is ranked sixth amongst the food products subjected to fraud in Europe [9]. Therefore, strict monitoring and quality assurance are needed to prevent adulterated products from entering the consumer market. Adulteration is only one practice by which the quality of honey can be compromised. Other factors include, but are not limited to, the application of a heat treatment or the use of improper storage conditions [10,11].

Due to the high variability in honey composition, the detection of adulteration is not an easy task. Honey adulteration can be carried out by adding water (dilution) or cheap sugar solutions, such as high-fructose corn syrup [12]. High-quality honey can also be mixed with honey of low quality, and sold at a higher price [13]. Adulteration may compromise consumer experience and expectations, which may lead to the reduced demand for honey and its products. Moreover, legitimate honey producers are often unable to compete with the low-priced adulterated honey. Consequently, there is growing interest in screening honey for adulteration and quality before market distribution.

Various methods have been developed to check for honey adulteration, with each method having its advantages [14]. Physicochemical analysis, such as of the sucrose content, the sum of glucose and fructose, moisture content, HMF, acidity, or diastase activity, are often used for this purpose [12,15]. Honey may be considered adulterated if one or more of these parameters does not meet international or domestic standards. Analysis of the sucrose content is also often used to identify honey adulteration. Authentic honey must not contain more than 5% sucrose [16]. Honey is mainly composed of glucose and fructose in varying concentrations (55 to 75 g/100 g), with a minimum acceptable limit of 60%. The second main component of honey is water (15 to 25 g/100 g), with an acceptable moisture level being below 17% [17]. HMF is an indicator of honey freshness and is commonly used in quality analysis. It is a compound formed during the acid-catalyzed dehydration of hexoses [18]. Generally, it should be present in very low amounts in fresh honey, with its complete absence being indicative of high-quality honey. As per the Codex Alimentarius honey standard [3] and the United Arab Emirates (UAE) standard, HMF content must not be higher than 80 mg/kg. The EC Council directive 2001/110/EC from the European community has set an HMF limit of 40 mg/kg, with an exceptional value of 80 mg/kg for honey coming from countries with tropical temperatures [4]. Honey with an HMF value of >80 mg/kg is characterized as being of very low quality. HMF quantities are affected by pH, heat treatment [19], and the conditions used for honey storage [20], with warm environments increasing the HMF concentration [21]. In addition, mixing honey with invert syrups can also increase HMF values [22]. Honey contains small amounts of different enzymes, and one of the most important ones is amylase (diastase). This enzyme is sensitive to heat and is, therefore, able to indicate the overheating of honey and its degree of thermal preservation [23]. Thus, diastase activity is considered to be a quality indicator used for the freshness of honey, set by Codex [3]. The diastase activity is usually expressed in Schade units, also known as the diastase number (DN), which is defined as the amount of enzyme that will convert 0.01 g of starch to the prescribed end-point in 1 h at 40 °C under the conditions of the test [24]. According to the Honey Quality and International Regulatory Standards, the diastase activity must not be less than or equal to eight, determined after processing and blending for all retail honey.

Besides the above stated parameters, free acidity is also used to identify fraudulent honey. The Codex Alimentarius [3] has set a permitted range of 50 meq acid/kg. A high value of free acidity in honey is an indicator that glucose and fructose fermentation by yeasts has occurred, converting the sugars to alcohol and carbon dioxide. In the presence of oxygen, alcohol is hydrolyzed to acetic acid, consequently increasing the free acidity [21]. On an obvious note, physical contaminants, such as hair and insects, are unacceptable. Recent research work has evaluated the quality of honey. A study by Kazeminia, Mahmoudi, Aali, and Ghajarbygi (2021) [25] assessed 43 honey samples collected from Qazvin province, Iran, and showed that the pH and acidity values conformed perfectly (100%) with the Iranian honey standard. However, 44.2% of the samples did not meet the acceptable quality level regarding HMF. For moisture content, 2.3% of the samples were above the acceptable limit. There was also a high percentage of samples that were not compliant with sucrose (53.5%), and glucose and fructose (25.6%) content requirements. Another study conducted by Gürbüz et al. (2020) [26] reported that all 68 honey samples collected from the Southeastern Anatolia region of Turkey were in compliance with the international standard in Turkey for sucrose content, the sum of glucose and fructose, moisture content and free acidity. However, 20.6% of samples were non-compliant concerning diastase number. For HMF, 7 of 68, or 10.3% of honey samples, had a higher HMF content than the legally permitted EU maximum level of 40 mg/kg. A study conducted by Yayinie, Atlabachew, Tesfaye, Hilluf, and Reta (2021) [27] found that all 47 honey samples collected from different geographical areas within the Amhara region, Ethiopia, met the Codex Alimentarius [3] standard. Boussaid et al. (2018) [28] reported that all 9 honey samples collected from southern Tunisia met the standards of the Codex Alimentarius for pH, free acidity, water activity and HMF.

Since there has not been any investigation available that has examined the imported honey quality in the UAE, the present study was undertaken to assess the honey imported to the UAE via ports in the Dubai Emirates over a 5 year period.

2. Materials and Methods

2.1. Sample Collection

A total of 1330 honey samples were collected from Dubai ports in the UAE between 2017 and 2021. They were classified according to honey type: honey (1180), blended honey (58), honeycomb (47), acacia honey (25), and forest honey (20). Blended honey included honey mixed with spices, pollen, verbena, lemon, mint, chili, ginger, pepper, cinnamon, or hibiscus. Comb honey is produced from traditional hives [29] and it contains honeycomb, which is the wax structure in which honey bees store honey and pollen in these hives. Acacia honey is a monofloral honey produced by *Apis mellifera*, a cultured bee that harvests the extra-floral nectar from the forest mangrove or mangium tree (*Acacia mangium*) [30]. Forest honey is produced by bees from oak, holm oak, and cork oak forests.

The sampling of honey was conducted by trained, authorized food inspectors from 5 sites located in the Dubai ports, municipality of Dubai. The collected samples were sent to the food analysis laboratory where they were stirred to yield uniformity within each sample, and then analyzed.

2.2. Chemicals and Reagents

All the solvents and chemicals used in extraction procedures and in the preparation of mobile phases were of LC-MS/MS reagent grade and were obtained from Sigma-Aldrich Chemie GmbH (Taufkirchen, Germany). Milli-Q ultra-pure water was used for all analyses (Merck, Milli-Q® IQ Element Water Purification, Burlington, MA, USA).

2.3. Determination of Sugar Composition

Determination of sugars (fructose, glucose, and sucrose) was carried out by High Performance Liquid Chromatography (Agilent Infinity 1260 II, Agilent, Santa Clara, CA, USA), using a refractive index detector following the AOAC official method 977.20-1977. Sample preparation was carried out by dissolving 2 mg honey in 25 mL deionized water. Standard solutions of 1% fructose, 1% glucose, and 0.5% sucrose (Sigma-Aldrich, St. Louis, MO, USA) were prepared in distilled water. Mixed standards were prepared at 1.0% for fructose and glucose and 0.5% for sucrose. The mixed standard was diluted to yield 0.2, 0.4, 0.6 and 0.8%. The chromatographic separation of sugars was achieved using an Agilent Zorbax carbohydrate column maintained at 35 °C with acetonitrile/water (75:25, *v/v*), which was used as the mobile phase. Then 10 µL of sample was injected at a flow rate of 1.0 mL/min. The temperature of the column was maintained at 27 °C during the entire run.

2.4. Moisture Analysis

The moisture content of the honey samples was determined using the refractive index (RI). Water content was obtained from a Chataway table [31]. An automatic digital refractometer, Atago RX-5000α (Bashumi Instr. Control Services, Northriding, Randburg, SA), calibrated with distilled water, was used for the measurement. A drop of honey was placed on the surface of the prism and a refractive index; the reading was taken at 20 °C and converted to a percentage (g/100 g) using the Chataway table [31].

2.5. Determination of Hydroxymethylfurfural (HMF)

The HMF content was determined based on the UV absorbance of HMF at 284 nm using a UV-Visible spectrophotometer (Thermo Fisher Scientific, Waltham, MA, USA) [32]. In order to avoid the interference of other components at this wavelength, the difference between the absorbance of a clear aqueous honey solution and the same solution after the addition of bisulfite was determined. The HMF content was calculated after the subtraction of the background absorbance at 336 nm. In a 50 mL volumetric flask containing 2 mg of honey dissolved in 25 mL of water, 0.5 mL of Carez I solution was added, followed by 0.5 mL of Carez II solution. Water was added to the flask to make up a volume of 50 mL, and the resulting solution was filtered. After discarding the first 5 mL of the filtrate, 5 mL of 0.2% sodium bisulfite solution was added to the test tube. In another test tube, 5 mL of pure water was added as a blank. A UV-visible spectrophotometer was used to measure the solution's absorbance at 284 and 336 nm in 10 mm quartz cells within 1 h. The calculation was performed using the formula below:

$$\text{HMF (mg/kg)} = (A284 - A336) \times 149.7 \times 5 \times \frac{D}{W}$$

It should be noted that 149.7 is a constant, A284 is the absorbance at 284 nm, A336 is the absorbance at 336 nm, 5 is the theoretical nominal sample weight, D is the dilution factor (in case dilution is necessary), and W is the weight of honey taken.

2.6. Determination of Honey Acidity

A titrimetric method was used to determine free acidity following the AOAC official method 962.19-1977. In a 100 mL beaker, a 3 mg sample of homogenized honey was dissolved in water. The solution was then titrated against a 0.1 N NaOH solution until the formation of a pink color and the titer value was noted. The results were reported as milliequivalents (meq) per kg of honey [33].

2.7. Diastase Activity

Diastase activity was calculated according to the AOAC method 958.09 [34]. The honey sample (5 g) was diluted in 10 mL of deionized water and 2.5 mL of acetate buffer (1.59 M, pH 5.3). The diluted sample was then transferred to a 25 mL volumetric flask containing 1.5 mL of 0.5 M NaCl solution. Ten ml of honey solution was mixed with 100 mL of 1% (w/v) starch solution and incubated in a water bath at 40 °C for 5 min. After that, 1 mL of the treated sample was mixed with 10 mL of 0.0007 M diluted iodine solution and measured at 660 nm in a spectrophotometer (Thermo-scientific Evolution 60S model, Waltham, MA, USA).

2.8. Physical Contaminants

The determination of extraneous matter was conducted by visual inspection. Honey samples were checked for the presence of hair, other foreign material, or insect parts.

3. Results and Discussion

3.1. Compliance of Imported Honey with Standards

The conformity assessment of imported honey samples, based on the UAE honey standard [35] in Table 1, is presented in Table 2. The data presented are arranged to identify the types of honey that are more likely to be adulterated. An evaluation of 1330 samples collected revealed that 79.2% conformed with UAE legislation. The level of non-conformity was lower than the 64.4% reported by Al-Farsi et al. (2018) [6] for 58 honey samples collected from 18 geographical regions in Oman, following a comparison with Gulf Standardization Organization (GSO) standards. It is notable that Boussaid et al. (2018) [28] found that all 9 honey samples examined in Tunisia complied with the Codex Alimentarius standards.

Table 1. UAE standards of honey [35].

	UAE.S 147, 2019
Sucrose content	Max 5%
Sum of glucose and fructose	Min 60%
Moisture *	Max limit 20%
HMF	Max 80 mg/kg
Acidity	Max 50 meq/kg
Physical hazards (hair, insects)	Absent
Diastase activity	$8\,^0$ Goth

* The permissible moisture content in honey according to the UAE standard was previously 17% [36].

Table 2. Summary of imported honey compliance with UAE standards from 2017 to 2021.

Type of Honey	No. Samples	No. Compliant (%)	No. Non-Compliant (%)
Honey	1180	929 (79)	251 (21)
Blended honey	58	46 (79.5)	12 (20.5)
Honeycomb	47	37 (79)	10 (21)
Acacia honey	25	23 (92)	2 (8)
Forest honey	20	19 (95)	1 (5)
Total	1330	1054 (79.3)	276 (20.8)

In the current study, the proportion of non-conformity ranged from 5.0 to 21.0% for the various types of honey, with the lowest values noted for forest honey. Honey, blended honey, and honeycomb had non-conformity levels that were similar, at 20.5 to 21.0%.

Table 3 presents the honey compliance across the years from 2017 to 2021. For the individual years, statistical significance was noted between the conforming and non-conforming samples. However, across the years, only 2017 was shown to be significantly different from the other years; here, 174, or 90.6%, of 1054 samples were compliant, while 18, or 9.4%, of 276 were non-compliant. None of the other years (2018, 2019, 2020 or 2021) had significantly different numbers of non-conforming samples, but in 2019, the proportion of conforming samples was the lowest.

Table 3. Honey compliance across the years from 2017 to 2021 using Chi-square.

	Results		
	Conforming (n = 1054) n (%)	Non-Conforming (n = 276) n (%)	*p*-Value
2017 (n = 192)	174(90.6) [aB]	18(9.4) [bB]	
2018 (n = 268)	223(83.2) [aA]	45(16.8) [bA]	
2019 (n = 297)	201(67.7) [aA]	96(32.3) [bA]	<0.001
2020 (n = 263)	225(85.6) [aA]	38(14.4) [bA]	
2021 (n = 310)	231(74.5) [aA]	79(25.5) [bA]	

a, b: Different letters indicate significant differences in the proportions of conforming and non-conforming samples across the individual years ($p < 0.001$). A, B: Different letters indicate significant differences in the proportions of conforming and non-conforming samples across the different years ($p < 0.001$).

3.2. Compliance of Imported Honey with Recognized Standards

The honey samples were assessed for compliance with the Emirates standard [35] for moisture, total sucrose content, the sum of glucose and fructose, HMF, acidity limit, diastase activity and the presence of physical contaminants (Table 4).

Table 4. Identification of non-compliant quality parameters among the 276 imported, non-compliant honey samples.

Type of Non-Compliant Honey		Sucrose Content (Max 5%)	Sum of Glucose and Fructose (Min 60%)	Moisture Max Limit		HMF (Max 80 mg/kg)	Acidity Max 50 meq/kg	Physical Hazards (Hair, Insects)	Diastase Activity (^{0} Goth)	Violations Involving Multiple Non-Compliant Parameters
				17%	20%					
Honeycomb (n = 10)	No. (%)		8 (80)	1 (10)						1 (10)
	Mean (%)		43.86	18.1						
	Range (%)		22.5−56.4	18−18.2						
Acacia honey (n = 2)	No. (%)		1 (50)	1 (50)						
	Mean (%)		50.4	18.6						
	Range (%)									
Blended honey (n = 12)	No. (%)		3 (25)	1 (8.)		5 (41.)				3 (25)
	Mean (%)		51.95	17.8		236.3				
	Range (%)		41.9−57.5			107−458				
Forest honey (n = 1)	No. (%)			1 (100)						
	Mean (%)			17.7						
	Range (%)									
Honey (n = 251)	No. (%)	6 (2.4)	101 (40.2)	26 (10.4)	5 (2)	67 (26.7%)	5 (2.0)	5 (2.0)	1 (0.4)	35 (13.9)
	Mean (%)	14.1	52.5	18.3	22	154.7	74	Present	2.1	
	Range (%)	5.1−33.4	19.6−59.1	17.7−19.3	20.6−24.6	83.2−663	52−85	Present		
Number of non-compliant samples (276)	No (%)	6 (2.2%)	113 (40.9)	35 (12.7%)		72 (26.1%)	5 (1.8%)	5 (1.8)	1 (0.4)	39 (14.1)

3.2.1. Total Sugar Content in Honey Samples
Sucrose Content

In 2.4%, or 6/251 samples of honey, non-compliance with the UAE honey standard was only due to sucrose content being higher than the 5% limit permitted. This level of non-compliance was lower than the 53.5% non-conformity, due to the high sucrose found among the 43 honey samples that originated in Iran [25]. During another study conducted by Gürbüz et al. (2020), sucrose was not detected in 55.9% or 38/68 honey samples collected from the Southeastern Anatolia region of Turkey [26]. The sucrose content of the remaining 30 samples (44.1%) was less than the legally permissible maximum value of 5%. In other work involving 9 honey samples collected from southern Tunisia it was found that the sucrose content ranged from 2.3 to 4.5% [28]. In the present study, the mean sucrose content of the 6 non-conforming samples was 14.1%, with a range of 5.1 to 33.4%. This high amount of sucrose could have been due to overfeeding the bees with sugar in spring [37] or to the early harvesting of honey before the full transformation of sugar into glucose and fructose [38]. The high sucrose content could be also an indication of possible adulteration by the direct addition of sugar to honey [39].

Sum of Glucose and Fructose

The determination of reducing sugars (the sum of glucose and fructose) in honey is also a quality criterion used to indicate honey freshness. Of the total number of non-conforming samples (n = 276), 113 (40.9%) did not meet the UAE criterion for acceptable total sugar content. While all 20 forest honey samples met the UAE standard, blended honey, honeycomb and acacia honey did not. Honey showed the highest percentage of non-conformities at 40.2%. The mean total reducing sugar content of these samples was 52.5%, with a range of 19.6 to 59.1%. An amount of fructose and glucose below 60% is taken to indicate honey adulteration. It should be noted that the ratio of fructose to glucose in any particular honey depends largely on the source of the nectar [37]. The mean total glucose and fructose content obtained in the present study for non-conforming honey samples at 52.5% was similar to the 54.3% obtained for 29 samples of Sidr honey collected from Oman. The acceptable total glucose and fructose content as per GSO honey standards is a minimum of 45% [6]; however, the UAE legislation is more strict and accepts a minimum value of only 60%.

The sugar composition results found during the current study are not in accordance with the values reported in a study by Geană, Ciucure, Costinel, and Ionete (2020) [40] where the total amount of fructose and glucose in 48 honey samples that originated in Romania was higher than the 60% specified in the EU standards [4]. Another study conducted in Turkey showed that the mean combined glucose and fructose content of 68 honey samples ranged from 62.6 to 77.3%, with a mean concentration of 71.0% [26].

3.2.2. Moisture Content

The non-conformity concerning the moisture content was 12.7%. The moisture content of the samples was assessed based on the sampling year, since the permissible moisture content in honey according to the UAE standard was recently increased from 17% [36] to 20% [35]. In honey samples, 26 of the 31 samples did not meet the old criterion for moisture (UAE.S 147, 2017) [36], with a mean moisture content of 18.3% (17.7 to 19.3%). Using the updated criterion of 20% moisture, 5 of the 31 samples did not conform [38], and had a mean moisture content of 22% (20.6 to 24.6%). Similar values were reported for Tanzanian honey [41], with a moisture range of 21.6 to 22.8%, and Philippine honey [41], with a moisture range between 22.0 to 23.1%. In contrast, Australian honey samples were fully compliant with the 20% requirement, having mean sample moisture contents ranging from 10.6 to 17.8% [21]. High moisture content may have resulted from honey harvest under high humidity conditions or an early seasonal honey extraction [42]. The variance in moisture content can also be attributed to the botanical source, the season, as well as the geographic conditions.

3.2.3. HMF Content

HMF is commonly used as a parameter for honey freshness and authenticity. Normally, fresh honey contains low amounts of HMF, with the HMF content depending on the rate of honey monosaccharide decomposition. According to the UAE honey standard [35], HMF content should not exceed 80 mg/kg in honey from countries with tropical temperatures. Of the 276 samples that did not conform to the standards, 72, or 26.1%, had HMF values above 80 mg/kg. For blended honey, 5 of the 12 non-conforming samples had an average value of 236.4 mg/kg (107 to 458 mg/kg) of HMF. For honey samples, 67 of 251 non-compliant samples had a mean HMF content of 154.7 mg/kg (83.2 to 663 mg/kg). The results of the present study are in agreement with those from a study conducted by Gürbüz et al. (2020) [26], where it was found that 10.3% of the honey samples had a higher HMF content than the legally permitted maximum of 40 mg/kg. The results of the current study are also similar to those from a study conducted by Al-Farsi et al. (2018) [6], where the values observed were 16.2 and 1062 mg/kg in Sumer and multiflora honey, respectively. Ajlouni and Sujirapinyokul (2010) [21] observed that HMF in two Australian honey samples (2.22 and 17.7 mg/kg) was within the international limit of 40 mg/kg. The variation in HMF values could result from the influence of factors such as pH, heating, storage conditions and floral type [43]. High HMF values are a reflection of overheating honey, use of inappropriate storage conditions or its mixture with an invert syrup made by acid or enzymatic inversion [37].

3.2.4. Acidity

The legislation of the UAE [35] and Codex regulations [3] do not accept an acidity of >50 meq/kg in honey. This criterion was not met by 2% of the samples in the current study. The free acidity in the studied honey samples was 74 meq/kg (52 to 85 meq/kg). Similarly, in Oman, mean acidity values of 84.9 meq/kg were reported for Sumer honey, with 18 of 21 samples exceeding the Codex standard [6]. Research studies on honey collected from Turkey [44], Portugal [45], Argentina [46], and Ethiopia [39] showed lower values of free acidity (25.0, 40.3, 11.0, and 45.0 meq/kg, respectively). The high acidity observed in 5 samples in the present study could have been due to the fermentation of honey sugars to form organic acid [30]. This could be controlled by adopting more modern techniques for producing honey that enable reducing the moisture or by pasteurizing the honey to control microorganisms. The variation in acidity between the samples may also be attributed to the presence of different acids of varying floral origin or harvest season [44,45].

3.2.5. Diastase Activity

It is well known that the natural enzyme diastase is an indicator of freshness in honey. The activity of the enzyme can be a measure of honey exposure to heat and/or inappropriate storage conditions [15,46]. All the honey samples in the present study were observed to have diastase activity, which was in accord with UAE standards [35]. The minimum standard value for the diastase index observed was 8 in all the samples, except 1 (honey, diastase value = 2). It is probable that this sample was not fresh or was inadequately pasteurized. This is in contrast with results from a study conducted on honey in Oman, where 16 of 58 samples did not conform to the GSO standard. In Iran, Kazeminia et al. (2021) [25] observed that 46.5% of 43 honey samples were non-compliant with standards for diastase activity. Compliant values for diastase were reported for honey in Ethiopia [47] and Argentina [48].

3.2.6. Physical Contaminants

As per UAE legislation, samples are required to be free from any form of physical contaminants. In the present study, only 5 of 251 honey samples did not meet the criterion. Two of them had hair, and 3 were found to contain insect parts. In a study conducted by Brasil da Silva (2021) [49] on 14 honey samples collected from the north of Brazil, it was reported, following microscopic examination, that 50% (7/14) of the samples contained what appeared to be dirt. The presence of physical contaminants could result from improper processing practices including poor hygiene, inadequate storage conditions, poor pest control practices, as well as inadequate packaging.

3.2.7. Combination of Non-Conforming Elements

Of the 276 non-conforming honey samples, 39 (14.1%) had more than one instance of non-compliance with the quality criteria. Approximately 35/251 samples, or 13.9%, of the honey samples did not meet the quality criteria on more than one count, with the distribution being as follows: glucose, fructose, and sucrose (9 samples); glucose, fructose, and HMF (8 samples); HMF and moisture (8 samples); glucose, fructose and moisture (4 samples); glucose, fructose, HMF, and moisture (2 samples); glucose, fructose, and acidity (2 samples); sucrose and HMF (1 sample); and HMF, diastase content, and sucrose (1 sample).

The combination of non-conformities in 9 samples (glucose, fructose and sucrose) may have resulted from the addition of sucrose or table sugar, which will reduce the percent of glucose and fructose present in total sugars detected. The combination of non-conformities in one sample (HMF, diastase content and sucrose) was probably due to the exposure of this sample to overheating; published work has shown that an increased temperature increases HMF and reduces diastase activity, especially for temperatures over 60 °C [50]. Therefore, the sample that was non-compliant due to a combination of HMF and diastase levels was not acceptably fresh.

The majority of the combined infractions included the violation of glucose and fructose content, which is highly affected by climate, honeybee flora, and honey handling practices [51].

3.2.8. Classification of Honey Non-Conformities According to the Country of Origin

Samples that did not conform with international and local standards, as per the country of origin, and honey type are presented in Table 5. Of the 251 non-conforming honey samples, 137, or 32.5% of samples, originated in India. This high percentage of non-conformity is in agreement with a report published by the Centre for Science and Environment, which indicated a high percentage of non-compliance in honey samples from India, with 77% identified as being adulterated with sugar syrups.

In total, 10 of the 13 honey brands failed the purity test used during a study by Dhingra (2020) [52]. However, it should be noted that the highest number of samples examined had been imported from India in the study because India greatly increased honey exports recently [53]. This could potentially introduce bias in the observations of the study. India exports honey to more than 65 countries. The United States is the biggest importer of honey, largely from India, with approximately 80% of the total imported honey being of Indian origin. The UAE is the third largest importer of honey from India with a value of USD 2.66 million, representing 3.31% of the total Indian exports [54]. In the current study, 33, or 19.5%, of 170 samples from Australia did not conform to the standards. This percentage of non-conformity is in accord with the 18% found during a study conducted by Zhou, Taylor, Salouros, and Prasad (2018) [55] on Australian honey samples. In the current study, 12 samples from New Zealand, 9 samples from Pakistan, and 8 from Turkey did not meet the legislative criteria. For blended honey, 5 of the 12 non-compliant samples originated from India. For honeycomb, out of the 10 non-conforming samples, 4 samples originated from Turkey.

Table 5. Number of non-conforming samples (%) according to the country of origin and type of honey.

Country (No. of Imported Samples)	No. of Imported Samples (No. of Rejected Samples, %)				
	Honey	**Blended Honey**	**Honey Comb**	**Acacia Honey**	**Forest Honey**
India (430) [$]	422 [$$] (137 [$$$], 32.5 [$$$$])	6 (5, 83.4)	2 (2, 100)	0 (0, 0)	0 (0, 0)
Australia (177)	170 (33, 19.5)	2 (0, 0)	3 (0, 0)	0 (0, 0)	2 (0, 0)
Germany (94)	67 (3, 4.5)	9 (1, 11.1)	8 (2, 25)	7 (1, 14.3)	3 (0, 0)
New Zealand (74)	69 (12, 17.4)	3 (0, 0)	2 (0, 0)	0 (0, 0)	0 (0, 0)
France (58)	44 (5, 11.5)	8 (0, 0)	3 (0, 0)	3 (0, 0)	0 (0, 0)
Pakistan (53)	53 (9, 17)	0 (0, 0)	0 (0, 0)	0 (0, 0)	0 (0, 0)
Switzerland (44)	16 (0, 0)	4 (1, 25)	0 (0, 0)	10 (1, 0.1)	14 (1, 7)
Turkey (44)	23 (8, 35)	7 (4, 57)	13 (4, 30.7)	1 (0, 0)	0 (0, 0)
United Kingdom (40)	40 (5, 12.5)	0 (0, 0)	0 (0, 0)	0 (0, 0)	0 (0, 0)
Spain (38)	37 (2, 5.4)	0 (0, 0)	0 (0, 0)	0 (0, 0)	1 (0, 0)
Egypt (24)	22 (5, 22.7)	2 (0, 0)	0 (0, 0)	0 (0, 0)	0 (0, 0)
Saudi Arabia (24)	24 (2, 8.3)	0 (0, 0)	0 (0, 0)	0 (0, 0)	0 (0, 0)
Kyrgyzstan (23)	21 (2, 9.5)	1 (1, 100)	0 (0, 0)	1 (0, 0)	0 (0, 0)
Italy (20)	18 (0, 0)	2 (0, 0)	0 (0, 0)	0 (0, 0)	0 (0, 0)
Lebanon (19)	17 (3, 17.6)	2 (0, 0)	0 (0, 0)	0 (0, 0)	0 (0, 0)
Honk Kong (18)	17 (3, 17.6)	1 (0, 0)	0 (0, 0)	0 (0, 0)	0 (0, 0)
Hungary (17)	8 (3, 37.5)	0 (0, 0)	9 (0, 0)	0 (0, 0)	0 (0, 0)
Iran (16)	14 (5, 35.7)	0 (0, 0)	2 (1, 50)	0 (0, 0)	0 (0, 0)
Yemen (15)	12 (7, 58.4)	1 (0, 0)	0 (0, 0)	0 (0, 0)	0 (0, 0)
Kazakhstan (12)	12 (2, 16.7)	0 (0, 0)	0 (0, 0)	0 (0, 0)	0 (0, 0)
Bulgaria (11)	7 (0, 0)	2 (0, 0)	0 (0, 100)	0 (0, 0)	2 (0, 0)
Ireland (9)	6 (1, 16.7)	1 (0, 0)	1 (0, 0)	1 (0, 0)	0 (0, 0)
Chile (7)	6 (1, 16.7)	1 (0, 0)	0 (0, 0)	0 (0, 0)	0 (0, 0)
United states (6)	5 (0, 0)	1 (0, 0)	0 (0, 0)	0 (0, 0)	0 (0, 0)
Greece (6)	6 (0, 0)	0 (0, 0)	0 (0, 0)	0 (0, 0)	0 (0, 0)
Uzbekistan (5)	4 (0, 0)	1 (0, 0)	0 (0, 0)	0 (0, 0)	0 (0, 0)
Russia Federation (4)	3 (0, 0)	0 (0, 0)	1 (0, 0)	0 (0, 0)	0 (0, 0)

Table 5. *Cont.*

Country (No. of Imported Samples)	No. of Imported Samples (No. of Rejected Samples, %)				
	Honey	Blended Honey	Honey Comb	Acacia Honey	Forest Honey
China (4)	4 (0, 0)	0 (0, 0)	0 (0, 0)	0 (0, 0)	0 (0, 0)
United Arab Emirates (4)	4 (0, 0)	0 (0, 0)	0 (0, 0)	0 (0, 0)	0 (0, 0)
Canada (3)	3 (0, 0)	0 (0, 0)	0 (0, 0)	0 (0, 0)	0 (0, 0)
Malaysia (3)	2 (1, 50)	1 (0, 0)	0 (0, 0)	0 (0, 0)	0 (0, 0)
Mexico (3)	3 (1, 33.3)	0 (0, 0)	0 (0, 0)	0 (0, 0)	0 (0, 0)
Ukraine (2)	1 (0, 0)	1 (0, 0)	0 (0, 0)	0 (0, 0)	0 (0, 0)
Belgium (2)	1 (0, 0)	1 (0, 0)	0 (0, 0)	0 (0, 0)	0 (0, 0)
Bosnia (2)	2 (0, 0)	0 (0, 0)	0 (0, 0)	0 (0, 0)	0 (0, 0)
Afghanistan (2)	2 (0, 0)	0 (0, 0)	0 (0, 0)	0 (0, 0)	0 (0, 0)
Mauritius (2)	2 (0, 0)	0 (0, 0)	0 (0, 0)	0 (0, 0)	0 (0, 0)
Armenia (2)	2 (0, 0)	0 (0, 0)	0 (0, 0)	0 (0, 0)	0 (0, 0)
Poland (2)	2 (0, 0)	0 (0, 0)	0 (0, 0)	0 (0, 0)	0 (0, 0)
Oman (2)	2 (0, 0)	0 (0, 0)	0 (0, 0)	0 (0, 0)	0 (0, 0)
Sudan (1)	1 (0, 0)	0 (0, 0)	0 (0, 0)	0 (0, 0)	0 (0, 0)
Nigeria (1)	1 (1, 100)	0 (0, 0)	0 (0, 0)	0 (0, 0)	0 (0, 0)
Indonesia (1)	0 (0, 0)	0 (0, 0)	1 (1, 100)	0 (0, 0)	0 (0, 0)
Korea (1)	0 (0, 0)	1 (0, 0)	0 (0, 0)	0 (0, 0)	0 (0, 0)
Thailand (1)	1 (0, 0)	0 (0, 0)	0 (0, 0)	0 (0, 0)	0 (0, 0)
Kuwait (1)	1 (0, 0)	0 (0, 0)	0 (0, 0)	0 (0, 0)	0 (0, 0)
Morocco (1)	1 (0, 0)	0 (0, 0)	0 (0, 0)	0 (0, 0)	0 (0, 0)
Argentina (1)	1 (0, 0)	0 (0, 0)	0 (0, 0)	0 (0, 0)	0 (0, 0)
Austria (1)	1 (0, 0)	0 (0, 0)	0 (0, 0)	0 (0, 0)	0 (0, 0)
Total	1180 (251, 21)	58 (12, 20.7)	47 (10, 21.3)	25 (2, 8)	20 (1, 5)

[$] represents the total number of honey samples originating from the mentioned country. [$$] represents the total number of samples of each honey type originating from the mentioned country. [$$$] represents the rejected samples from the total number of samples from this honey type and [$$$$] the percentage of non-conformity out of the total samples from this honey type.

To better understand the results, a detailed conformity assessment is presented in Figure 1, according to the country of origin of honey samples (n = 1180). This approach was chosen only for the honey samples because of its greater sample size; it represented 88% of the total 1330 samples examined.

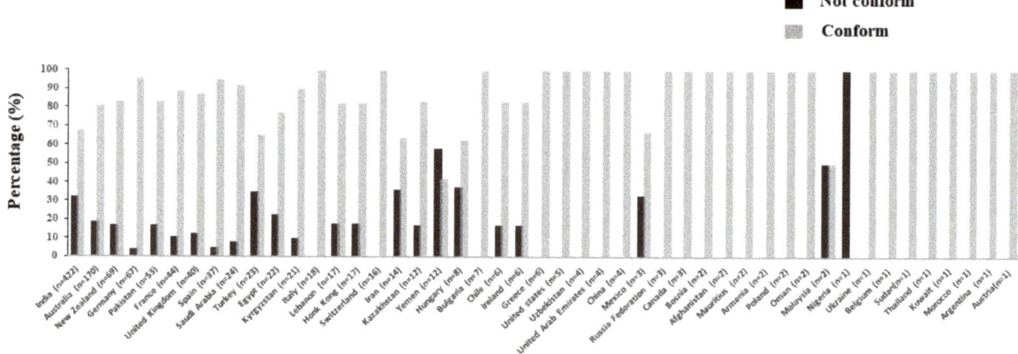

Figure 1. Origin and percent of imported honey samples (n = 1180) tested that were not compliant with UAE honey quality standards.

The honey samples originated from 5 of the 49 countries that supplied honey in this study, namely India (422), Australia (170), New Zealand (69), Germany (67), and Pakistan (53). India had the highest proportion of non-compliant samples (32.5%), while Germany had the lowest at 4.5%.

The most frequent non-compliant parameters were the sum of glucose and fructose in samples originating from India, Australia, Pakistan, and Turkey (Figure 2). On the other hand, for samples from New Zealand, the moisture content and HMF were the major non-compliant parameters.

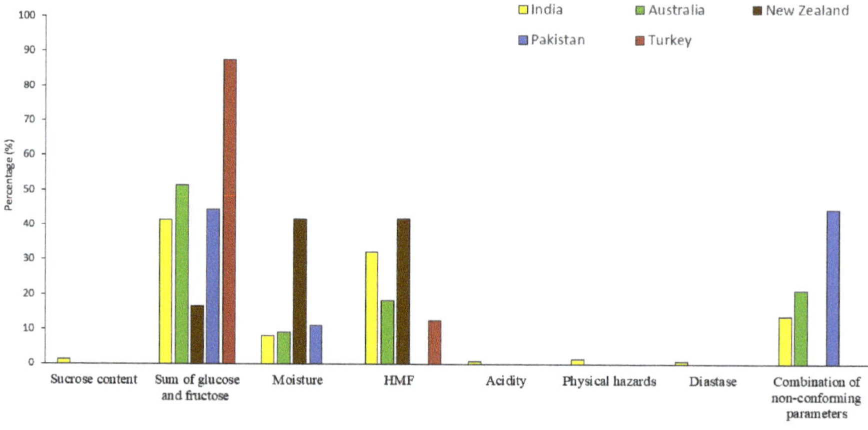

Figure 2. Occurrence of non-compliant parameters in imported honey samples from countries having the greatest levels of non-compliance.

4. Conclusions

Assessment of the physicochemical parameters of honey is necessary for quality assurance purposes. This study evaluated the physicochemical quality characteristics of honey imported to the UAE. Of the 1330 honey samples tested, 1054 complied with the UAE honey standard. Examination of the country of origin showed that India supplied the highest proportion of non-compliant samples, at 32.5%, while Germany had the lowest, at 4.5%. However, it is important to note that the number of samples from each country differed and hence this may have introduced bias. The most frequent source of non-compliance involved the sum of glucose and fructose. This study emphasizes the need for continued border inspection, in conjunction with the physicochemical analysis of honey samples, in order to prevent the entry of adulterated honey into the country.

Author Contributions: T.M.O., R.S.O.: Conceived and designed the experiments; wrote the paper. M.S.A.S., W.A.M.B.O., A.A.S.A.A.A.: Conceived and designed the experiments. V.G., F.S.B.B.: Performed the experiments; contributed reagents, materials, analysis tools or data. N.E.D., H.H., R.H.: Analyzed and interpreted the data; wrote the paper. All authors have read and agreed to the published version of the manuscript.

Funding: This research received no external funding.

Data Availability Statement: The data are available from the corresponding author.

Acknowledgments: The authors would like to thank University of Sharjah, Sharjah, United Arab Emirates and Dubai Municipality, Dubai, United Arab Emirates.

Conflicts of Interest: The authors declare no conflict of interest.

References

1. Viuda-Martos, M.; Ruiz-Navajas, Y.; Fernández-López, J.; Pérez-Álvarez, J.A. Functional Properties of Honey, Propolis, and Royal Jelly. *J. Food Sci.* **2008**, *73*, R117–R124. [CrossRef] [PubMed]
2. Mehryar, L.; Esmaiili, M. Honey & Honey Adulteration Detection: A Review. In Proceedings of the 11th International Congress on Engineering and Food, Athens, Greece, 22–26 May 2011; Saravacos, G., Ed., Elsevier Procedia: Amsterdam, The Netherlands, 2011.
3. Alimentarius, C. Revised Codex Standart for Honey. *Codex Stan.* **2001**, *12*, 1981.
4. Council, E. Council Directive 2001/110/EC of 20 December 2001 Relating to Honey. *Off. J. Eur. Communities L10 47–52 Off J. Eur. Commun.* **2002**, *110*, 47–50.
5. Bogdanov, S.; Lüllmann, C.; Martin, P.; von der Ohe, W.; Russmann, H.; Vorwohl, G.; Oddo, L.P.; Sabatini, A.-G.; Marcazzan, G.L.; Piro, R. Honey Quality and International Regulatory Standards: Review by the International Honey Commission. *Bee World* **1999**, *80*, 61–69. [CrossRef]
6. Al-Farsi, M.; Al-Belushi, S.; Al-Amri, A.; Al-Hadhrami, A.; Al-Rusheidi, M.; Al-Alawi, A. Quality Evaluation of Omani Honey. *Food Chem.* **2018**, *262*, 162–167. [CrossRef]
7. Kumar, A.; Gill, J.P.S.; Bedi, J.S.; Manav, M.; Ansari, M.J.; Walia, G.S. Sensorial and Physicochemical Analysis of Indian Honeys for Assessment of Quality and Floral Origins. *Food Res. Int.* **2018**, *108*, 571–583. [CrossRef]
8. Brar, D.S.; Pant, K.; Krishnan, R.; Kaur, S.; Rasane, P.; Nanda, V.; Saxena, S.; Gautam, S. A Comprehensive Review on Unethical Honey: Validation by Emerging Techniques. *Food Control* **2023**, *145*, 109482. [CrossRef]
9. Parliament, E. *Report on the Food Crisis, Fraud in the Food Chain and the Control Thereof (2013/2091 (INI))*; European Parliament: Strasbourg, France, 2013.
10. Morales, V.; Sanz, M.L.; Martín-Álvarez, P.J.; Corzo, N. Combined Use of HMF and Furosine to Assess Fresh Honey Quality. *J. Sci. Food Agric.* **2009**, *89*, 1332–1338. [CrossRef]
11. Al-Diab, D.; Jarkas, B. Effect of Storage and Thermal Treatment on the Quality of Some Local Brands of Honey from Latakia Markets. *J. Entomol. Zool. Stud.* **2015**, *3*, 328–334.
12. Cotte, J.-F.; Casabianca, H.; Chardon, S.; Lheritier, J.; Grenier-Loustalot, M.-F. Application of Carbohydrate Analysis to Verify Honey Authenticity. *J. Chromatogr. A* **2003**, *1021*, 145–155. [CrossRef]
13. Li, S.; Shan, Y.; Zhu, X.; Zhang, X.; Ling, G. Detection of Honey Adulteration by High Fructose Corn Syrup and Maltose Syrup Using Raman Spectroscopy. *J. Food Compos. Anal.* **2012**, *28*, 69–74. [CrossRef]
14. Soares, S.; Amaral, J.S.; Oliveira, M.B.P.P.; Mafra, I. A Comprehensive Review on the Main Honey Authentication Issues: Production and Origin. *Compr. Rev. Food Sci. Food Saf.* **2017**, *16*, 1072–1100. [CrossRef]
15. Crăciun, M.E.; Parvulescu, O.C.; Donise, A.C.; Dobre, T.; Stanciu, D.R. Characterization and Classification of Romanian Acacia Honey Based on Its Physicochemical Parameters and Chemometrics. *Sci. Rep.* **2020**, *10*, 20690. [CrossRef] [PubMed]
16. Guo, W.; Zhu, X.; Liu, Y.; Zhuang, H. Sugar and Water Contents of Honey with Dielectric Property Sensing. *J. Food Eng.* **2010**, *97*, 275–281. [CrossRef]
17. Seraglio, S.K.T.; Schulz, M.; Gonzaga, L.V.; Fett, R.; Costa, A.C.O. Current Status of the Gastrointestinal Digestion Effects on Honey: A Comprehensive Review. *Food Chem.* **2021**, *357*, 129807. [CrossRef]
18. Belitz, H.D.; Grosch, W. *Química de Los Alimentos [Food Chemistry]*; Acribia S.A.: Zaragoza, Spain, 1997.
19. Tosi, E.; Martinet, R.; Ortega, M.; Lucero, H.; Ré, E. Honey Diastase Activity Modified by Heating. *Food Chem.* **2008**, *106*, 883–887. [CrossRef]
20. Zappala, M.; Fallico, B.; Arena, E.; Verzera, A. Methods for the Determination of HMF in Honey: A Comparison. *Food Control* **2005**, *16*, 273–277. [CrossRef]
21. Ajlouni, S.; Sujirapinyokul, P. Hydroxymethylfurfuraldehyde and Amylase Contents in Australian Honey. *Food Chem.* **2010**, *119*, 1000–1005. [CrossRef]

22. Tosun, M.; Keles, F. Investigation Methods for Detecting Honey Samples Adulterated with Sucrose Syrup. *J. Food Compos. Anal.* **2021**, *101*, 103941. [CrossRef]
23. Ahmed, M.; Djebli, N.; Aissat, S.; Khiati, B.; Meslem, A.; Bacha, S. In Vitro Activity of Natural Honey Alone and in Combination with Curcuma Starch against Rhodotorula Mucilaginosa in Correlation with Bioactive Compounds and Diastase Activity. *Asian Pac. J. Trop. Biomed.* **2013**, *3*, 816–821. [CrossRef]
24. Huang, Z.; Liu, L.; Li, G.; Li, H.; Ye, D.; Li, X. Nondestructive Determination of Diastase Activity of Honey Based on Visible and Near-Infrared Spectroscopy. *Molecules* **2019**, *24*, 1244. [CrossRef] [PubMed]
25. Kazeminia, M.; Mahmoudi, R.; Aali, E.; Ghajarbygi, P. Evaluation of Authenticity in Honey Samples from Qazvin, Iran. *J. Chem. Health Risks* **2021**, *13*, 73–83.
26. Gürbüz, S.; Çakıcı, N.; Mehmetoğlu, S.; Atmaca, H.; Demir, T.; Arıgül Apan, M.; Atmaca, Ö.F.; Güney, F. Physicochemical Quality Characteristics of Southeastern Anatolia Honey, Turkey. *Int. J. Anal. Chem.* **2020**, *2020*, 8810029. [CrossRef]
27. Yayinie, M.; Atlabachew, M.; Tesfaye, A.; Hilluf, W.; Reta, C. Quality Authentication and Geographical Origin Classification of Honey of Amhara Region, Ethiopia Based on Physicochemical Parameters. *Arab. J. Chem.* **2021**, *14*, 102987. [CrossRef]
28. Boussaid, A.; Chouaibi, M.; Attouchi, S.; Hamdi, S.; Ferrari, G. Classification of Southern Tunisian Honeys Based on Their Physicochemical and Textural Properties. *Int. J. Food Prop.* **2018**, *21*, 2590–2609. [CrossRef]
29. Belay, A.; Solomon, W.K.; Bultossa, G.; Adgaba, N.; Melaku, S. Botanical Origin, Colour, Granulation, and Sensory Properties of the Harenna Forest Honey, Bale, Ethiopia. *Food Chem.* **2015**, *167*, 213–219. [CrossRef]
30. Moniruzzaman, M.; Sulaiman, S.A.; Azlan, S.A.M.; Gan, S.H. Two-Year Variations of Phenolics, Flavonoids and Antioxidant Contents in Acacia Honey. *Molecules* **2013**, *18*, 14694–14710. [CrossRef]
31. Cunnif, P. *Official Methods of Analysis Method 969.38B, P21*, 16th ed.; AOAC Internacional: Washington, DC, USA, 1995; Volume II.
32. White, J.W., Jr. Spectrophotometric Method for Hydroxymethylfurfural in Honey. *J. Assoc. Off. Anal. Chem.* **1979**, *62*, 509–514. [CrossRef]
33. Horowitz, W.; Latimer, G.W. *Official Methods of Analysis of AOAC International*; AOAC International: Gaithersburg, MD, USA, 2006; Volume 18.
34. AOAC. *Official Methods of Analysis of the Association of Official Analytical Chemists*, 19th ed.; AOAC International: Gaithersburg, MD, USA, 2012.
35. *UAE.S 147; Honey*. Emirates Authority for Standards & Metrology (ESMA): Abu Dhabi, United Arab Emirates, 2019.
36. *UAE.S 147: 2017; Honey*. Emirates Authority for Standards & Metrology (ESMA): Abu Dhabi, United Arab Emirates, 2017.
37. Wu, L.; Du, B.; Vander Heyden, Y.; Chen, L.; Zhao, L.; Wang, M.; Xue, X. Recent Advancements in Detecting Sugar-Based Adulterants in Honey–A Challenge. *TrAC Trends Anal. Chem.* **2017**, *86*, 25–38. [CrossRef]
38. Da, C.; Azeredo, L.; Azeredo, M.A.A.; De Souza, S.R.; Dutra, V.M.L. Protein Contents and Physicochemical Properties in Honey Samples of Apis Mellifera of Different Floral Origins. *Food Chem.* **2003**, *80*, 249–254.
39. Gebremariam, T.; Brhane, G. Determination of Quality and Adulteration Effects of Honey from Adigrat and Its Surrounding Areas. *Int. J. Technol. Emerg. Engin. Res.* **2014**, *2*, 71–76.
40. Geană, E.-I.; Ciucure, C.T.; Costinel, D.; Ionete, R.E. Evaluation of Honey in Terms of Quality and Authenticity Based on the General Physicochemical Pattern, Major Sugar Composition and Δ13C Signature. *Food Control* **2020**, *109*, 106919. [CrossRef]
41. Gidamis, A.B.; Chove, B.E.; Shayo, N.B.; Nnko, S.A.; Bangu, N.T. Quality Evaluation of Honey Harvested from Selected Areas in Tanzania with Special Emphasis on Hydroxymethyl Furfural (HMF) Levels. *Plant Foods Hum. Nutr.* **2004**, *59*, 129–132. [CrossRef] [PubMed]
42. Karabagias, I.K.; Vlasiou, M.; Kontakos, S.; Drouza, C.; Kontominas, M.G.; Keramidas, A.D. Geographical Discrimination of Pine and Fir Honeys Using Multivariate Analyses of Major and Minor Honey Components Identified by 1H NMR and HPLC along with Physicochemical Data. *Eur. Food Res. Technol.* **2018**, *244*, 1249–1259. [CrossRef]
43. Fallico, B.; Zappala, M.; Arena, E.; Verzera, A. Effects of Conditioning on HMF Content in Unifloral Honeys. *Food Chem.* **2004**, *85*, 305–313. [CrossRef]
44. Derebaşı, E.; Bulut, G.; Col, M.; Güney, F.; Yaşar, N.; Ertürk, Ö. Physicochemical and Residue Analysis of Honey from Black Sea Region of Turkey. *Fresenius Environ. Bull.* **2014**, *23*, 10–17.
45. Estevinho, L.M.; Feás, X.; Seijas, J.A.; Vázquez-Tato, M.P. Organic Honey from Trás-Os-Montes Region (Portugal): Chemical, Palynological, Microbiological and Bioactive Compounds Characterization. *Food Chem. Toxicol.* **2012**, *50*, 258–264. [CrossRef]
46. Aloisi, P.V. Determination of Quality Chemical Parameters of Honey from Chubut (Argentinean Patagonia). *Chil. J. Agric. Res.* **2010**, *70*, 640–645. [CrossRef]
47. Getu, A.; Birhan, M. Chemical Analysis of Honey and Major Honey Production Challenges in and around Gondar, Ethiopia. *Acad. J. Nutr.* **2014**, *3*, 6–14.
48. Cantarelli, M.A.; Pellerano, R.G.; Marchevsky, E.J.; Camiña, J.M. Quality of Honey from Argentina: Study of Chemical Composition and Trace Elements. *J. Argentine Chem. Soc.* **2008**, *96*, 33–41.
49. da Silva, J.B. Evaluation of Physic Contamiants and Contamination with Coliforms, Molds and Yeasts of Honey from the Northern Brazil. *Rev. Bras. Ciênc. Vet.* **2021**, *28*, 117–123.
50. Pasias, I.N.; Kiriakou, I.K.; Proestos, C. HMF and Diastase Activity in Honeys: A Fully Validated Approach and a Chemometric Analysis for Identification of Honey Freshness and Adulteration. *Food Chem.* **2017**, *229*, 425–431. [CrossRef] [PubMed]

51. Tigistu, T.; Worku, Z.; Mohammed, A. Evaluation of the Physicochemical Properties of Honey Produced in Doyogena and Kachabira Districts of Kembata Tambaro Zone, Southern Ethiopia. *Heliyon* **2021**, *7*, e06803. [CrossRef] [PubMed]
52. Dhingra, S. Available online: https://policycommons.net/artifacts/2233584/httpcdncseindiaorgattachments093428900_1606892 722_report_laboratory-results-of-honey-testing/2991517/ (accessed on 24 January 2023).
53. Jamwal, S.; Sharma, N.; Dhiman, A.; Kumari, S. Current Status and Future Strategies to Increase Honey Production in India. In *Honey*; CRC Press: Boca Raton, FL, USA, 2021; pp. 191–206. ISBN 1003175961.
54. Singh, R. Current Honey Market in India-Volume and Value. *Int. J. Ayurveda Pharma Res.* **2021**, *9*, 82–88. [CrossRef]
55. Zhou, X.; Taylor, M.P.; Salouros, H.; Prasad, S. Authenticity and Geographic Origin of Global Honeys Determined Using Carbon Isotope Ratios and Trace Elements. *Sci. Rep.* **2018**, *8*, 14639. [CrossRef] [PubMed]

 *foods*

Article

Physicochemical Composition of Local and Imported Honeys Associated with Quality Standards

Hael S. A. Raweh [1], Ahmed Yacine Badjah-Hadj-Ahmed [2], Javaid Iqbal [1] and Abdulaziz S. Alqarni [1,*]

[1]   Melittology Research Lab, Department of Plant Protection, College of Food and Agriculture Sciences, King Saud University, Riyadh 11451, Saudi Arabia; hraweh@ksu.edu.sa (H.S.A.R.); jiqbal@ksu.edu.sa (J.I.)

[2]   Department of Chemistry, College of Science, King Saud University, Riyadh 11451, Saudi Arabia; ybadjah@ksu.edu.sa

*   Correspondence: alqarni@ksu.edu.sa

Abstract: The compliance with honey standards is crucial for its validity and quality. The present study evaluated the botanical origin (pollen analysis) and physicochemical properties: moisture, color, electrical conductivity (EC), free acidity (FA), pH, diastase activity, hydroxymethylfurfural (HMF), and individual sugar content of forty local and imported honey samples. The local honey exhibited low moisture and HMF (14.9% and 3.8 mg/kg, respectively) than imported honey (17.2% and 23 mg/kg, respectively). Furthermore, the local honey showed higher EC and diastase activity (1.19 mS/cm and 11.9 DN, respectively) compared to imported honey (0.35 mS/cm and 7.6 DN, respectively). The mean FA of local honey (61 meq/kg) was significantly naturally higher than that of imported honey (18 meq/kg). All local nectar honey that originated from *Acacia* spp. exhibited naturally higher FA values that exceeded the standard limit ($\leq$50 meq/kg). The Pfund color scale ranged from 20 to 150 mm in local honey and from 10 to 116 mm in imported honey. The local honey was darker, with a mean value of 102.3 mm, and was significantly different from imported honey (72.7 mm). The mean pH values of local and imported honey were 5.0 and 4.5, respectively. Furthermore, the local honey was more diverse in pollen grain taxa compared to imported honey. Local and imported honey elicited a significant difference regarding their sugar content within individual honey type. The mean content of fructose, glucose, sucrose, and reducing sugar of local honey (39.7%, 31.5%, 2.8%, and 71.2%, respectively) and imported honey (39.2%, 31.8%, 0.7%, and 72.0%, respectively) were within the permitted quality standards. This study indicates the necessity of increasing the awareness regarding quality investigations for healthy honey with good nutritional value.

Keywords: honey; physicochemical properties; melissopalynological analysis; quality control

Citation: Raweh, H.S.A.; Badjah-Hadj-Ahmed, A.Y.; Iqbal, J.; Alqarni, A.S. Physicochemical Composition of Local and Imported Honeys Associated with Quality Standards. *Foods* **2023**, *12*, 2181. https://doi.org/10.3390/foods12112181

Academic Editors: Liming Wu and Qiangqiang Li

Received: 27 April 2023
Revised: 24 May 2023
Accepted: 25 May 2023
Published: 29 May 2023

1. Introduction

Honey is a natural substance produced by honey bees, and it originates from floral nectar and some plants exudates. Honey composition and quality characteristics are variable and are mainly affected by different factors such as soil composition, nectar source, climatic conditions, beekeeping practices, processing type, and storage conditions. Floral origin is quite influential on the physicochemical properties of honey such as electrical conductivity, color, moisture, pH, mineral content, and acidity level; conversely, other parameters, e.g., hydroxymethylfurfural (HMF) content and purity, are related to the manufacturing process [1]. The main constitutes of honey are carbohydrates (80–85%) and water (15–17%). Fructose and glucose are the most dominant sugars responsible for the majority of the physical and nutritional properties of honey. Small quantities of other sugars (disaccharides, trisaccharides, and oligosaccharides) are also present in honey, in addition to minerals, free amino acids, flavonoids, vitamins, enzymes, and phenolic and organic acids. The acidity level indicates the maturity of honey and characterizes its stability and changes of quality during storage [2]. There is an increasing interest in alternative medicine for public health, and honey is a main element in this regard. Therefore, honey consumption rates are increasing in Saudi Arabia [3]. Moreover, limited production of honey and inappropriate agricultural practices have led to an increase in honey adulteration [4]. Honey becomes an easy target of adulteration due to the high demand for its therapeutic and healing properties. Adulteration of honey could be direct by adding a substance to the honey such as cane sugar, beet sugar, and molasses, or indirect by feeding honeybee colonies with adulterating substances. Excessive heat used for pasteurization and packing, on the other hand, can have negative effects on honey quality, such as loss of enzyme activity [5]. Honey fermentation and spoilage may occur when honey is harvested with high humidity [6]. However, authenticity of honey is not only restricted to adulteration; indeed, post-harvest quality alterations are also possible during the flow season or in the store. In the recent past, various tools were developed to assess the quality and authenticity of honey as desired by the consumers, as well as to provide fair competition to honey producers. The international standards of honey quality parameters are available in different standards such as the Codex Alimentarius Standard and European Honey Directive [7]. In recent years, the importance of the physicochemical properties of honey has been increased, because these parameters are vital for issuing honey quality certificates [8]. Honey quality is generally accessed quantitatively by analyzing its composition, as described in international standards and legislations of honey (sucrose content $\leq$ 5%, fructose 31–42%, glucose 23–32%, reducing sugar $\geq$ 60%, moisture content $\leq$ 21%, water-insoluble content $\leq$ 0.1%, electrical conductivity $\leq$ 0.8 mS/cm, mineral content (ash) $\leq$ 0.6%, free acidity $\leq$ 50 meq/kg, diastase activity $\geq$ 8 DN (Schade units), and hydroxymethylfurfural (HMF) content $\leq$ 40 mg/kg) [7,9]. These specifications of honey were also adopted in Saudi Arabia by the Gulf Standardization Organization [10]. Honey is of special importance due to religious and cultural reasons in many Muslim countries, including Saudi Arabia. It is not only used as a sweetening additive but also as a healing agent. On the basis of official statistics, about 27,347 tons of honey were brought to the Saudi market last year (2021), of which, 3233 tons were produced locally and 24,114 tons were imported from different countries. In Saudi Arabia, the import of honey has increased during the last five years (2016–2020), from 13,568 to 16,441, 17,099, 18,526, and 24,114 tons, respectively [11]. Nevertheless, most of the locally produced honey is processed and marketed without a verified quality check and assessment of its origin information. This has led to increased honey adulteration and its marketing without verified quality. This study is of particular importance in order to add a comprehensive database of characterizing Saudi Arabian honey, as well as imported honey, which may contribute, if available, positively in terms of reformulating a proper and nationally accepted honey quality standard. Thus, the current study investigated the different physicochemical attributes of honey composition associated with quality standards from honey of local and imported origins.

2. Materials and Methods

The botanical origin (pollen analysis) and physicochemical characteristics such as moisture content, color, electrical conductivity (EC), free acidity (FA), pH, hydroxymethylfurfural (HMF), sugar content, and diastase activity (DN) of local and imported honey samples were evaluated according to the recommended methods [12–22]. The honey analyses were performed at the Honey Quality Research Laboratory, Department of Plant Protection, King Saud University, Riyadh.

2.1. Honey Samples

Twenty samples of each honey type (local and imported) of diversified botanical origin were collected from different sources in Saudi Arabia during 2020–2021 (Table 1). The honey samples were kept in the dark at room temperature until the subsequent analysis.

Table 1. Detail of tested honey samples.

Origin of Honey	No.	Sample Code	Detail of Honey Type and Location
Native honey (Kingdom of Saudi Arabia, KSA)	1	ACS1	*Acacia gerardii* honey (Huraymila: 25°12″ N, 46°10″ E)
	2	ACS2	*Acacia gerardii* honey (Hail: 27°31″ N, 41°41″ E)
	3	ACS3	*Acacia gerardii* honey (Riyadh: 24°24″ N 46°71″ E)
	4	ACS4	*Acacia gerardii* honey (Al Qassim: 25°49″ N, 42°51″ E)
	5	ACS5	*Acacia gerardii* honey (Al Taif: 21°16″ N, 44°25″ E)
	6	ACS6	*Acacia gerardii* honey (Asir: 18°13″ N, 42°23″ E)
	7	ACS7	*Acacia gerardii* honey (Al Ahsa: 25°17″ N, 49°29″ E)
	8	SDS1	Sidr, *Ziziphus* sp. honey (Huraymila: 25°12″ N, 46°10″ E)
	9	SDS2	Sidr, *Ziziphus* sp. honey (Hail: 27°31″ N, 41°41″ E)
	10	SDS3	Sidr, *Ziziphus* sp. honey (Riyadh: 24°24″ N 46°71″ E)
	11	SDS4	Sidr, *Ziziphus* sp. honey (Al Qassim: 25°49″ N, 42°51″ E)
	12	SDS5	Sidr, *Ziziphus* sp. honey (Al Taif: 21°16″ N, 44°25″ E)
	13	SDS6	Sidr, *Ziziphus* sp. honey (Asir: 18°13″ N, 42°23″ E)
	14	SDS7	Sidr, *Ziziphus* sp. honey (Al Ahsa: 25°17″ N, 49°29″ E)
	15	ALS	Alfalfa honey (Al Qassim: 25°49″ N, 42°51″ E)
	16	SES	*Vachellia seyal* honey (Huraymila: 25°12″ N, 46°10″ E)
	17	SMS1	*Acacia tortilis* honey (Huraymila: 25°12″ N, 46°10″ E)
	18	SMS2	*Acacia tortilis* honey (Riyadh: 24°24″ N 46°71″ E)
	19	SMS3	*Acacia tortilis* honey (Al Taif: 21°16″ N, 44°25″ E)
	20	SHS	Shafallah–caper bush honey, *Capparis spinose* (Al Taif: 21°16″ N, 44°25″ E)
Imported Honey (from different countries of the world)	1	SMF	Multifloral honey, Spain
	2	IMF	Multifloral honey, India
	3	PAS1	Honey imported from different countries but packed in KSA
	4	PAS2	Honey imported from different countries but packed in KSA
	5	PAS3	Honey imported from different countries but packed in KSA
	6	PAS4	Honey imported from different countries but packed in KSA
	7	PAS5	Honey imported from different countries but packed in KSA
	8	PAS6	Honey imported from different countries but packed in KSA
	9	MKN1	Manuka honey, New Zealand
	10	MKN2	Manuka honey, New Zealand
	11	BFG	Black forest honey, Germany
	12	CMF	Multifloral honey, China
	13	CSD	Sidr, *Ziziphus* sp. honey, China
	14	PAG	*Robinia pseudoacacia* Black locust honey, Germany
	15	SWMF	Multifloral honey, Switzerland
	16	BMF	Multifloral honey, United Kingdom
	17	FMF	Multifloral honey, France
	18	AMF	Multifloral honey, Australia
	19	CTE	Citrus honey, Egypt
	20	KSD	Sidr, *Ziziphus* sp. honey, Pakistan

2.2. Melissopalynological Analysis

The pollen presence is fundamental for the melissopalynological analysis of honey [9]. The presence of pollen in the honey samples and the botanical origins of honey samples was tested according to the recommended protocols [15–17]. Briefly, ten grams of honey was mixed in 20 mL of warm distilled water (40 °C), centrifuged for 10 min at 2500 rpm, poured into a small tube, and centrifuged again for 10 min. The entire sediment was put on a slide, spread out over an area of 20 mm^2, and dried by slight heating at 40 °C. The sediment was mounted with glycerin gelatin and liquefied by heating in a water bath at 40 °C [18]. The identification of pollen grain in the treated honey samples was performed according to the pollen atlas [19].

2.3. Physicochemical Analysis

The color, moisture content, EC, FA, pH, HMF, DN, and sugar content of local and imported honey samples were determined as per the recommended protocol [15]. Every honey sample was tested three times for every parameter, and the data were expressed as mean values.

2.3.1. Color Analysis

The Pfund scale was used to measure the color intensity of honey samples according to the recommended protocol [15,21]. Half of the cuvette was filled with homogenous honey (without air bubbles) using a 10 mm light path. Color grades (0–150 mm) were determined using a color photometer (HI 96785, Hanna® Instruments, Nusfalau, Romania), in which the cuvette was inserted. The analytical-grade glycerol standard was used to compare the Pfund grades of honey according to the United States Department of Agriculture [15,21].

2.3.2. Moisture Content

The refractometric method [12,15] was used to measure the moisture content in terms of refractive indices with the help of a refractometer (Hammann® honey refractometer, Hassloch, Germany) at ambient temperature. The refractive index directly increased with increases in the solid content of the honey sample. A drop of thoroughly mixed honey was put on the lens, and the lid of the refractometer was carefully closed for the even spreading of honey without any air bubbles. The refractometer was held towards the light to record the interface position. Before the testing of every honey sample, the instrument was thoroughly cleaned and dried.

2.3.3. Electrical Conductivity (EC)

The EC was measured using an EC meter (Hanna® pH PPM Meter HI-9813-6N, Nusfalau, Romania). It was first calibrated with deionized water, and the conductance cell was dipped into 10% honey solution (10.0%). The reading of EC was recorded after the stabilization of the instrument [12,15].

2.3.4. pH

Ten grams of honey was mixed in 75 mL deionized water. Honey solution was transferred into a beaker, and a pH meter (Hanna® pH PPM Meter HI-9813-6N, Nusfalau, Romania) was put in the solution. The stable readings of pH were recorded from the pH meter [12,15].

2.3.5. Free Acidity (FA)

FA was measured using the titrimetric method. Ten grams of honey was dissolved in 75 mL of deionized water. The honey solution was titrated with sodium hydroxide (NaOH 0.05 N) until the pH value reached at 8.5. The final acidity number was expressed in meq/kg [12,15].

2.3.6. Hydroxymethylfurfural (HMF)

HMF was recorded by determining the absorbance of the solutions at 284 and 336 nm, which was done using a GenesysTM10S UV-visible spectrometer (Thermo Fisher Scientific, Shanghai, China) [18]. The following equation was used to calculate the HMF content:

$$\text{HMF}(\text{mg/kg}) = (A284) - (A336) \times 149.7$$

where A284: absorbance value at 284 nm, A336: absorbance at 336 nm, and 149.7: a factor calculated by the molecular weight of HMF and the mass of the sample [12,15].

2.3.7. Diastase Activity (DN)

The diastase number (DN) displaying the diastase activity and the DN of the honey samples was measured using the recommended protocol [7,15,18,22]. The absorbance of samples was recorded, and a calibration curve was formulated.

2.3.8. Sugar Content

The percentages of sugar contents (fructose, glucose, sucrose, and reducing sugar) in honey samples were measured using high-performance liquid chromatography HPLC (Agilent Technologies®, Santa Clara, CA, USA) with RID detector and carbohydrate column). Sample preparation for HPLC was performed according to Raweh et al. [18]. The chromatogram peaks of the sugars were identified by comparison with those of previously injected standard sugars [15,20].

2.4. Statistical Analysis

The data of different physicochemical properties of honey are expressed as mean ± SE. The quantified variables of the honey samples were compared using the analysis of variance

(ANOVA) and Duncan's multiple range test. The statistical significance ($p < 0.05$) for the parameter values was calculated using SAS® 9.2 software.

3. Results

3.1. Presence of Pollen Grains

The melissopalynological studies revealed the presence of different types of pollen grains in all tested local and imported honey samples that originated from diverse topographical origins (Table 2). A great diversity of pollens was observed in the pollen spectra. The majority of pollens belonged to four families (Fabaceae, Asteraceae, Rhamnaceae, and Capparaceae), which were detected from local honey samples. In imported honey samples, the majority of pollens that belonged to three families (Fabaceae, Asteraceae, and Rhamnaceae) were detected, but the pollen diversity was lower than those of local honey (Table 2).

Table 2. Microscopic analyses of the pollen grain types present in the local and imported honey samples.

Origin of Honey	No.	Sample Code	Detail of Pollen Grains
	1	ACS1	Fabaceae, and others.
	2	ACS2	Fabaceae, and others.
	3	ACS3	Fabaceae, Rhamnaceae, and others.
	4	ACS4	Fabaceae, Asteraceae, and others.
	5	ACS5	Fabaceae, Capparaceae, Malvaceae, Asteraceae.
	6	ACS6	Fabaceae, and others.
	7	ACS7	Fabaceae, Asteraceae, and others.
	8	SDS1	Rhamnaceae, and others.
	9	SDS2	Rhamnaceae, Fabaceae, Asteraceae, and others.
Native honey	10	SDS3	Rhamnaceae, Tamaricaceae, Capparaceae, Asteraceae, Combretaceae, and others.
	11	SDS4	Rhamnaceae, Fabaceae, Combretaceae, Capparaceae, and others.
	12	SDS5	Rhamnaceae, Capparaceae, Fabaceae, and others.
	13	SDS6	Rhamnaceae, Capparaceae, and others.
	14	SDS7	Rhamnaceae, Fabaceae, Combretaceae, Capparaceae, and others.
	15	ALS	Fabaceae, Capparaceae, and others.
	16	SES	Fabaceae, Tamaricaceae, Asteraceae, and Capparaceae.
	17	SMS1	Fabaceae, Asteraceae, and others.
	18	SMS2	Fabaceae, Rhamnaceae, Asteraceae, Capparaceae, and others.
	19	SMS3	Fabaceae, Capparaceae, and others.
	20	SHS	Capparaceae, Fabaceae, Rhamnaceae, Asteraceae, and others.
	1	SMF	Malvaceae, Asteraceae, Fabaceae, Santalaceae, and others.
	2	IMF	Fabaceae, and others.
	3	PAS1	Fabaceae, and others.
	4	PAS2	Fabaceae, Asteraceae, and others.
	5	PAS3	Myrtaceae, Fabaceae, Asteraceae, and others.
	6	PAS4	Rhamnaceae, Fabaceae, Asteraceae, Malvaceae, Solanaceae, and others.
	7	PAS5	Fabaceae, Asteraceae, Malvaceae, and others.
	8	PAS6	Fabaceae, Asteraceae, and others.
	9	MKN1	Myrtaceae, Lamiaceae, and others.
Imported honey	10	MKN2	Myrtaceae, Solanaceae, Asteraceae, and others.
	11	BFG	Rhamnaceae, and others.
	12	CMF	Rosaceae, and others.
	13	CSD	Rhamnaceae, Fabaceae, Rosaceae, and Asteraceae.
	14	PAG	Fabaceae, Convolvulaceae, and others.
	15	SWMF	Rhamnaceae, Fabaceae, Ericaceae, Fabaceae, and others.
	16	BMF	Asteraceae, and others.
	17	FMF	Pinaceae, and others.
	18	AMF	Rutaceae, and others.
	19	CTE	Rutaceae, Fabaceae, Rhamnaceae, Asteraceae, Solanaceae, and others.
	20	KSD	Rhamnaceae, Fabaceae, Asteraceae, and others.

Others: pollen families that were not identified.

3.2. Physicochemical Analysis of Honey

The physicochemical properties (moisture content, color, EC, pH, FA, HMF, DN, and sugar contents) were determined from the local and imported honey samples. The majority of the local and imported honey exhibited adequate quality physicochemical properties that were compatible with international regulation of honey quality [9,10]. Local honey samples showed certain physicochemical properties (Tables 3 and 4) that were marked as relatively better (low moisture and HMF; high EC and DN) than that of imported honey samples, with a few exceptions. The low moisture content depicts the maturation of honey without any fermentation and long shelf life, low HMF with high DN illustrates the freshness and proper handling, and higher EC in the local exceptional nectar honey shows the presence of more mineral elements due to their botanical origin.

Table 3. Physicochemical properties of local honey samples.

No.	Sample Code	Color	* Pfund Color (mm)	Moisture (%)	EC (mS/cm)	pH	Free Acidity (meq/kg)	HMF (mg/kg)	DN
1	ACS1	Dark amber	144 ± 0.0 b	13.8 ± 0.0 k	1.91 ± 0.0 b	5.2 ± 0.0 e	106 ± 0.3 b	1.6 ± 0.2 efg	8.7 ± 0.3 gh
2	ACS2	Amber	113 ± 0.0 g	14.3 ± 0.0 j	1.51 ± 0.0 e	4.5 ± 0.0 l	95 ± 0.0 e	1.6 ± 0.2 efg	10.0 ± 0.0 f
3	ACS3	Dark amber	150 ± 0.0 a	15.6 ± 0.0 c	1.35 ± 0.0 h	4.5 ± 0.0 l	100 ± 0.0 d	6.3 ± 0.1 c	11.0 ± 0.0 e
4	ACS4	Dark amber	125 ± 0.0 c	14.4 ± 0.0 j	1.55 ± 0.0 d	4.9 ± 0.0 h	90 ± 0.0 f	0.7 ± 0.3 fg	10.0 ± 0.0 f
5	ACS5	Dark amber	120 ± 0.0 d	14.4 ± 0.0 j	1.27 ± 0.0 i	4.5 ± 0.0 l	66 ± 0.0 i	10.9 ± 0.3 b	8.0 ± 0.0 i
6	ACS6	Dark amber	150 ± 0.0 a	14.5 ± 0.0 ij	1.83 ± 0.0 c	5.1 ± 0.0 f	105 ± 0.0 b	0.9 ± 0.2 fg	10.9 ± 0.1 e
7	ACS7	Dark amber	150 ± 0.0 a	13.1 ± 0.0 l	2.00 ± 0.0 a	4.8 ± 0.0 i	110 ± 0.0 a	0.2 ± 0.0 fg	10.9 ± 0.1 e
8	SDS1	Dark amber	115 ± 0.0 e	15.2 ± 0.0 d	1.43 ± 0.0 f	5.0 ± 0.0 g	102 ± 0.3 c	0.3 ± 0.3 fg	28.7 ± 0.3 a
9	SDS2	Extra light amber	44 ± 0.0 n	14.9 ± 0.0 fgh	0.62 ± 0.0 m	4.9 ± 0.0 h	15 ± 0.0 n	1.1 ± 0.2 fg	29.0 ± 0.0 a
10	SDS3	Light amber	60 ± 0.0 m	15.03 ± 0.0 def	0.55 ± 0.0 n	5.1 ± 0.0 f	15 ± 0.0 n	0.8 ± 0.1 fg	17.4 ± 0.0 b
11	SDS4	Amber	103 ± 0.0 i	15.1 ± 0.0 de	1.39 ± 0.0 g	6.4 ± 0.0 c	23 ± 0.0 l	1.3 ± 0.2 efg	11.0 ± 0.0 e
12	SDS5	Amber	86 ± 0.0 j	15.0 ± 0.0 efg	1.33 ± 0.0 h	7.1 ± 0.0 a	11 ± 0.0 o	0.7 ± 0.2 fg	12.0 ± 0.0 d
13	SDS6	Light amber	81 ± 0.0 k	15.6 ± 0.0 c	1.45 ± 0.0 f	5.4 ± 0.0 d	20 ± 0.0 m	0.0 ± 0.0 g	10.3 ± 0.0 f
14	SDS7	Light amber	75 ± 0.0 l	14.7 ± 0.1 hi	0.66 ± 0.0 l	6.6 ± 0.0 b	11 ± 0.0 o	3.4 ± 0.3 de	8.0 ± 0.0 i
15	ALS	White	31 ± 0.0 o	16.0 ± 0.0 b	0.23 ± 0.0 o	3.5 ± 0.0 n	30 ± 0.0 k	2.2 ± 0.4 efg	9.0 ± 0.0 g
16	SES	Amber	103 ± 0.0 i	13.8 ± 0.0 k	1.24 ± 0.0 i	4.6 ± 0.0 k	67 ± 0.0 h	2.4 ± 0.4 ef	8.3 ± 0.0 hi
17	SMS1	Amber	114 ± 0.0 f	16.0 ± 0.0 b	1.19 ± 0.0 j	4.6 ± 0.0 k	75 ± 0.0 g	5.6 ± 0.1 cd	9.0 ± 0.0 g
18	SMS2	Amber	112 ± 0.0 h	14.8 ± 0.1 gh	0.96 ± 0.0 k	4.6 ± 0.0 k	65 ± 0.0 j	29.2 ± 1.7 a	5.6 ± 0.0 j
19	SMS3	Dark amber	150 ± 0.0 a	15.0 ± 0.0 efg	1.19 ± 0.0 j	4.7 ± 0.0 j	90 ± 0.0 f	7.2 ± 0.1 c	5.2 ± 0.0 j
20	SHS	White	20 ± 0.0 p	17.1 ± 0.1 a	0.23 ± 0.0 o	3.8 ± 0.0 m	29 ± 0.3 k	0.2 ± 0.1 fg	15.3 ± 0.0 c

Means with the same letters are not significantly different from each other ($p < 0.05$, Tukey's test). * Color was determined in mm on the Pfund scale according to the U.S. Department of Agriculture classifications (water white: <9, extra white: 9–17, white: 18–34, extra light amber: 35–50, light amber: 51–85, amber: 86–114, dark amber: >114). Codex Alimentarius Standard (moisture: ≤20%, Pfund color: 0–150 mm; EC: ≤0.8 mS/cm; pH: 3.4–6.1; FA: ≤50 meq/kg; HMF: ≤40 mg/kg (in tropical regions: 80 mg/kg); DN: ≥8) [9,10].

Table 4. Physicochemical properties of imported honey samples.

No.	Sample Code	Color	* Pfund Color	Moisture %	EC mS/cm	pH	Free Acidity (meq/kg)	HMF (mg/kg)	DN
1	SMF	Light amber	70 ± 0.0 k	17.2 ± 0.0 h	0.26 ± 0.0 j	4.0 ± 0.0 j	20 ± 0.0 d	20 ± 0.0 h	11.0 ± 0.0 d
2	IMF	Light amber	72 ± 0.0 j	17.5 ± 0.0 ef	0.13 ± 0.0 m	4.4 ± 0.0 g	14 ± 0.0 g	38 ± 0.0 d	0.0 ± 0.0 m
3	PAS1	Light amber	68 ± 0.0 i	17.7 ± 0.0 de	0.25 ± 0.0 j	4.1 ± 0.0 i	20 ± 0.0 d	40 ± 0.0 cd	0.0 ± 0.0 m
4	PAS2	Light amber	60 ± 0.0 m	16.8 ± 0.0 i	0.05 ± 0.0 o	4.0 ± 0.0 j	7 ± 0.0 l	67 ± 0.0 b	0.0 ± 0.0 m
5	PAS3	Amber	93 ± 0.0 f	14.7 ± 0.0 m	0.88 ± 0.0 a	7.2 ± 0.0 a	8 ± 0.0 k	2 ± 0.0 k	8.0 ± 0.0 f
6	PAS4	Light amber	85 ± 0.0 g	18.4 ± 0.0 c	0.10 ± 0.0 n	3.9 ± 0.0 k	11 ± 0.0 i	42 ± 0.6 c	6.6 ± 0.0 gh
7	PAS5	White	28 ± 0.0 q	17.9 ± 0.0 d	0.17 ± 0.0 l	4.5 ± 0.0 f	7 ± 0.0 l	24 ± 0.3 g	8.3 ± 0.0 ef
8	PAS6	Amber	109 ± 0.0 b	18.7 ± 0.1 b	0.30 ± 0.0 h	3.8 ± 0.0 l	27 ± 0.0 c	85 ± 2.1 a	5.0 ± 0.0 i
9	MKN1	Amber	99 ± 0.0 d	19.4 ± 0.0 a	0.48 ± 0.0 f	4.0 ± 0.0 j	30 ± 0.0 b	10 ± 0.0 j	7.0 ± 0.0 fg
10	MKN2	Extra light amber	50 ± 0.0 n	16.7 ± 0.1 i	0.59 ± 0.0 d	4.2 ± 0.0 h	30 ± 0.0 b	14 ± 0.1 i	2.5 ± 0.0 jk
11	BFG	Light amber	78 ± 0.0 i	15.8 ± 0.0 k	0.73 ± 0.0 c	4.5 ± 0.0 f	37 ± 0.0 a	2 ± 0.1 k	14.0 ± 0.0 c
12	CMF	White	33 ± 0.0 p	18.2 ± 0.0 c	0.28 ± 0.0 i	4.7 ± 0.0 d	10 ± 0.0 j	10 ± 0.2 j	2.0 ± 0.0 kl
13	CSD	Light amber	82 ± 0.0 h	16.9 ± 0.0 i	0.38 ± 0.0 g	5.3 ± 0.0 c	10 ± 0.0 j	2 ± 0.3 k	9.4 ± 0.0 e
14	PAG	Extra white	10 ± 0.0 r	17.46 ± 0.1 fg	0.16 ± 0.0 l	3.8 ± 0.0 l	15 ± 0.0 f	27 ± 0.0 f	5.5 ± 1.1 hi
15	SWMF	Amber	98 ± 0.0 e	17.3 ± 0.1 gh	0.39 ± 0.0 g	3.9 ± 0.0 k	30 ± 0.0 b	3 ± 0.0 k	7.5 ± 0.0 fg
16	BMF	Amber	100 ± 0.0 c	18.2 ± 0.1 c	0.06 ± 0.0 o	4.6 ± 0.0 e	7.0 ± 0.0 l	9 ± 0.0 j	20.0 ± 0.0 b
17	FMF	Light amber	82 ± 0.0 h	17.4 ± 0.0 fgh	0.54 ± 0.0 e	4.4 ± 0.0 g	30 ± 0.0 b	9 ± 0.1 j	10.9 ± 0.0 d
18	AMF	Extra light amber	43 ± 0.0 o	16.2 ± 0.0 j	0.25 ± 0.0 j	4.7 ± 0.0 d	10 ± 0.0 j	20 ± 0.7 h	3.6 ± 0.0 j
19	CTE	Light amber	78 ± 0.0 i	14.9 ± 0.0 l	0.20 ± 0.0 k	3.9 ± 0.0 k	16 ± 0.0 e	32 ± 1.3 e	0.7 ± 0.0 lm
20	KSD	Drak amber	116 ± 0.0 a	16.9 ± 0.0 i	0.84 ± 0.0 b	6.1 ± 0.0 b	13 ± 0.3 h	3 ± 0.0 k	29.0 ± 0.0 a

Means with the same letters are not significantly different from each other ($p < 0.05$, Tukey's test). * Color was determined in mm on the Pfund scale according to the U.S. Department of Agriculture classifications (water white: <9, extra white: 9–17, white: 18–34, extra light amber: 35–50, light amber: 51–85, amber: 86–114, dark amber: >114). Codex standard (moisture: ≤20%, Pfund color: 0–150 mm; EC: ≤0.8 mS/cm; pH: 3.4–6.1; FA: ≤50 meq/kg; HMF: ≤40 mg/kg (in tropical regions: 80 mg/kg); DN: ≥8) [9,10].

3.2.1. Color

The honey color is dependent on their botanical origins, and it was significantly different among local honey (Table 3) and imported honey (Table 4) samples. The color ranged from white to dark amber for local honey, and extra white to dark amber for the imported honey. The Pfund scale ranged from 20 to 150 mm and 10 to 116 mm in local and imported honey, respectively (Tables 3 and 4). The mean value for the Pfund color (102.3 ± 5.1) of local honey was significantly different from the imported honey (72.7 ± 3.59) (Table 5). The Pfund color scale of local and imported honey was within the suggested range of the International Codex [9,10,23].

Table 5. Comparison among mean values of tested physicochemical properties of local and imported honey.

Physicochemical Properties	Local Honey (Mean ± SE)	Imported Honey (Mean ± SE)	*p*-Value
Moisture (%)	14.9 ± 0.2	17.2 ± 0.3	0.018 *
Color	102.3 ± 5.1	72.7 ± 3.59	0.004 *
EC (mS/cm)	1.19 ± 0.1	0.35 ± 0.1	<0.000 *
pH	5.0 ± 0.2	4.5 ± 0.2	0.424
FA (meq/kg)	61 ± 8.3	18 ± 2.1	<0.000 *
HMF (mg/kg)	3.8 ± 1.5	23 ± 5.0	<0.000 *
DN	11.9 ± 1.4	7.6 ± 1.6	0.040 *

Asterisk represents a significant difference between local and imported honey for each single attribute ($p < 0.05$, *t*-test). Codex Alimentarius Standard (moisture: ≤20%, Pfund color: 0–150 mm; EC: ≤0.8 mS/cm; pH: 3.4–6.1; FA: ≤50 meq/kg; HMF: ≤40 mg/kg (in tropical regions: 80 mg/kg); DN: ≥8) [9,10].

3.2.2. Moisture Content

The moisture content (%) was significantly different within the individual honey type, namely, local and imported honey samples (Tables 3 and 4). The moisture content ranged from 13.1 ± 0.0 to 17.1 ± 0.1%, with a mean value of 14.9 ± 0.2% among local honey, (Table 3), and from 14.7 ± 0.0 to 19.4 ± 0.0%, with a mean value of 17.2 ± 0.3% for imported honey (Table 4).

Table 5 shows the moisture content values after comparison among local and imported honey samples. The imported honey possessed a significantly higher moisture content (17.2 ± 0.3%) than local honey (14.9 ± 0.2%) (Table 5), but these were within the permitted range (>20%) according to the international standards for honey [9,10].

3.2.3. Electrical Conductivity (EC)

The values of EC ranged from 0.23 ± 0.0 to 2.00 ± 0.0 mS/cm, with a mean value of 1.19 ± 0.1 mS/cm (Table 3) for the local honey samples and from 0.05 ± 0.0 to 0.88 ± 0.0 mS/cm, with a mean value of 0.35 ± 0.1 mS/cm for the imported honey samples (Table 4). The mean EC value of local honey (1.19 ± 0.1 mS/cm) was significantly higher than that of imported honey (0.35 ± 0.1 mS/cm) (Table 5), which showed values within the permitted range (≤0.8 mS/cm) of international standards for blossom honey [9,10].

3.2.4. pH

The pH values of the local honey (Table 3) and imported honey (Table 4) samples were acidic, and within the standard limit (3.4–6.1) of the international standard [9]. The pH values ranged from 3.5 ± 0.0 to 7.1 ± 0.0, with a mean of 5.0 ± 0.2 for local honey samples (Table 3), and 3.8 ± 0.0 to 6.1 ± 0.0, with a mean value of 4.5 ± 0.2 for imported honey samples (Table 4). Three local Sidr (*Ziziphus* spp.) honey samples (SDS4, SDS5, and SDS7) had pH values of 6.4 ± 0.0, 7.1 ± 0.0, and 6.6 ± 0.0, respectively (Table 3), which exceeded the standard limit (3.4–6.1). There was a significant difference within the samples of local (Table 3) and imported honey (Table 4). The mean pH values of local and imported honey did not show any significant differences (Table 5).

3.2.5. Free Acidity (FA)

The results indicated a significant difference for FA level among honey samples of individual honey types (Tables 3 and 4). The FA value ranged from 11 ± 0.0 to 110 ± 0.0 meq/kg (mean = 61 ± 8.3 meq/kg) for local honey samples (Table 3). The local honey samples (ACS1–7: originated from *Acacia* spp. plant), SDS1, SES, and SMS1–3 were characterized with high FA that exceeded the permitted limit (≤50 meq/kg) of honey standards (Table 3). The FA value for imported honey samples was within the permitted limit and ranged from 7 ± 0.0 to 37 ± 0.0 meq/kg (mean = 18 ± 2.1 meq/kg) (Table 4). The mean FA value (61 ± 8.3 meq/kg) of local honey was significantly higher as compared to the FA value (18 ± 2.1 meq/kg) of imported honey (Table 5).

3.2.6. Hydroxymethylfurfural (HMF)

The HMF values of local honey (Table 3) and imported honey (Table 4) samples were lower than the standard limit ($\leq$40 mg/kg) provided in the international standards [9] GSO [10]. The HMF values were significantly different among samples of individual honey type (Tables 3 and 4). The HMF value of local honey ranged from 0.0 ± 0.0 to 29.2 ± 1.7 mg/kg (mean = 3.8 ± 1.5 mg/kg). The HMF value of imported honey ranged from 2.0 ± 0.0 mg/kg to 85 ± 2.1 mg/kg (mean = 23.0 ± 5.0 mg/kg) (Tables 3 and 5). Three imported honey samples (PAS2, PAS4, and PAS6) had exceptional HMF values (67 ± 0.0, 42 ± 0.6, and 85 ± 2.1, respectively) (Table 4) that exceeded the standard limit ($\leq$40 mg/kg). The mean HMF value was significantly lower in local honey samples (3.8 ± 1.5 mg/kg) than that of imported honey samples (23 ± 5.0 mg/kg) (Table 5).

3.2.7. Diastase Activity (DN)

The diastase activity of honey is an important feature that is closely associated with the freshness of honey. The data of diastase activity were measured in terms of diastase number (DN). The values of local honey ranged from 5.2 ± 0.0 to 29.0 ± 0.0 DN, with mean of 11.9 ± 1.4 DN (Table 3), and were within the Codex standard limits ($\geq$8). Out of twenty local honey samples, only two samples (SMS2 and SMS3) showed DN lower than the Codex standard limit (Table 3). The values of imported honey ranged from 0.0 ± 0.0 to 29.0 ± 0.0 DN, with a mean of 7.6 ± 1.6 DN. Out of twenty imported honey samples, eight samples (SMF, PAS3, PAS5, BFG, CSD, BMF, FMF, and KSD) showed DN values within the Codex standard limits ($\geq$8), and the rest of all the samples had DN lower than the Codex standard limits (Table 4). The mean values of local honey (11.9 ± 1.4 DN) were higher and were significantly different from that of the imported honey samples (7.6 ± 1.6 DN) (Table 5).

3.3. Sugar Content of Honey

The HPLC analysis revealed the percentage of sugar (fructose, glucose, and sucrose) detected in the tested local honey (Figure 1A) and imported honey samples (Figure 1B). The sequence pattern of sugar content was similar in local and imported honey samples, being within the permitted quality range (fructose: 31–42%, glucose: 23–32%, sucrose: $\leq$5%). The level of reducing sugar (fructose + glucose) was also within the permitted quality standard ($\geq$60%) in both local and imported honey (Figure 1A,B).

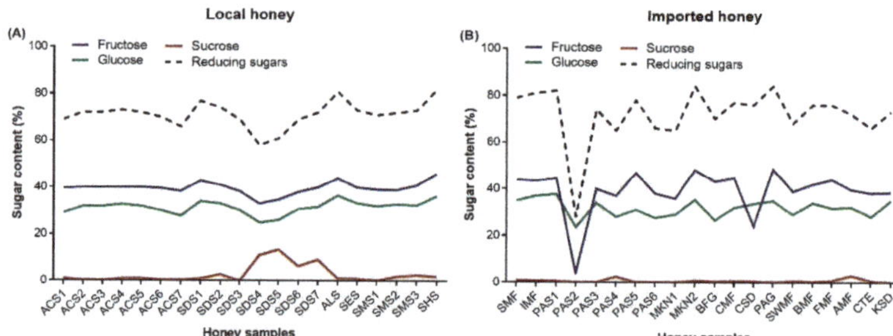

Figure 1. HPLC-based sugar profile of local honey and imported honey. (**A**) Local honey: ACS1–7 (*Acacia gerardii* honey from seven locations), SDS1–7 (Sidr, *Ziziphus* sp. honey from seven locations), ALS (alfalfa honey), SES (*Vachellia seyal* honey), SMS (*Acacia tortilis* honey), and SHS (Shafallah–caper bush honey, *Capparis spinose*). (**B**) Imported honey: SMF (multifloral honey, Spain), IMF (multifloral honey, India), PAS1–6 (honey imported from different countries but packed in KSA), MKN1–2 (manuka honey, New Zealand), BFG (black forest honey, Germany), CMF (multifloral honey, China), CSD (Sidr, *Ziziphus* sp. honey, China), PAG (*Robinia pseudoacacia* black locust honey, Germany), SWMF (multifloral honey, Switzerland), BMF (multifloral honey, United Kingdom), FMF (multifloral honey, France), AMF (multifloral honey, Australia), CTE (citrus honey, Egypt), and KSD (Sidr, *Ziziphus* sp. Pakistan).

The sugar content (fructose, glucose, sucrose, and reducing sugars) indicated significant differences within the individual type of local (Table 6) and imported honey (Table 7). Fructose and glucose were the two main carbohydrates that were detected in all the analyzed local and imported honey samples. Fructose percentage was relatively higher than that of glucose in local honey (Table 6) and imported honey, with an exception of two imported honey samples (PAS2 and CSD) that indicated a higher glucose percentage than that of fructose (Table 7).

Table 6. Analysis of sugar content (%) present in local honey samples.

No.	Sample Code	Sugar Content (%)			
		Fructose	Glucose	Sucrose	Reducing Sugars
1	ACS1	39.6 ± 0.0 ef	29.0 ± 0.0 h	1.0 ± 0.0 gh	69 ± 0.0 ef
2	ACS2	40.0 ± 0.0 e	31.8 ± 0.2 f	0.3 ± 0.0 i	72 ± 0.2 cde
3	ACS3	40.0 ± 0.0 e	31.9 ± 0.0 f	0.2 ± 0.0 i	72 ± 0.0 cde
4	ACS4	40.0 ± 0.0 e	32.7 ± 0.0 cde	1.0 ± 0.0 gh	73 ± 0.0 cd
5	ACS5	40.0 ± 0.0 e	31.9 ± 0.1 ef	1.0 ± 0.0 gh	72 ± 0.1 cde
6	ACS6	39.6 ± 0.0 ef	30.0 ± 0.0 g	0.5 ± 0.0 hi	70 ± 0.0 def
7	ACS7	38.4 ± 0.2 gh	27.7 ± 0.1 i	0.5 ± 0.0 hi	66 ± 0.2 f
8	SDS1	38.3 ± 0.4 h	30.2 ± 0.3 g	0.1 ± 0.0 i	69 ± 0.8 ef
9	SDS2	42.9 ± 0.1 c	34.0 ± 0.0 b	1.0 ± 0.0 gh	77 ± 0.1 b
10	SDS3	41.0 ± 0.0 d	33.0 ± 0.0 cd	2.8 ± 0.0 e	74 ± 0.0 bc
11	SDS4	33.0 ± 0.0 j	24.9 ± 0.1 k	10.9 ± 0.1 b	58 ± 0.1 g
12	SDS5	34.8 ± 0.1 i	26.0 ± 0.0 j	13.3 ± 0.3 a	61 ± 0.0 g
13	SDS6	38.0 ± 0.0 h	30.7 ± 0.3 g	6.3 ± 0.3 d	69 ± 0.3 ef
14	SDS7	40.0 ± 0.0 e	31.7 ± 0.1 f	9.0 ± 0.0 c	72 ± 0.1 cde
15	ALS	44.0 ± 0.0 b	36.6 ± 0.0 a	1.2 ± 0.2 g	81 ± 0.0 a
16	SES	40.0 ± 0.0 e	33.3 ± 0.3 bc	1.0 ± 0.0 gh	73 ± 0.3 bcd
17	SMS1	39.2 ± 0.2 f	32.0 ± 0.0 ef	0.2 ± 0.0 i	71 ± 0.2 cde
18	SMS2	39.0 ± 0.0 fg	32.9 ± 0.0 cd	2.0 ± 0.0 f	72 ± 0.1 cde
19	SMS3	41.0 ± 0.0 d	32.4 ± 0.2 def	2.5 ± 0.0 ef	73 ± 0.2 bc
20	SHS	46.0 ± 0.0 a	36.4 ± 0.2 a	2.0 ± 0.0 f	82 ± 0.2 a
	Mean	39.7 ± 0.6	31.5 ± 0.7	2.8 ± 0.9	71.2 ± 1.3

Means with the same letters are not significantly different from each other ($p < 0.05$, Tukey's test). Codex Alimentarius Standard (fructose: 31–42%, glucose: 23–32%, sucrose: ≤5%, reducing sugar: ≥60%) [9,10].

Table 7. Analysis of sugar content (%) present in imported honey samples.

No.	Sample Code	Sugar Content (%)			
		Fructose	Glucose	Sucrose	Reducing Sugars
1	SMF	43.9 ± 0.0 abc	35.0 ± 0.0 c	0.9 ± 0.0 c	79 ± 0.0 abc
2	IMF	43.5 ± 0.0 abc	37.0 ± 0.0 b	0.8 ± 0.0 cd	81 ± 0.0 abc
3	PAS1	44.6 ± 0.4 abc	37.8 ± 0.2 a	0.7 ± 0.0 cd	82 ± 0.5 ab
4	PAS2	4.0 ± 0.0 e	23.6 ± 0.0 j	0.0 ± 0.0 f	28 ± 0.0 j
5	PAS3	40.0 ± 0.0 abc	34.0 ± 0.0 d	0.1 ± 0.0 ef	74 ± 0.0 abcdefgh
6	PAS4	37.0 ± 0.0 bc	28.0 ± 0.0 h	2.5 ± 0.0 b	65 ± 0.0 ghi
7	PAS5	46.9 ± 0.0 ab	31.0 ± 0.0 f	0.2 ± 0.0 ef	78 ± 0.0 abcd
8	PAS6	38.0 ± 0.0 abc	27.6 ± 0.1 h	0.1 ± 0.0 ef	66 ± 0.1 fghi
9	MKN1	35.9 ± 0.0 c	29.0 ± 0.0 g	0.2 ± 0.2 ef	65 ± 0.0 hi
10	MKN2	48.2 ± 0.2 a	35.4 ± 0.2 c	0.9 ± 0.0 c	84 ± 0.3 a
11	BFG	43.2 ± 0.2 abc	26.6 ± 0.3 i	0.6 ± 0.0 d	70 ± 0.4 cdefgh
12	CMF	44.7 ± 0.1 abc	32.0 ± 0.0 e	0.8 ± 0.0 cd	77 ± 0.1 abcde
13	CSD	23.9 ± 9.0 d	33.7 ± 0.1 d	0.7 ± 0.0 cd	76 ± 0.6 i
14	PAG	48.6 ± 0.1 a	35.0 ± 0.0 c	0.3 ± 0.0 e	84 ± 0.0 a
15	SWMF	38.9 ± 0.0 abc	29.0 ± 0.0 g	0.8 ± 0.0 cd	68 ± 0.1 defghi
16	BMF	42.0 ± 0.0 abc	34.0 ± 0.0 d	0.6 ± 0.0 d	76 ± 0.0 abcdef
17	FMF	44.0 ± 0.0 abc	31.8 ± 0.0 e	0.9 ± 0.0 c	76 ± 0.0 abcdefg
18	AMF	39.7 ± 0.1 abc	32.2 ± 0.2 e	3.0 ± 0.0 a	72 ± 0.3 bcdefgh
19	CTE	38.2 ± 0.2 abc	28.0 ± 0.0 h	0.8 ± 0.0 cd	66 ± 0.2 efghi
20	KSD	38.5 ± 0.0 abc	35.0 ± 0.0 c	0.0 ± 0.0 f	73 ± 0.0 abcdefgh
	Mean	39.2 ± 2.3	31.8 ± 0.8	0.7 ± 0.2	72 ± 2.8

Means with the same letters are not significantly different from each other ($p < 0.05$, Tukey's test). Codex Alimentarius Standard (fructose: 31–42%, glucose: 23–32%, sucrose: ≤5%, reducing sugar: ≥60%) [9,10].

In local honey (Table 6), the range of sugar content was 33 ± 0.0 to 46.0 ± 0.0% (mean: 39.7 ± 0.6%) for fructose, 24.9 ± 0.1 to 36.6 ± 0.0% (mean: 31.5 ± 0.7%) for glucose, 0.1 ± 0.0

to 13.3 ± 0.3% (mean: 2.8 ± 0.9%) for sucrose, and 58 ± 0.1 to 82 ± 0.2% (mean: 71.2 ± 1.3%) for reducing sugars. Four local honey samples (SDS4, SDS5, SDS6, and SDS7) showed sucrose contents that exceeded the permitted limits (≤5%) of International Codex and GSO standards [9,10].

In imported honey (Table 7), the range of sugar content was 4.0 ± 0.0 to 48.6 ± 0.1% (mean: 39.2 ± 2.3%) for fructose, 23.6 ± 0.0% to 37.8 ± 0.2 (mean: 31.8 ± 0.8%) for glucose, 0.0 ± 0.0% to 3.0 ± 0.0% (mean: 0.7 ± 0.2%) for sucrose, and 28 ± 0.0 to 84 ± 0.0% (mean: 72.0 ± 2.8%) for reducing sugars. All sugar contents were within the permitted limits (fructose, glucose, sucrose, and reducing sugar: 31–42%, 23–32%, ≤5%, and ≥60%, respectively) of the International Codex and GSO standard [9,10]. One local honey (SDS4: 58 ± 0.1%) (Table 6) and one imported honey (PAS2: 28 ± 0.0%) (Table 7) possessed lower percentages of reducing sugar than the permitted range (≥60%).

Table 8 revealed the mean values of various sugar contents in local and imported honey. Only the percentage of sucrose was significantly different among local (2.8%) and imported honey (0.7%). The contents of fructose, glucose, and reducing sugar were similar in both local and imported honey. The mean percentages of fructose, glucose, sucrose, and reducing sugar of the local and imported honey were within the permitted limits (fructose: 31–42%, glucose: 23–32%, sucrose: ≤5%, reducing sugar: ≥60%) of the International Codex and GSO standard [9,10].

Table 8. Comparison among sugar contents of local and imported honey in Saudi Arabia.

Physicochemical Properties (Sugar Content)	Local Honey (Mean ± SE)	Imported Honey (Mean ± SE)	*p*-Value
Fructose (%)	39.7 ± 0.6	39.2 ± 2.3	0.833
Glucose (%)	31.5 ± 0.7	31.8 ± 0.8	0.780
Sucrose (%)	2.8 ± 0.9	0.7 ± 0.2	0.020 *
Reducing sugar (%)	71.2 ± 1.3	72.0 ± 2.8	0.078

Asterisk represents a significant difference between local and imported honey for each single attribute (*p* < 0.05, *t*-test). The mean percentages of fructose, glucose, sucrose, and reducing sugar of local and imported honey were within the permitted limits (fructose: 31–42%, glucose: 23–32%, sucrose: ≤5%, reducing sugar: ≥60%) of the International Codex Alimentarius and GSO standard [9,10].

4. Discussion

Pollen is a fundamental element in the analysis and quality evaluation of honey [24]. Melissopalynological analysis of honey provides the identification of the pollen types and potential plant source of honey [25]. This knowledge of pollen species is expedient in elucidating the sources of floral nectar that bees forage to produce honey of specific geographical and botanical sources [26]. In the present study, the pollen spectra from honey samples revealed that local honey had a relatively wide variety of botanical families than the imported honey. The possible explanation for the diversity in pollen content taxa between local and imported honey is because of different geographical regions, as well as the treatment of fine filtration. The local honey was without any fine filtration; unlike the imported honey, which might be commonly exposed to fine filtration to remove most of its pollen content before commercialization. According to USDA standards, commercial honey is filtered to remove suspended particles, including pollen grains [27]. Ponnuchamy et al. [28] reported the diversity of pollen spectra in the honey collected from one area at different times of the year.

We found that the color diversity among the honey samples (local and imported) ranged from white to dark amber, which is in accordance with the Pfund scale [21]. The diversity in honey color is common, and a previous study also reported the diversity ranged from colorless to amber and dark amber to black [14]. It is evident that the commercially available honey varied greatly in quality due to its color, flavor, and density over the globe [29]. Honey color is closely connected with botanical origin, and is an imperative to assess the honey quality [30]. Light-colored honey has a mild flavor, while dark honey has a more concentrated and rich flavor [31]. Furthermore, darker honey also has a high content of manganese, iron, phenolic compounds, and copper [14,32]. Many factors such as the environment, season, mineral, Maillard reaction, phenolic content, pollen, wax used, floral origin, and length of storage can affect the color of honey [25,32–34]. The Pfund color scale of local and imported honey was within the suggested range (0–150 mm) of the International Codex [9,10,23] and depended on the botanical origin.

The moisture content in the honey is important to determine honey quality, stability, resistance to spoilage, resistance to fermentation, and granulation during storage [25]. We found a relatively higher mean moisture content in imported honey than local honey, but both were within the acceptable limit (<20%) of international standards [9,10,23]. The prevalent subtropical climate conditions of high temperature and low humidity in Saudi Arabia could be the reason for the low moisture level in local honey. Moisture level is also vulnerable to geographical moisture conditions (temperature and humidity) during honey production, level of honey maturity in the hive, content of floral nectar, harvesting time, processing techniques, storage conditions, and apiary management [35–37]. The low moisture content would be an advantage for long storage with the prolonged shelf life of honey [25,38]. Other studies also found comparable findings of low moisture content in Saudi honey [24,39,40].

The level of EC is an important indicator of the quality of honey [41]. Our result showed that the EC value of local nectar honey exceeded the permitted limit (0.8 mS/cm) of international and Gulf standards [9,10]. The level of EC depends on the presence of mineral contents, storage time, floral origin, proteins, and organic acids in honey [14,40,42]. The higher level of these contents resulted in the higher EC, and vice versa [43]. EC is the most appropriate parameter for differentiating the geographical source and identification of flora of honey [41]. The level of EC showed great variation depending on the floral origin of honey [38]. The storage, floral sources, and color of honey also affect the EC values, as dark honey provides a higher EC than light-colored honey due to differences in the levels of minerals [43]. The local Saudi honey was exceptional nectar honey, which is characterized by naturally higher EC, and likewise, previous studies also presented higher EC in Saudi honey [25,40,43].

The pH value is linked with the number of organic acids present in the honey. It can also be influenced by various other factors such as the presence of inorganic ions, as well as extraction and storage conditions, which affect the structure, stability, and shelf life of honey, as well as the fermentation process [14,37]. In the present study, the mean pH values of local and imported honey (5.0 ± 0.2 and 4.5 ± 0.2, respectively) were within the permitted limit (3.4–6.1) of standards [9,10,23]. Generally, our results regarding pH values are in agreement with those described in the literature from different countries [38,39]. We also found that few samples of local Sidr honey (*Ziziphus sp.*) exhibited higher pH (>6.1) than standard limits (3.40 to 6.10), which is in line with previous studies where Sidr honey revealed high pH [40].

FA is a characteristic that originally depends on the floral source, geographical origin, and climatic conditions. We demonstrated that the mean FA (18 ± 2.1 meq/kg) of imported honey samples was within the permitted range (≤50 meq/kg) of standards) and was in agreement with previous studies [44]. Conversely, the mean FA of local honey samples (61 ± 8.3 meq/kg) was higher than that of imported honey and exceeded the permitted limits (≤50 meq/kg). The high mean FA value in local honey was due to the honey samples that originated from *Acacia* spp. plants, which had high FA because of the nature of floral source [45]. The high FA value exceeding the permitted standards in honey originated from *Acacia* plants were in agreement with previous studies conducted in different Gulf countries, such as Oman, Saudi Arabia, and Yemen [33,40,46]. Irrespective of geographical origin, *Acacia* honey has distinctive acidic properties. The nature of the *Acacia* nectar and the effect of the honey harvest season of hot summer, as well as high-salinity soils, could be the possible reason for the acidity of *Acacia* honey [47]. In honey, essential acid gluconic acid is produced by oxidation of glucose with an enzyme glucose oxidase, which makes honey slightly acidic [20]. Thus, the increase in FA may be due the presence of a high level of gluconic acid in *Acacia* flowers as a rich source of nectar [48]. The variations in FA among local and imported honey samples might be due to differences in geographical conditions; the presence of organic acids, particularly gluconic acid; inorganic ions (phosphate and chloride); floral sources; the fermentation process; and the bee species [33]. A high FA value in *Acacia* honey samples could therefore be a feature of honey related to the floral origin of honey.

HMF is one of the most important criteria to monitor the freshness of honey, beekeeping practices, honey exposure to high temperature, and storage conditions [49]. In fresh honey, the level of HMF is naturally in small quantities, but its concentration increases with storage duration and prolonged heating [50]. HMF is an indicator for poor storage conditions at high temperature [18]. Our result revealed that mean HMF content of the local honey samples (3.8 ± 1.5 mg/kg) was lower than that of the imported honey samples (23.0 ± 5.0 mg/kg), but both were within the permitted limits (≤40 or 80 mg/kg) of honey standards [9,10,23]. Only in a few imported honey samples did HMF exceed these standards. These high HMF values in the present results might have been due to storage time and honey exposure to heating [20]. The accepted level of HMF in honey differs among countries, i.e., being greater in hot tropical countries, and should not exceed 80 mg/kg, whereas in other countries, 40 mg/kg is the maximum accepted level [9,10,23]. Our results are in agreement with the findings of previous studies [24,51]. The production of HMF can be increased with the presence of simple sugars (glucose and fructose), many acids and minerals in honey, in addition to honey processing practices or long storage [52].

The diastase enzyme is a significant enzyme secreted by bees during the conversion of nectar into honey, in addition to its floral source. It is greatly affected by the floral origin, climate, poor storage, and exposure of honey to heating; the activity of the diastase enzyme indicates the freshness of honey [53]. The storage duration and honey exposed to heating can modify the diastase activity of honey [54,55]. Our results exhibited that the diastase activity (11.9 ± 1.4 DN) of local honey met the requirements of international and local standards (≥8), with the exception of only two samples with lower diastase activity than the standard limits. Comparable values for diastase activity have been reported for Ethiopian, Argentinian, and Omani honey [56]. The mean diastase activity of imported honey was 7.6 ± 1.6 DN, and the majority of samples were out of the standard limits of international and local standards [9,10,23]. These results indicated that imported samples were either older, stored in poor conditions, or exposed to heating [55] that degraded the enzyme and resulted in decreased diastase activity. The diastase activity values are in agreement with the findings of Mesallam and El-Shaarawy [51].

The level of sugar content in honey is an important for its quality assessment [24,25]. Our results revealed that fructose, glucose, and sucrose were the most important sugars found in the analyzed honey samples, and the levels of these sugars were significantly different among local and imported honey samples. These sugars were sourced from the floral nectar that bees forage and consume during honey production, and the floral source can be identified from sugar analysis [57]. The mean levels of reducing sugar in local (71.2 ± 1.3%) and imported honey (72 ± 2.8%) were within the permitted range (≥60) of honey standards [9,10,23], and these outcomes were in confirmation with previous findings [24,51]. In the present study, fructose was the main sugar in the honey samples compared to glucose and sucrose. The mean fructose level was 39.7 ± 0.6% in local honey and 39.2 ± 2.3% in imported honey, and they were within the permitted range (31–42%) of honey standards [9,10,23]. Only two samples of imported honey had less fructose but higher glucose, indicating that these samples were possibly adulterated [24,58]. The mean percentage of glucose in our data was 31.5 ± 0.7% in local honey and 31.8 ± 0.8% in imported honey, being within the permitted range (23–32%) of honey standards [9,10,23]. These values are in agreement with findings of previous studies [56].

In our data, the local honey had a higher sucrose percentage (2.8 ± 0.9%) than that of imported honey (0.7 ± 0.2%) and was within the permitted range (≤5%) of honey standards [9,10,23]. Tigistu et al. [33] also found high sucrose (2.54 ± 0.40%) in Ethiopian honey. The normal levels of sucrose in most samples indicate that these honey samples were highly matured [50]. Some Sidr honey samples in our data had a higher level of sucrose, which could be attributed to the fact that some beekeepers harvest their honey before the complete sealing of honeycombs. This early harvest is related to the two short peaks of honey flow during Sidr flowering season [45,59,60].

The adulteration of commercial honey is a continued concern worldwide. Generally, adulteration of honey involves the addition of different sugary syrups such as C3 and C4 sugars and certain oligosaccharides. The common source of C4 sugar is sugarcane and corn, with C3 sugar coming from rice and beetroot, while starch-based polysaccharides come from rice and corn [61,62]. Resin technology, a new kind of adulteration, is also being used to produce adulterated honey, hide its origin, and eliminate any trace of contamination and antibiotics. The FDA has notified that the honey going through resin technology should not be labelled as honey. Resin technology can eliminate/alter the chemical components of honey color, flavor, and aroma; pollen; antibiotics; and residues. It also helps the commercial companies to customize the color, aroma, and flavor of honey [63]. The modified sugar syrups are difficult to catch because they are designed not to be detected by the regular testing sugar methods. An advanced global standard specialized testing using nuclear magnetic resonance (NMR) is under debate, which might be needed in order to verity the quality of honey, as well as to analyze the presence of modified sugars in the honey. NMR is a powerful analytical tool that can detect the presence and structure of different substances in honey [64].

We proposed a recommendation for the implication of NMR testing at the country level for the export and import of honey in order to authenticate the honey. A thorough surveillance with solid custom regulations could help to alleviate the import of honey adulterated with common sugars. A regular inspection of honey processing units for sampling honey and testing with NMR is also recommended because there is a high probability that adulterated honey with C3, C4, polysaccharides, and fructose syrups could bypass the normal purity tests.

5. Conclusions

The quality of local and imported honey was determined by evaluating their physicochemical properties such moisture, color, electrical conductivity, pH, diastase activity, free acidity, sugar content, and HMF. The majority of these tested parameters of local and imported honey complied with the different quality standards. The local honey showed lower moisture content and HMF, as well as higher diastase enzyme activity and EC, than the imported honey. The free acidity level was higher than the quality standards in the local exceptional nectar honey. The pollen analysis identified different types of pollen present in the honey samples, as well as their plant sources of floral nectar. The characterization and estimation of the physicochemical parameters of local and imported honey is crucial in order to monitor the quality of honey, prepare certification marks for validity, produce high-quality honey in Saudi Arabia, and propose new standards that are based on the characteristics of Saudi honey.

Author Contributions: Conceptualization, H.S.A.R., A.S.A. and J.I.; methodology, H.S.A.R., A.Y.B.-H.-A. and A.S.A.; validation, A.S.A., A.Y.B.-H.-A. and J.I.; formal analysis, H.S.A.R., J.I., and A.S.A.; investigation, H.S.A.R., A.S.A. and J.I.; resources, A.S.A.; writing—original draft preparation, H.S.A.R. and J.I.; writing—review and editing, J.I. and A.S.A.; visualization, A.S.A.; supervision, A.Y.B.-H.-A. and A.S.A.; project administration, A.S.A. and J.I.; funding acquisition, A.S.A. All authors have read and agreed to the published version of the manuscript.

Funding: This research was funded by the Deputyship for Research and Innovation, Ministry of Education, Saudi Arabia, through the project number IFKSURG-2-383.

Data Availability Statement: Data is contained within the article.

Acknowledgments: The authors extend their appreciation to the Deputyship for Research and Innovation, Ministry of Education in Saudi Arabia, for funding this research work through the project number IFKSURG-2-383.

Conflicts of Interest: The authors declare no conflict of interest.

References

1. Vorlova, L.; Pridal, A. Invertase and diastase activity in honeys of Czech provenience. *Acta Univ. Agric. Silvic. Mendel. Brun.* **2002**, *5*, 57–66.
2. Dimins, F.; Kuka, P.; Kuka, M.; Cakste, I. The criteria of honey quality and its changes during storage and thermal treatment. *Proc. Latv. Univ. Agric.* **2006**, *16*, 73–78.
3. Alqarni, A. *Beekeeping in Saudi Arabia: Current and Future (in Arabic)*, 1st ed.; Saudi Society for Agricultural Sciences, King Saud University: Riyadh, Saudi Arabia, 2011; Volume 21, p. 40.
4. Wang, Y.; Rodolfo Juliani, H.; Simon, J.E.; Ho, C.-T. Amino acid-dependent formation pathways of 2-acetylfuran and 2,5-dimethyl-4-hydroxy-3[2H]-furanone in the Maillard reaction. *Food Chem.* **2009**, *115*, 233–237. [CrossRef]
5. Zábrodská, B.; Vorlová, L. Adulteration of honey and available methods for detection—A review. *Acta Vet. Brno* **2014**, *83*, 85–102. [CrossRef]
6. Bogdanov, S.; Martin, P. Honey Authenticity: A Review. *Mitt. Aus Dem Geb. Der Lebensm. Hyg.* **2002**, *93*, 232–254.
7. Bogdanov, S.; Lüllmann, C.; Martin, P.; von der Ohe, W.; Russmann, H.; Vorwohl, G.; Oddo, L.P.; Sabatini, A.-G.; Marcazzan, G.L.; Piro, R.; et al. Honey quality and international regulatory standards: Review by the International Honey Commission. *Bee World* **1999**, *80*, 61–69. [CrossRef]
8. Ur-Rehman, S.; Farooq Khan, Z.; Maqbool, T. Physical and spectroscopic characterization of Pakistani honey. *Cienc. E Investig. Agrar.* **2008**, *35*, 199–204. [CrossRef]
9. Codex, A.C. Codex Alimentarius Commission: Revised Codex Standard for Honey Codex Stan 12-1981, Rev. 1 (1987), Rev. 2 (2001). *Codex Stand.* **2001**, *12*, 1–7.
10. *GSO 5/FDS*; Final Draft for Honey. Gulf Cooperation Council, Standardization Organization (GSO): Riyadh, Saudi Arabia, 2014; pp. 1–12.
11. MEWA. *Statistical Yearbook of 2020, Information and Statistics*; Ministry of Environment Water and Agriculture: Riyadh, Saudi Arabia, 2020; p. 77.
12. AOAC. *Official Methods of Analysis*, 17th ed.; The Association of Official Analytical Chemists International: Gaithersburg, MD, USA, 2000; Volume 2, pp. 22–33.
13. AOAC. *Official Methods of Analysis*, 15th ed.; AOAC International: Association of Official Analytical Chemist: Washington, DC, USA, 1990; p. 870.
14. Terrab, A.; González-Miret, L.; Heredia, F.J. Colour characterisation of thyme and avocado honeys by diffuse reflectance spectrophotometry and spectroradiometry. *Eur. Food Res. Technol.* **2004**, *218*, 488–492. [CrossRef]
15. AOAC. *Officials Methods of Analysis*, 17th ed.; Association of Official Analytical Chemists: Washington, DC, USA, 2010.
16. Louveaux, J.; Maurizio, A.; Vorwohl, G. Methods of Melissopalynology. *Bee World* **1978**, *59*, 139–157. [CrossRef]
17. Ohe, W.V.D.; Oddo, L.P.; Piana, M.L.; Morlot, M.; Martin, P. Harmonized methods of melissopalynology. *Apidologie* **2004**, *35*, S18–S25.
18. Raweh, H.S.A.; Badjah-Hadj-Ahmed, A.Y.; Iqbal, J.; Alqarni, A.S. Impact of different storage regimes on the levels of physico-chemical characteristics, especially free acidity in Talh (*Acacia gerrardii* Benth.) honey. *Molecules* **2022**, *27*, 5959. [CrossRef]
19. Gosling, W.D.; Miller, C.S.; Livingstone, D.A. Atlas of the tropical West African pollen flora. *Rev. Palaeobot. Palynol.* **2013**, *199*, 1–135. [CrossRef]
20. Bogdanov, S. Harmonised methods of the international honey commission. *Bee Prod. Sci.* **2009**, *5*, 1–63.
21. USDA. *United States Standards for Grades of Extracted Honey*; US Department of Agriculture: Washington, DC, USA, 1985; pp. 1–14.
22. AOAC. Diastatic activity of honey. In *Official Methods of Analysis*; Association of Official Analytical Chemists: Washington, DC, USA, 2010.
23. EU Council. European Union Council, Council Directive 2001/110/EC of 20 December 2001 relating to honey. *Off. J. Eur. Communities* **2002**, *10*, 47–52.
24. Aljohar, H.I.; Maher, H.M.; Albaqami, J.; Al-Mehaizie, M.; Orfali, R.; Orfali, R.; Alrubia, S. Physical and chemical screening of honey samples available in the Saudi market: An important aspect in the authentication process and quality assessment. *Saudi Pharm. J.* **2018**, *26*, 932–942. [CrossRef] [PubMed]
25. El Sohaimy, S.A.; Masry, S.H.D.; Shehata, M.G. Physicochemical characteristics of honey from different origins. *Ann. Agri. Sci.* **2015**, *60*, 279–287. [CrossRef]
26. Aronne, G.; Micco, V.D.; Guarracino, M.R. Application of support vector machines to melissopalynological data for honey classification. *Int. J. Agric. Environ. Inf. Syst.* **2010**, *1*, 85–94. [CrossRef]
27. Bryant, V. Filtering honey almost every filter removes some pollen. *Bee Cult. Mag. Am. Beekeep.* **2017**, *145*, 75–76.
28. Ponnuchamy, R.; Bonhomme, V.; Prasad, S.; Das, L.; Patel, P.; Gaucherel, C.; Pragasam, A.; Anupama, K. Honey pollen: Using Melissopalynology to understand foraging preferences of bees in tropical Ssouth India. *PLoS ONE* **2014**, *9*, e101618. [CrossRef]

29. González-Miret, M.L.; Terrab, A.; Hernanz, D.; Fernández-Recamales, M.A.; Heredia, F.J. Multivariate correlation between color and mineral composition of honeys and by their botanical origin. *J. Agric. Food Chem.* **2005**, *53*, 2574–2580. [CrossRef]
30. Naab, O.A.; Tamame, M.A.; Caccavari, M.A. Palynological and physicochemical characteristics of three unifloral honey types from central Argentina. *Span. J. Agric. Res.* **2008**, *6*, 566–576. [CrossRef]
31. Kayode, J.; Oyeyemi, S.D. Physico-chemical investigation of honey samples from bee farmers in Ekiti State, Southwest Nigeria. *J. Plant Sci.* **2014**, *2*, 246. [CrossRef]
32. Moniruzzaman, M.; Sulaiman, S.A.; Azlan, S.A.; Gan, S.H. Two-year variations of phenolics, flavonoids and antioxidant contents in acacia honey. *Molecules* **2013**, *18*, 14694–14710. [CrossRef]
33. Tigistu, T.; Worku, Z.; Mohammed, A. Evaluation of the physicochemical properties of honey produced in Doyogena and Kachabira Districts of Kembata Tambaro zone, Southern Ethiopia. *Heliyon* **2021**, *7*, e06803. [CrossRef] [PubMed]
34. Beretta, G.; Granata, P.; Ferrero, M.; Orioli, M.; Maffei Facino, R. Standardization of antioxidant properties of honey by a combination of spectrophotometric/fluorimetric assays and chemometrics. *Anal. Chim. Acta* **2005**, *533*, 185–191. [CrossRef]
35. Saeed, M.A.; Jayashankar, M. Physico-chemical characteristics of some Indian and Yemeni Honey. *J. Bioenergy Food Sci.* **2020**, *7*, 2832019. [CrossRef]
36. Corbella, E.; Cozzolino, D. Classification of the floral origin of Uruguayan honeys by chemical and physical characteristics combined with chemometrics. *LWT Food Sci. Technol.* **2006**, *39*, 534–539. [CrossRef]
37. Fuad, A.M.A.; Anwar, N.Z.R.; Zakaria, A.J.; Shahidan, N.; Zakaria, Z. Physicochemical characteristics of Malaysian honeys influenced by storage time and temperature. *J. Fundam. Appl. Sci.* **2017**, *9*, 841–851. [CrossRef]
38. Terrab, A.; Díez, M.J.; Heredia, F.J. Palynological, physico-chemical and colour characterization of Moroccan honeys: I. River red gum (*Eucalyptus camaldulensis* Dehnh) honey. *Int. J. Food Sci. Technol.* **2003**, *38*, 379–386. [CrossRef]
39. Al-Doghairi, M.A.; Al-Rehiayani, S.; Ibrahim, G.H.; Osman, K. Physicochemical and Antimicrobial Properties of Natural Honeys Produced in Al-Qassim Region, Saudi Arabia. *J. King Abdulaziz Univ. Meteorol. Environ. Arid. Land Agric. Sci.* **2007**, *18*, 3–18. [CrossRef]
40. Alqarni, A.S.; Owayss, A.A.; Mahmoud, A.A. Physicochemical characteristics, total phenols and pigments of national and international honeys in Saudi Arabia. *Arab. J. Chem.* **2016**, *9*, 114–120. [CrossRef]
41. Sousa, J.M.B.d.; Souza, E.L.d.; Marques, G.; Benassi, M.d.T.; Gullón, B.; Pintado, M.M.; Magnani, M. Sugar profile, physicochemical and sensory aspects of monofloral honeys produced by different stingless bee species in Brazilian semi-arid region. *LWT Food Sci. Technol.* **2016**, *65*, 645–651. [CrossRef]
42. Habib, H.M.; Al Meqbali, F.T.; Kamal, H.; Souka, U.D.; Ibrahim, W.H. Physicochemical and biochemical properties of honeys from arid regions. *Food Chem.* **2014**, *153*, 35–43. [CrossRef]
43. Alqarni, A.S.; Owayss, A.A.; Mahmoud, A.A.; Hannan, M.A. Mineral content and physical properties of local and imported honeys in Saudi Arabia. *J. Saudi Chem. Soc.* **2014**, *18*, 618–625. [CrossRef]
44. Makhloufi, C.; Schweitzer, P.; Azouzi, B.; Oddo, L.P.; Choukri, A.; Hocine, L.; D'Albore, G.R. Some Properties of Algerian Honey. *Apiacta* **2007**, *42*, 73–80.
45. Raweh, H.S.A.; Ahmed, A.Y.B.H.; Iqbal, J.; Alqarni, A.S. Monitoring and evaluation of free acidity levels in Talh honey originated from Talh tree *Acacia gerrardii* Benth. *J. King Saud Univ. Sci.* **2022**, *34*, 101678. [CrossRef]
46. Sajwani, A.; Farooq, S.; Eltayeb, E. Differentiation of Omani Acacia and white Acacia honey by botanical and physicochemical analysis. *Ukr. J. Food Sci.* **2019**, *7*, 264–285. [CrossRef]
47. Afifi, H.S.; Abu-Alrub, I.; Ma\\sry, S. Discrimination of Samar and Talh honey produced in the Gulf Cooperation Council (GCC) region using multivariate data analysis. *Ann. Agric. Sci.* **2021**, *66*, 31–37. [CrossRef]
48. Awad, A.M.; Owayss, A.A.; Alqarni, A.S. Performance of two honey bee subspecies during harsh weather and *Acacia gerrardii* nectar-rich flow. *Sci. Agric.* **2017**, *74*, 474–480. [CrossRef]
49. Mouhoubi-Tafinine, Z.; Ouchemoukh, S.; Bey, M.; Louaileche, H.; Tamendjari, A. Effect of storage on hydroxymethylfurfural (HMF) and color of some Algerian honey. *Int. Food Res. J.* **2018**, *25*, 1044–1050.
50. Tesfaye, B. Evaluation of Physico-chemical properties of honey produced in Bale Natural Forest, Southeastern Ethiopia. *Int. J. Agric. Sci. Food Technol.* **2016**, *2*, 21–27. [CrossRef]
51. Mesallam, A.S.; El-Shaarawy, M.I. Quality attributes of honey in Saudi Arabia. *Food Chem.* **1987**, *25*, 1–11. [CrossRef]
52. Bastos, D.H.M.; Monaro, É.; Siguemoto, É.S.; Séfora, M. Maillard Reaction Products in Processed Food: Pros and Cons. In *Food Industrial Processes—Methods and Equipment*; IntechOpen: London, UK, 2012. [CrossRef]
53. Juan-Borrás, M.; Domenech, E.; Hellebrandova, M.; Escriche, I. Effect of country origin on physicochemical, sugar and volatile composition of acacia, sunflower and tilia honeys. *Food Res. Int.* **2014**, *60*, 86–94. [CrossRef]
54. Bogdanov, S.; Martin, P.; Lullmann, C.; Borneck, R.; Flamini, C.; Morlot, M.; Lheritier, J.; Vorwohl, G.; Russmann, H.; Persano, L.; et al. Harmonised methods of the European Honey Commission. *Apidologie* **1997**, *28*, 1–59.
55. Iftikhar, F.; Masood, M.A.; Waghchoure, E.S. Comparison of *Apis cerana*, *Apis dorsata*, *Apis florea* and *Apis mellifera*: Honey from different areas of Pakistan. *Asian J. Exp. Biol. Sci.* **2011**, *2*, 399–403.
56. Al-Farsi, M.; Al-Belushi, S.; Al-Amri, A.; Al-Hadhrami, A.; Al-Rusheidi, M.; Al-Alawi, A. Quality evaluation of Omani honey. *Food Chem.* **2018**, *262*, 162–167. [CrossRef]
57. Val, A.; Huidobro, J.F.; Sánchez, M.P.; Muniategui, S.; Fernández-Muiño, M.A.; Sancho, M.T. Enzymatic determination of galactose and lactose in honey. *J. Agric. Food Chem.* **1998**, *46*, 1381–1385. [CrossRef]

58. Aparna, A.R.; Rajalakshmi, D. Honey—Its characteristics, sensory aspects, and applications. *Food Rev. Int.* **1999**, *15*, 455–471. [CrossRef]
59. Alqarni, A.S.; Iqbal, J.; Raweh, H.S.; Hassan, A.M.A.; Owayss, A.A. Beekeeping in the Desert: Foraging activities of honey bee during major honeyflow in a hot-arid ecosystem. *Appl. Sci.* **2021**, *11*, 9756. [CrossRef]
60. Raweh, H.S. Honey Bee, *Apis mellifera* L., Activities on Sidr, *Ziziphus nummularia* (Burm.F.) Wight & Arn., and Honey Production under the Environmental Conditions of Riyadh Province. Master's Thesis, King Saud University, Riyadh, Saudi Arabia, 2015.
61. Fakhlaei, R.; Selamat, J.; Khatib, A.; Razis, A.F.A.; Sukor, R.; Ahmad, S.; Babadi, A.A. The toxic impact of honey adulteration: A review. *Foods* **2020**, *9*, 1538. [CrossRef]
62. Guler, A.; Kocaokutgen, H.; Garipoglu, A.V.; Onder, H.; Ekinci, D.; Biyik, S. Detection of adulterated honey produced by honeybee (*Apis mellifera* L.) colonies fed with different levels of commercial industrial sugar (C_3 and C_4 plants) syrups by the carbon isotope ratio analysis. *Food. Chem.* **2014**, *155*, 155–160. [CrossRef] [PubMed]
63. Wang, Q.; Zhao, H.; Zhu, M.; Zhang, J.; Cheng, N.; Cao, W. Method for identifying acacia honey adulterated by resin absorption: HPLC-ECD coupled with chemometrics. *Lebensm. Wiss. Technol.* **2020**, *118*, 108863. [CrossRef]
64. Rachineni, K.; Rao Kakita, V.M.; Awasthi, N.P.; Shirke, V.S.; Hosur, R.V.; Chandra Shukla, S. Identifying type of sugar adulterants in honey: Combined application of NMR spectroscopy and supervised machine learning classification. *Curr. Res. Food Sci.* **2022**, *5*, 272–277. [CrossRef] [PubMed]

Article

Rapid Identification of Corn Sugar Syrup Adulteration in Wolfberry Honey Based on Fluorescence Spectroscopy Coupled with Chemometrics

Shengyu Hao [1], Jie Yuan [2], Qian Wu [2], Xinying Liu [3], Jichun Cui [4] and Hongzhuan Xuan [2,*]

[1] School of Physical Science and Information Technology, Liaocheng University, Liaocheng 252059, China; haoshengyu@lcu.edu.cn
[2] School of Life Sciences, Liaocheng University, Liaocheng 252059, China; 2010150207@stu.lcu.edu.cn (J.Y.); 2110150205@stu.lcu.edu.cn (Q.W.)
[3] Animal Product Quality and Safety Center of Shandong Province, Jinan 250010, China; sdfengjian@sina.com
[4] School of Chemistry and Chemical Engineering, Liaocheng University, Liaocheng 252059, China; cuijichun@lcu.edu.cn
* Correspondence: hongzhuanxuan@163.com

Abstract: Honey adulteration has become a prominent issue in the honey market. Herein, we used the fluorescence spectroscopy combined with chemometrics to explore a simple, fast, and non-destructive method to detect wolfberry honey adulteration. The main parameters such as the maximum fluorescence intensity, peak positions, and fluorescence lifetime were analyzed and depicted with a principal component analysis (PCA). We demonstrated that the peak position of the wolfberry honey was relatively fixed at 342 nm compared with those of the multifloral honey. The fluorescence intensity decreased and the peak position redshifted with an increase in the syrup concentration (10–100%). The three-dimensional (3D) spectra and fluorescence lifetime fitting plots could obviously distinguish the honey from syrups. It was difficult to distinguish the wolfberry honey from another monofloral honey, acacia honey, using fluorescence spectra, but it could easily be distinguished when the fluorescence data were combined with a PCA. In all, fluorescence spectroscopy coupled with a PCA could easily distinguish wolfberry honey adulteration with syrups or other monofloral honeys. The method was simple, fast, and non-destructive, with a significant potential for the detection of honey adulteration.

Keywords: wolfberry honey; adulteration; fluorescence spectroscopy; fluorescence lifetime; principal component analysis

Citation: Hao, S.; Yuan, J.; Wu, Q.; Liu, X.; Cui, J.; Xuan, H. Rapid Identification of Corn Sugar Syrup Adulteration in Wolfberry Honey Based on Fluorescence Spectroscopy Coupled with Chemometrics. *Foods* **2023**, *12*, 2309. https://doi.org/10.3390/foods12122309

Academic Editor: Mircea Oroian

Received: 12 May 2023
Revised: 6 June 2023
Accepted: 6 June 2023
Published: 8 June 2023

1. Introduction

Honey has been a functional food for centuries due to its nutrients and therapeutic effects [1]. Honey is rich in various chemical components, including monosaccharides (fructose and glucose) and other types of sugars; these are the dominant constituents, accounting for 70–80% of honey [2]. It also contains small amounts of amino acids (including tryptophan, tyrosine, and phenylalanine), proteins (enzymes), vitamins (especially vitamin B6, thiamine, niacin, riboflavin, and pantothenic acid), minerals (including calcium, copper, iron, magnesium, manganese, phosphorus, potassium, sodium, and zinc), phenolic acids (caffeic acid, ferulic acid, chlorogenic acid, and vanillin acid), flavonoids (galangin, chrysin, apigenin, pinobanksin, naringenin, and quercetin) and royal jelly aliphatic acids [3–5]. It is because honey contains these minor materials that it differs from syrups and other sweeteners; thus, these substances are also the basis for the authentication identification of honey adulteration.

Honey adulteration with various syrups such as corn syrup, sugarcane syrup, beet syrup, rice syrup, wheat syrup, inverted syrup, and inulin syrup has become a prominent issue that harms the interests of consumers and beekeepers [2,3,6]. With the increase in the types of adulterated honey, other technologies are needed to detect adulterated honey [2]. The traditional methods mainly include pollen identification, which identifies the sources of the nectar and plants through the characteristics and quantity of the honey pollen, and characteristic parameters such as the water content, Brix, electrical conductivity, amylase value, and 5-hydroxymethylfurfural content [7–9]. However, these methods are often time-consuming and labor-intensive, and have advanced requirements for technicians. Modern analytical techniques, including ultra-performance liquid chromatography-quadrupole time-of-flight mass spectrometry (UPLC-QToF MS) [10], gas chromatography (GC-MS) [11], a stable carbon isotopic ratio analysis (SCIRA) [12], high-performance anion exchange chromatography (HPAEC) [13], nuclear magnetic resonance (NMR) [14], and e-noses and e-tongues have also been employed for the detection of syrup in adulterated honey [15,16], but these methods or techniques are either complicated to operate, or the detection process and sample pre-processing are time-consuming or expensive. For example, SCIRA can be used for the C4 plant sugar corn syrup adulteration in honey, but the instrument is expensive and complicated to operate.

Fluorescence spectroscopy is a type of spectral detection technology that has rapidly developed in recent years because of its simple operation as well as it being fast and non-destructive [17,18]. The theory basis for fluorescence spectroscopy is that when a fluorescent molecule absorbs photons, it changes from an original ground state to an excited state. The excited-state molecule consumes part of its energy by colliding with the surrounding molecules and rapidly drops to the lowest vibration level of the first electron excited state, remaining there for about 10^{-9}–10^{-7} s. After that, the excess energy is directly released in the form of photon and drops to various vibration levels of the electronic ground state. At this time, the emitted light is fluorescent and can be detected using fluorescence spectrometer [19].

Honey is rich in vitamins, phenols, polypeptides, and amino acids (tryptophan, tyrosine, and phenylalanine) as well as other fluorophores, thus front-face fluorescence spectroscopy and three-dimensional (3D) synchronous fluorescence spectroscopy have been used in honey authentication of botanical origin and geographical origin [20–25]. However, few studies have been reported on the authenticity of honey by combining multiple fluorescence spectroscopy techniques, especially the authenticity of honey by measuring fluorescence lifetime. When a substance is excited by a laser beam, the molecules absorb energy and leap from the ground state to a certain excited state, and then fluoresce back to the ground state in the form of a radiative leap. When the excitation light is removed, the time required for the fluorescence intensity of the molecule to drop to 1/e of the maximum intensity of the fluorescence at the time of excitation is called the fluorescence lifetime.

Differences in the fluorophores between monofloral honeys of different botanical origin and geographical origin or between monofloral and multifloral honeys indicate different fluorescent characteristics. Similarly, the fluorescent characteristics of adulterated honey must change compared with authentic honey. Thus, it is possible to detect monofloral honey, multifloral honey, or adulterated honey by fluorescence spectroscopy.

Wolfberry honey is a typical monofloral honey that is mainly produced in Northwest China, including Ningxia, Qinghai, and Gansu provinces [14]. The nectar plant of wolfberry honey is wolfberry, a traditional health food with a variety of pharmacological activities. Customers have shown an increasing interest in wolfberry honey due to its potential health benefits [26]. Due to its high price, natural wolfberry honey, just like other monofloral honey, is easy to be adulterated in the honey market, and the common adulteration techniques involve sugar syrup adulteration, or adding multifloral honey to monofloral honey, or nectar adulteration. Thus, the rapid identification of wolfberry honey adulteration is an urgent need. Here, the authenticity of wolfberry honey was determined by combining multiple fluorescence spectroscopy techniques mainly from the maximum fluorescence intensity, peak positions, and fluorescence lifetime, and these data were further analyzed by a principal component analysis (PCA), which demonstrated that fluorescence spectroscopy was a simple, fast, and non-destructive method for the detection of honey adulteration.

2. Materials and Methods

2.1. Materials

A total of 23 wolfberry honey samples were collected from different apiaries in 2021 of Golmud in Qinghai Province, Northwest of China. Multifloral honey samples were obtained from different beekeepers in the counties of Meiyuan, Gangcha, and Datong of Qinghai Province.

An acacia honey, only as a monofloral honey control sample, was collected from an apiary in the Shandong Province of North China in 2021. Corn syrup was obtained from the Daesang Corporation of Korea, and corn maltose syrup (M50) was from the Hubei Hefeng Grain and Oil Group Co., Ltd., Wuhan, China. All samples were stored at 4 °C before use.

2.2. The Characteristic Parameters Analysis of Wolfberry Honey Samples

The water content, total sugar content, and Baume degree were determined using a honey refractometer (Shanghai Lichen Instrument Technology Co., Ltd., Shanghai, China). The pH values were measured by using a pH meter (Shanghai Yidian Instrument Technology Co., Ltd., Shanghai, China). The total protein content was tested using the standard Bradford's method. The diastase (α-amylase) activity was measured according to GB/T 18932 (China). The conductivity of dissolved honey solution was determined using an electrical conductivity instrument (Shanghai Leici Instrument Technology Co., Ltd., Shanghai, China).

2.3. Palynological Identification

Pollen grains in the wolfberry honey samples were obtained according to our previous publication [27]. In brief, sugar was first removed from the honey samples by centrifugation twice at 12,000 g for 15 min. The sediment was then resuspended with 1 mL glutaraldehyde solution (2.5%) overnight. After that, all samples were gradient dehydrated with ethanol and freeze-dried in a vacuum. The palynological identification was taken using a Hitachi S-750 SEM system (Hitachi Company, Tokyo, Japan).

2.4. Preparation of the Test Samples

All samples were diluted with distilled water to contain 30 mg per milliliter and were exhausted with an ultrasonic cleaner (40 kHz; 80 W; 30 s) prior to analysis.

Adulterated samples were prepared by adding 10%, 20%, 30%, 40%, 50%, 60%, 70%, 80%, and 90% acacia honey, corn syrup, and corn maltose syrup to the wolfberry honey.

2.5. Fluorescence Spectroscopy Detection

2.5.1. Fluorescence Spectra Measurement

The fluorescence spectra measurements of the different samples were determined with an Edinburgh F900 with a R928-P detector (Edinburgh Instruments, EI, Livingston, UK), and a 1 cm quartz cell was used in the measurement. The fluorescence intensity of each sample solution was directly measured with an excitation wavelength of 280 nm and an emission wavelength range from 300 to 540 nm with a 1 nm increment, and spectrometer slits were set for a 2.5 nm band-pass. The fluorescence intensity of all samples were recorded at room temperature.

2.5.2. Three-Dimensional Fluorescence Spectra Measurement

The 3D fluorescence spectra of the different samples were measured with an F-7000 FL spectrophotometer (Hitachi, Japan), and a 1 cm quartz cell was used in the measurement. The instrument parameters were an excitation wavelength from 200 to 450 nm and an emission wavelength from 260 to 560 nm; scan speed was 30,000 nm/min; excitation slit and emission slit were all set for 2.5 nm; PMT voltage was 600 V. The 3D fluorescence spectra of all samples were recorded at room temperature.

2.5.3. Fluorescence Lifetime Measurement

The fluorescence lifetime was tested using an Edinburgh F900 at an excitation wavelength of 280 nm and an emission wavelength of 350 nm. The light source was 900 µF. The average fluorescence lifetime was calculated according to our previous publication [19].

2.6. Statistical Analysis

The statistical analysis involved a paired Student's *t*-test, Tukey test, and an ANOVA using SPSS version 18.0. A *p*-value less than 0.05 was considered to indicate a statistically significant difference. Origin Pro8 software was also used for the fluorescence spectra data processing, mapping analysis, and principal component analysis.

3. Results and Discussion

3.1. Characteristic Parameters of Wolfberry Honey Samples

The characteristic parameters of wolfberry honey samples can be seen in Table 1, which preliminary indicated the quality of wolfberry honey used in the present study. The pH of the wolfberry honey samples was 4.17, which was within the typical blossom honey pH range of 3.5 to 4.5. The contents of water and total sugar as well as Baume degree were consistent with the standards of the European Regulations of quality. The protein content was about 27.62 mg per 100 g, which mainly derived from plant pollen or honey honeybee itself and contributed to the main fluorophores in honey. The conductivity was about 0.34 mS/cm, less than 0.8 mS/cm specified by relevant standards. Diastase activity is an important indicator to evaluate the quality of honey, and it is related to the freshness and processing and storage conditions. The diastase activity of wolfberry honey samples was 26.18 ± 1.91 mL/(g·h), obviously higher than the European quality standards, indicating that all the wolfberry honey samples were fresh and unprocessed.

Table 1. Characteristic parameters of wolfberry honey.

Variables	Units	Wolfberry Honey
pH	-	4.17 ± 0.08
Water	g per 100 g	20.48 ± 0.59
Total sugar	g per 100 g	77.74 ± 0.72
Baume degree	°Bé	41.42 ± 0.25
Protein content	mg per 100 g	27.62 ± 3.01
Conductivity	mS/cm	0.34 ± 0.02
Diastase activity	mL/(g·h)	26.18 ± 1.91

3.2. Palynology Characteristics of Wolfberry Honey Samples

In order to determine the authenticity of the botanic source of wolfberry honey, we observed the palynology characteristics of wolfberry honey samples using scanning electron microscopy. The morphology of the pollen grains from wolfberry honey samples showed typical palynological characteristics of *Lycium Linn.*, namely the morphology of pollen was subspherical, triple-grooved with grooves up to both poles, which can be found in Figure 1.

3.3. Fluorescence Spectra of Wolfberry Honey, Multifloral Honey and Syrups

The main components of honey are sugar and water, and it also contains small amounts of protein, free amino acids, phenolic acids, flavonoids, and minerals [4]. These minority components in honey showed fluorescent properties [22,23]. A peak with an excitation of 280 nm and emission of 340 nm can potentially suggest fluorescence from aromatic amino acids or protein in honey [28].

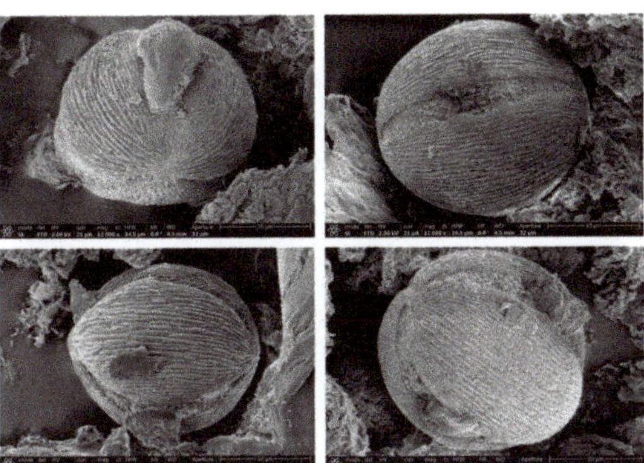

Figure 1. Scanning electron micrographs of pollen from wolfberry honey.

The fluorescence spectra of wolfberry honey, multifloral honey, and syrups at a fixed excitation wavelength of 280 nm can be found in Figure 2. The fluorescence spectra of all 23 wolfberry honey samples were consistent with similar fluorescence intensity and peak positions, indicating the typical fluorescence spectra of the wolfberry honey from the same botanical origin and geographical origin (Figure 2a), and the data were consistent with our previous study on acacia honey authenticity identification, and further confirm the accuracy of the methodology [19]. Conversely, the multifloral honey samples collected from different geographical origins including Meiyuan, Gangcha, and Datong of Qinghai province showed different fluorescence intensities and peak positions compared with those of wolfberry honey samples (Figure 2b). The main components of syrup are sugar with few fluorescent compounds such as proteins, amino acids, and phenolic acids, thus the fluorescence intensities of the two kinds of corn syrups were much lower than those of the authentic honey samples and the peak positions showed an obvious redshift (Figure 2c).

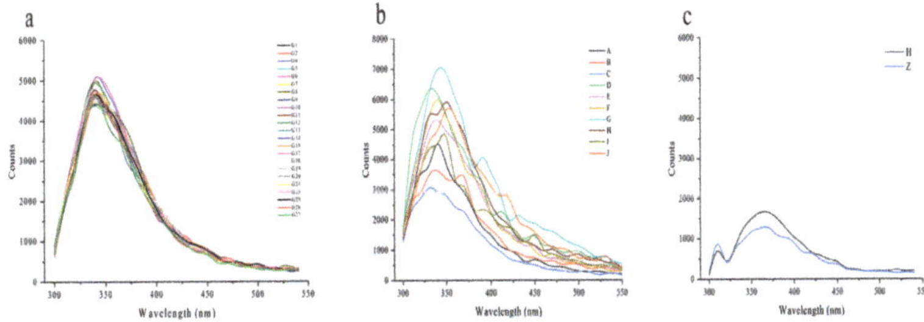

Figure 2. The fluorescence spectra of wolfberry honey, multifloral honey, and syrups. (**a**) The fluorescence spectra of all 23 wolfberry honey samples (G1–G27, 23 wolfberry honey samples). (**b**) The fluorescence spectra of the multifloral honey samples (A–J, 10 multifloral honey samples). (**c**) The fluorescence spectra of corn syrup and corn maltose syrup (H, corn syrup, Z, corn maltose syrup).

Monofloral honey, multifloral honey, and syrups showed different fluorescence spectra at a fixed excitation wavelength of 280 nm, indicating fluorescence spectroscopy is can differ the botanical origin and geographical origin of honey, and can easily distinguish honey from syrups by comparing the peak positions and fluorescence intensity.

Table 2 shows the differences in the maximum fluorescence intensity and the peak positions of the wolfberry honey samples, multifloral honey, corn syrups, and corn maltose syrup. The maximum fluorescence intensity of the wolfberry honey was 4838.49 ± 181.21, and that of multifloral honey was 4412.6 ± 305.49, both of which were significantly higher than that of the syrups. The peak position of the wolfberry honey samples was

342.65 ± 3.11 nm. However, the peak positions of the two types of syrups were completely different from those of the authentic wolfberry honey; these were 365 ± 1.41 nm and 363 ± 8.49 nm, respectively.

Table 2. The comparison of honey and syrups between the maximum fluorescence intensity and peak position.

	Wolfberry Honey	Multifloral Honey	Corn Syrup	Corn Maltose Syrup
Maximum fluorescence intensity	4838.49 ± 181.21 [a]	4412.6 ± 305.49 [b]	1813.25 ± 94.02 [b,c]	1400.9 ± 173.23 [b,c]
Peak position (nm)	342.65 ± 3.11 [b,c]	346.5 ± 2.99 [b]	365 ± 1.41 [a]	363 ± 8.49 [a]

Different letters (a, b, c) in each column indicate significant differences calculated using Tukey test.

The difference in chemical components or fluorophores among the wolfberry honey, multifloral honey, and two kinds of corn syrups lead to different fluorescence spectra at a fixed excitation wavelength of 280 nm, which can be used to distinguish the monofloral honey, multifloral honey and syrups.

3.4. Fluorescence Spectra Changes in the Adulterated Wolfberry Honey

When different proportions of corn syrup or corn maltose syrup (10–100%) were adulterated into wolfberry honey, the fluorescence intensity decreased with an increase in the syrup concentration, and the peak positions of the adulterated honey samples also redshifted (Figure 3a,b). However, if wolfberry honey was mixed with another monofloral honey, acacia honey, the characteristic properties of acacia honey can be found in Supplementary Table S1, the changes between the maximum fluorescence intensities and the peak positions were slight and not obvious compared with the genuine wolfberry honey (Figure 3c). Figure 3d showed the change trends in the peak positions and the fluorescence intensities of wolfberry honey adulterated with different proportions of syrups and acacia honey (0–100%). These results further demonstrated syrup adulteration in monofloral honey can be easily detected at a fixed excitation wavelength of 280 nm, and the maximum fluorescence intensities and the peak positions were two important parameters, but it was not easy to distinguish the mixture of two kinds of monofloral honey samples only by fluorescence spectra.

3.5. The 3D Fluorescence Spectra of the Wolfberry Honey Samples and Syrups

The 3D fluorescence spectroscopy is a matrix spectrum characterized by excitation wavelength (*Y*-axis), emission wavelength (*X*-axis), and fluorescence intensity (*Z*-axis), which is able to obtain the information of the excitation wavelength and emission wavelength as well as the information of fluorescence intensity. Therefore, 3D fluorescence spectroscopy is also called total luminescence spectra.

The 3D fluorescence spectra of the wolfberry honey, corn syrup, and corn maltose syrup at an excitation wavelength from 200 to 450 nm and an emission wavelength from 260 to 560 nm were obviously different and could easily be distinguished (Figure 4a–c). From the 3D contour map, we can obviously see that there were two peak values at the excitation wavelength of 280 nm and 230 nm and the corresponding emission wavelength at about 340 nm, which indicated the fluorophores might be aromatic amino acids and non-flavonoid phenolic compounds in the honey sample [28,29]. The 3D fluorescence spectra of two different corn syrups were totally different compared with the wolfberry honey sample without significant peak values and peak positions, which is consistent with the results from 280 nm excitation wavelength, and the 3D contour maps can visually distinguish honey from syrups.

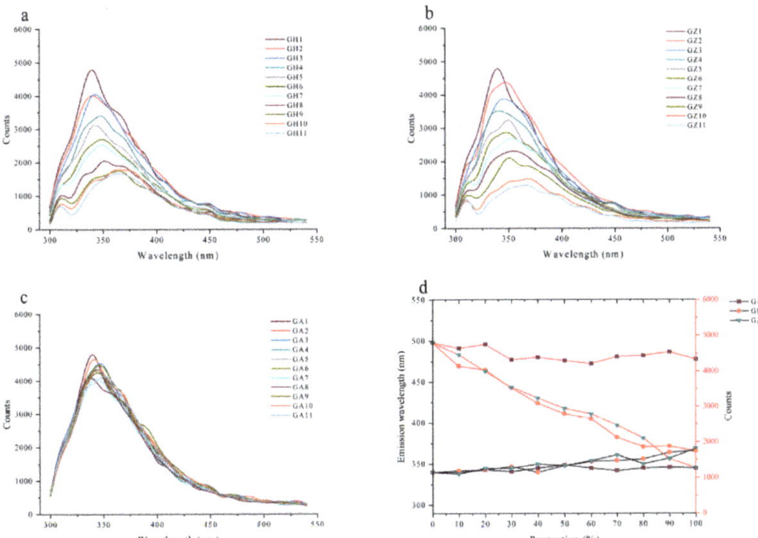

Figure 3. The fluorescence spectra of the adulterated wolfberry honey. (**a**) The fluorescence spectra of wolfberry honey adulterated with different proportions of corn syrup (0–100%). (**b**) The fluorescence spectra of wolfberry honey adulterated with different proportions of corn maltose syrup (0–100%). (**c**) The fluorescence spectra of wolfberry honey adulterated with different proportions of acacia honey (0–100%). (**d**) The change trends in emission wavelengths and fluorescence intensities of wolfberry honey adulterated with different proportions of syrups and acacia honey (0–100%). GA: wolfberry honey adulterated with different proportions of acacia honey; GH: wolfberry honey adulterated with different proportions of corn syrup; GZ: wolfberry honey adulterated with different proportions of corn maltose syrup.

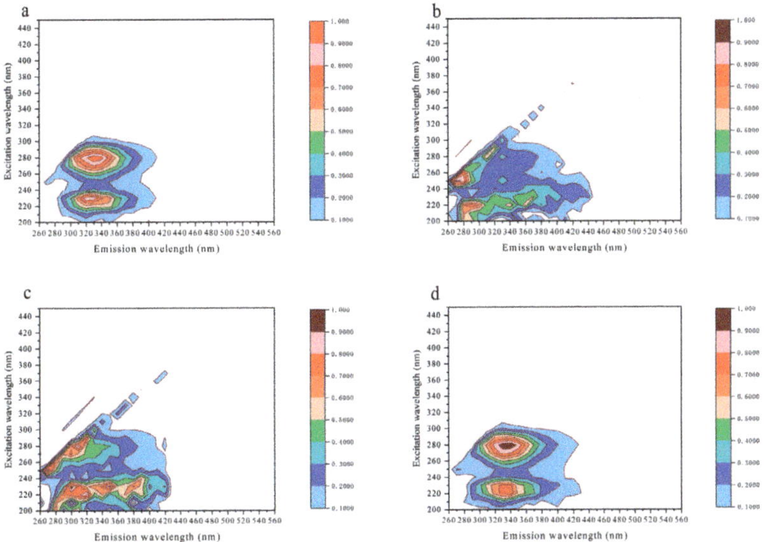

Figure 4. The three-dimensional (3D) fluorescence spectra of the wolfberry honey and syrups. (**a**) The 3D fluorescence spectra of wolfberry honey. (**b**) The 3D fluorescence spectra of corn syrup. (**c**) The 3D fluorescence spectra of corn maltose syrup. (**d**) The 3D fluorescence spectra of acacia honey.

Similarly, it was difficult to distinguish the wolfberry honey from the acacia honey by the 3D fluorescence spectra (Figure 4d), for two kinds of monofloral honey had similar 3D fluorescence spectra with similar two peak values and peak positions. The peak positions and the maximum fluorescence intensities of the wolfberry honey, corn syrup, corn maltose syrup, and acacia honey can be found in Supplementary Table S2.

3.6. The 3D Fluorescence Spectra Changes in the Adulterated Wolfberry Honey

The 3D fluorescence spectra demonstrated obvious changes in peak values and peak positions with an increase in the syrup concentration (10–100%) at an excitation wavelength from 200 to 450 nm and an emission wavelength from 260 to 560 nm, especially at excitation wavelength of 280 nm and 230 nm. However, the differences between the wolfberry honey and adulterated wolfberry honey with a different concentration of acacia honey (10–100%) were not obvious from the 3D fluorescence spectra. (Figure 5a–c), the change trends in peak position and peak value can be found in the Figure 5d,e.

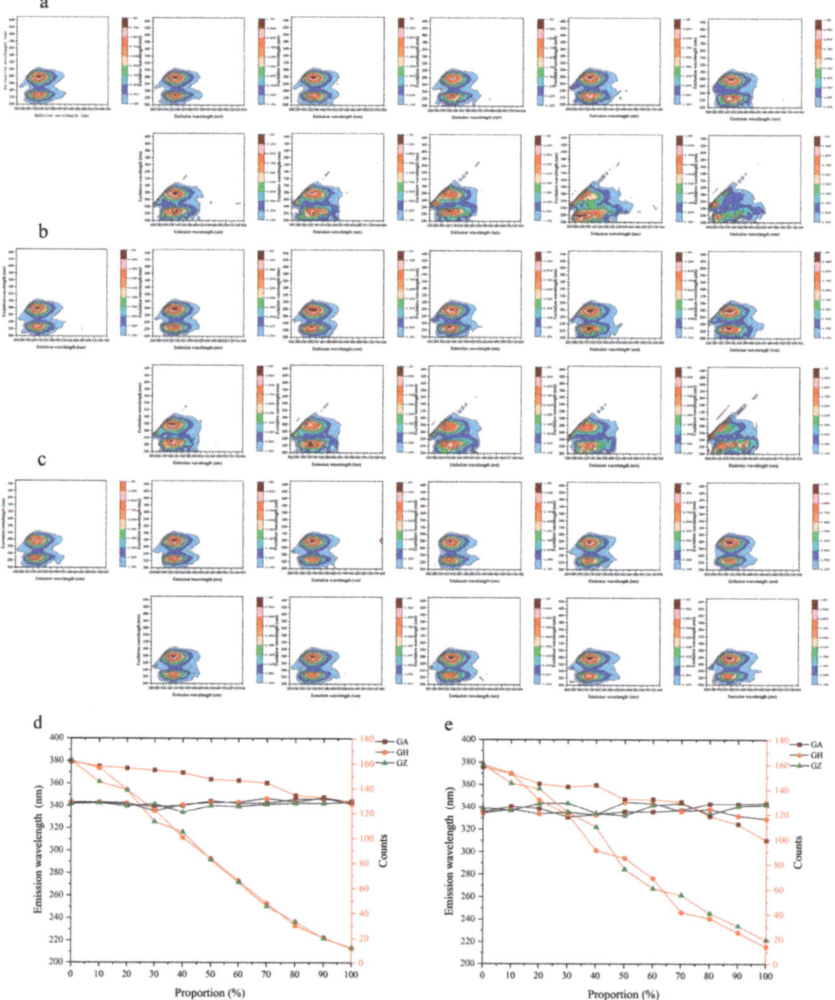

Figure 5. The three-dimensional (3D) fluorescence spectra of the wolfberry honey adulterated with different proportions of syrup or acacia honey. (**a–c**) The 3D fluorescence spectra of wolfberry honey

adulterated with different proportions of corn syrup (0–100%), corn maltose syrup (0–100%) and acacia honey (0–100%). (**d**) The change trends in emission wavelengths and fluorescence intensities of wolfberry honey adulterated with different proportions of syrups and acacia honey (0–100%) at the maximum peak value. (**e**) The change trends in emission wavelengths and fluorescence intensities of wolfberry honey adulterated with different proportions of syrups and acacia honey (0–100%) at the second peak value. GA: wolfberry honey adulterated with different proportions of acacia honey; GH: wolfberry honey adulterated with different proportions of corn syrup; GZ: wolfberry honey adulterated with different proportions of corn maltose syrup.

3.7. The Fluorescence Lifetime of Wolfberry Honey and Syrups

The fluorescence lifetime is an intrinsic parameter of fluorescent substances, which is not easily disturbed by the concentration of fluorescent molecules, stray light, fluorescence scattering angle, and can visually reflect the decay process of fluorescence signal of fluorescent substances [30].

The fluorescence lifetime of the wolfberry honey and the syrups can be determined from the decay of their fluorescence intensity as a function of time. The fluorescence lifetime fitting plots of the wolfberry honey samples almost overlapped together and had no significant difference (Figure 6a), indicating that authentic wolfberry honey samples had similar fluorescence lifetime decay plot. However, the fluorescence lifetime fitting plots of the adulterated wolfberry honey samples especially in the decay domain shifted obviously compared with the authentic honey samples with an increase in the syrup concentration (10–100%) (Figure 6b,c), which indicated that the difference in the fluorescence lifetime between wolfberry honey and adulterated honey. Similar to the results of fluorescence spectra at a fixed excitation wavelength of 280 nm and 3D fluorescence spectra, the fluorescence lifetime fitting plots of honey samples adulterated with acacia honey were slightly shifted with the increase in acacia honey incorporation ratio (10–100%) (Figure 6d). The mean fluorescence lifetime of the wolfberry honey, acacia honey, and syrups can be found in Supplementary Table S3.

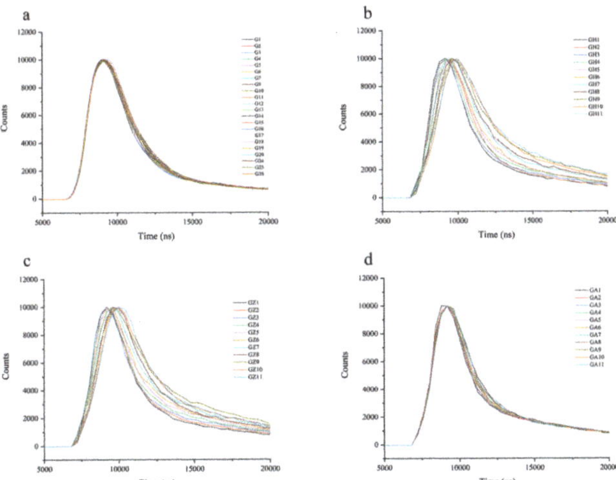

Figure 6. The fluorescence lifetime fitting plots of wolfberry honey and syrups. (**a**) The fluorescence lifetime fitting plots of wolfberry honey. (**b**) The fluorescence lifetime fitting plots of wolfberry honey adulterated with different proportions of corn syrup (0–100%). (**c**) The fluorescence lifetime fitting plots of wolfberry honey adulterated with different proportions of corn maltose syrup (0–100%). (**d**) The fluorescence lifetime fitting plots of wolfberry honey adulterated with different proportions of acacia honey (0–100%).

3.8. The PCA of Wolfberry Honey and Syrups

PCA is a dimensionality-reduction statistical method that is often used to reduce the dimensionality of large data sets. Seven main parameters, including the maximum fluorescence intensity and peak positions at the fixed excitation wavelength of 280 nm, two peak values, and two peak positions from the 3D fluorescence spectra, and fluorescence lifetime were used to perform the PCA. Two principal components, PC1 and PC2, were extracted according to the initial eigenvalues. The cumulative variance contribution rate reached 92.36%, which fully showed the difference between the wolfberry honey and the adulterated honey samples. In PC1, the maximum fluorescence intensity at 280 nm excitation wavelength and the peak values at 3D excitation wavelengths (200–450 nm excitation wavelength) had a positive effect, whereas the peak positions at 280 nm excitation wavelength and fluorescence lifetime had a negative effect. The fluorescence lifetime had higher loading factors in PC2, which had a positive effect on PC2. Although it was difficult to distinguish wolfberry honey from acacia honey using fluorescence spectra, they could be easily discriminated combined with PCA.

A PCA Biplot was used to depict the wolfberry samples and the adulterated honey samples. The dispersions were obvious and allowed us to easily distinguish the wolfberry honey from the adulterated honey at a 10% incorporation (Figure 7). From the PCA score plot we found the determination factor to distinguish acacia honey from wolfberry honey was mainly fluorescence lifetime. The maximum fluorescence intensity and peak positions at 280 nm excitation wavelength, the peak values at 3D excitation wavelengths, and fluorescence lifetime together determined the proportion of syrups in wolfberry honey. With an increase in the syrup concentration in wolfberry, the effects of peak positions at 280 nm excitation wavelength and fluorescence lifetime were more significant.

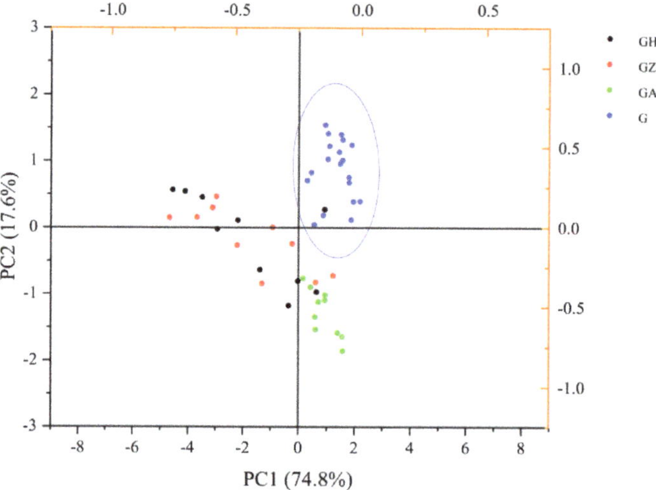

Figure 7. Principle components analysis of wolfberry honey and wolfberry honey adulterated with different proportions of syrups. GH: wolfberry honey adulterated with different proportions of corn syrup (10−100%); GZ: wolfberry honey adulterated with different proportions of corn maltose syrup (10−100%); GA: wolfberry honey adulterated with different proportions of acacia honey (10−100%); G: wolfberry honey samples.

4. Conclusions

We reported multiple fluorescence spectroscopy techniques combined with chemometrics can easily detect wolfberry honey adulteration with corn syrup, corn maltase syrup, and other pure monofloral honey such as acacia honey. The fluorescence intensities and peak positions are the easy-to-obtain characteristics to distinguish genuine honey from syrups because of their differences in the fluorophore concentration. The 3D fluorescence spectroscopy visually and comprehensively identified honey and syrup by obtaining the information of the excitation wavelength, emission wavelength, as well as fluorescence intensity. The fluorescence lifetime, independent of the fluorophore concentration, also can differentiate the adulterated honey samples. These results further confirmed the differences in their chemical constituents especially fluorophores of the monofloral honey, multifloral honey, syrups, and adulterated honey. Compared with other more expensive, sophisticated, time-consuming, and labor-intensive methods, the fluorescence spectroscopy detection time for each sample was only within several minutes and the pretreatment of the sample was simple. In all, fluorescence spectroscopy combined with chemometrics is a simple, fast, low-cost, and non-destructive method to quickly identify honey adulteration with corn syrup and corn maltose syrup.

Supplementary Materials: The following supporting information can be downloaded at: https://www.mdpi.com/article/10.3390/foods12122309/s1.

Author Contributions: Methodology, investigation, writing—original draft: S.H.; methodology: J.C.; data analysis: J.Y., Q.W. and X.L.; conceptualization, writing—review and editing, supervision, funding acquisition: H.X. All authors have read and agreed to the published version of the manuscript.

Funding: This work was supported by the grant from Shandong Province Modern Agricultural Technology System (SDAIT-24-05), Shandong Provincial Natural Science Foundation of China (ZR2021MC110), Taishan scholars (NO. tsqn202211172), and Liaocheng Key Research and Development Program (2022YDSF10).

Data Availability Statement: The data used to support the findings of this study can be made available by the corresponding author upon request.

Conflicts of Interest: The authors declare no conflict of interest.

References

1. Sarkar, S.; Chandra, S. Honey as a functional additive in yoghurt-a review. *Nutr. Food Sci.* **2019**, *50*, 168–178. [CrossRef]
2. Jaafar, M.B.; Othman, M.B.; Yaacob, M.; Talip, B.A.; Ilyas, M.A.; Ngajikin, N.H.; Fauzi, N.A.M. A Review on honey adulteration and the available detection approaches. *Int. J. Integr. Eng.* **2020**, *12*, 125–131.
3. Zulkiflee, N.S.B.; Zoolfakar, A.S.; Rani, R.A.; Aryani, D.; Zolkapli, M. Detection and classification of honey adulteration combined with multivariate analysis. *Int. J. Integr. Eng.* **2022**, *14*, 262–272. [CrossRef]
4. da Silva, P.M.; Gauche, C.; Gonzaga, L.V.; Costa, A.C.; Fett, R. Honey: Chemical composition, stability and authenticity. *Food Chem.* **2016**, *196*, 309–323. [CrossRef] [PubMed]
5. Isidorow, W.; Witkowski, S.; Iwaniuk, P.; Zambrzycka, M.; Swiecicka, I. Royal jelly aliphatic acids contribute to antimicrobial activity of honey. *J. Apic. Sci.* **2018**, *62*, 111–123. [CrossRef]
6. Wu, L.M.; Du, B.; Vander Heyden, Y.; Chen, L.Z.; Zhao, L.W.; Wang, M.; Xue, X.F. Recent advancements in detecting sugar-based adulterants in honey—A challenge. *TrAC-Trends Anal. Chem.* **2017**, *86*, 25–38. [CrossRef]
7. Ursulin-Trstenjak, N.; Puntaric, D.; Levanic, D.; Gvozdic, V.; Pavlek, Z.; Puntaric, A.; Puntaric, E.; Puntaric, I.; Vidosavljevic, D.; Lasic, D.; et al. Pollen, Physicochemical, and mineral analysis of croatian acacia honey samples: Applicability for identification of botanical and geographical origin. *J. Food Qual.* **2017**, *2017*, 8538693. [CrossRef]
8. Babaeva, E.Y.; Polevova, S.V.; Zugkiev, B.G. Identification of propolis and honey sources pollen analysis of the south of north Caucasus and the lower Volga. *Sci. Study Res.* **2021**, *22*, 437–451.
9. Truong, A.T.; Yoo, M.S.; Cho, Y.S.; Yoon, B. Identification of seasonal honey based on quantitative detection of typical pollen DNA. *Appl. Sci.* **2022**, *12*, 4846. [CrossRef]
10. Jandric, Z.; Frew, R.D.; Fernandez-Cedi, L.N.; Cannavan, A. An investigative study on discrimination of honey of various floral and geographical origins using UPLC-QToF MS and multivariate data analysis. *Food Control.* **2017**, *72*, 189–197.
11. Tahir, H.E.; Zou, X.B.; Huang, X.W.; Shi, J.Y.; Mariod, A.A. Discrimination of honeys using colorimetric sensor arrays, sensory analysis and gas chromatography techniques. *Food Chem.* **2016**, *206*, 37–43. [CrossRef]
12. Kalashnikova, D.A.; Simonova, G.V. Ratios of Stable Isotopes C-13/C-12 and N-15/N-14 in Samples of dead honey bees and beekeeping products. *J. Anal. Chem.* **2021**, *76*, 526–534. [CrossRef]
13. Tedesco, R.; Barbaro, E.; Zangrando, R.; Rizzoli, A.; Malagnini, V.; Gambaro, A.; Fontana, P.; Capodaglio, G. Carbohydrate determination in honey samples by ion chromatography-mass spectrometry (HPAEC-MS). *Anal. Bioanal. Chem.* **2020**, *412*, 5217–5227. [CrossRef]

14. Sun, T.; Zhang, Y.; Wang, X.; Zhang, Y.Y.; Liu, Z.; Liu, W.; Chen, P.A.; Zhang, Z.H.; Yu, Y.J. Chemometric strategy for aligning chemical shifts in 1H NMR to improve geographical origin discrimination: A case study for Chinese Goji honey. *Microchem. J.* **2022**, *174*, 107062. [CrossRef]
15. Romano, A.; Cuenca, M.; Makhoul, S.; Biasioli, F.; Martinello, L.; Fugatti, A.; Scampicchio, M. Comparison of E-Noses: The case study of honey. *Ital. J. Food Sci.* **2016**, *28*, 326–337.
16. Lozano-Torres, B.; Martinez-Bisbal, M.C.; Soto, J.; Borras, M.J.; Martinez-Manez, R.; Escriche, I. Monofloral honey authentication by voltammetric electronic tongue: A comparison with H-1 NMR spectroscopy. *Food Chem.* **2022**, *383*, 132460. [CrossRef] [PubMed]
17. Hassoun, A.; Mage, I.; Schmidt, W.F.; Temiz, H.T.; Li, L.; Kim, H.Y.; Nilsen, H.; Biancolillo, A.; Ait-Kaddour, A.; Sikorski, M. Fraud in Animal Origin Food Products: Advances in emerging spectroscopic detection methods over the past five years. *Foods* **2020**, *9*, 1069. [CrossRef]
18. Yan, S.; Sun, M.H.; Wang, X.; Shan, J.H.; Xue, X.F. A Novel, Rapid screening technique for sugar syrup adulteration in honey using fluorescence spectroscopy. *Foods* **2022**, *11*, 2316. [CrossRef]
19. Hao, S.Y.; Li, J.Y.; Liu, X.Y.; Yuan, J.; Yuan, W.Q.; Tian, Y.; Xuan, H.Z. Authentication of acacia honey using fluorescence spectroscopy. *Food Control.* **2021**, *130*, 108327. [CrossRef]
20. Soto, M.E.; Ares, A.M.; Bernal, J.; Nozal, M.J.; Bernal, J.L. Simultaneous determination of tryptophan, kynurenine, kynurenic and xanthurenic acids in honey by liquid chromatography with diode array, fluorescence and tandem mass spectrometry detection. *J. Chromatogr. A* **2011**, *1218*, 7592–7600. [CrossRef]
21. Bartolic, D.; Maksimovic, V.; Maksimovic, J.D.; Stankovic, M.; Krstovic, S.; Baosic, R.; Radotic, K. Variations in polyamine conjugates in maize (*Zea mays* L.) seeds contaminated with aflatoxin B1: A dose-response relationship. *J. Sci. Food Agric.* **2020**, *100*, 2905–2910.
22. Borges, C.V.; Nunes, A.; Costa, V.E.; Orsi, R.D.; Basilio, L.S.P.; Monteiro, G.C.; Maraschin, M.; Lima, G.P.P. Tryptophan and Biogenic Amines in the Differentiation and Quality of Honey. *Int. J. Tryptophan Res.* **2022**, *15*, 1–12.
23. Ruoff, K.; Karoui, R.; Dufour, E.; Luginbühl, W.; Bosset, J.O.; Bogdanov, S.; Amado, R. Authentication of the botanical origin of honey by front-face fluorescence spectroscopy. A Preliminary Study. *J. Agric. Food Chem.* **2005**, *53*, 1343–1347. [CrossRef] [PubMed]
24. Ruoff, K.; Luginbühl, W.; Künzli, R.; Bogdanov, S.; Bosset, J.O.; von der Ohe, K.; von der Ohe, W.; Amadò, R. Authentication of the botanical and geographical origin of honey by front-face fluorescence spectroscopy. *J. Agric. Food Chem.* **2006**, *54*, 6858–6866. [CrossRef] [PubMed]
25. Sergiel, I.; Pohl, P.; Biesaga, M.; Mironczyk, A. Suitability of three-dimensional synchronous fluorescence spectroscopy for fingerprint analysis of honey samples with reference to their phenolic profiles. *Food Chem.* **2014**, *145*, 319–326. [CrossRef]
26. Rotar, A.M.; Vodnar, D.C.; Bunghez, F.; Catunescu, G.M.; Pop, C.R.; Jimborean, M.; Semeniuc, C.A. Effect of Goji berries and honey on lactic acid bacteria viability and shelf life stability of yoghurt. *Not. Bot. Horti Agrobot. Cluj-Napoca* **2015**, *43*, 196–203.
27. Hao, S.Y.; Yuan, J.; Cui, J.C.; Yuan, W.Q.; Zhang, H.W.; Xuan, H.Z. The rapid detection of acacia honey adulteration by alternating current impedance spectroscopy combined with ^{1}H NMR profile. *LWT-Food Sci. Technol.* **2022**, *161*, 130.
28. Lenhardt, L.; Bro, R.; Zeković, I.; Dramićanin, T.; Dramićanin, M.D. Fluorescence spectroscopy coupled with PARAFAC and PLS DA for characterization and classification of honey. *Food Chem.* **2015**, *175*, 284–291.
29. Kečkeš, S.; Gašić, U.; Veličković, T.C.; Milojković-Opsenica, D.; Natić, M.; Tešić, Ž. The determination of phenolic profiles of Serbian unifloral honeys using ultra-high-performance liquid chromatography/high resolution accurate mass spectrometry. *Food Chem.* **2013**, *138*, 32–40.
30. Fišerová, E.; Kubala, M. Mean fluorescence lifetime and its error. *J. Lumin.* **2012**, *132*, 2059–2064. [CrossRef]

 *foods*

Article

Sunflower Honey—Evaluation of Quality and Stability during Storage

Milica Živkov Baloš, Nenad Popov, Sandra Jakšić, Željko Mihaljev, Miloš Pelić, Radomir Ratajac and Dragana Ljubojević Pelić *

Scientific Veterinary Institute "Novi Sad", 21000 Novi Sad, Serbia
* Correspondence: dragana@niv.ns.ac.rs

Abstract: Honey's unique qualities should last for several years when properly stored. Therefore, it is up to manufacturers to choose the right shelf life for their product while also considering the product's nature. Physicochemical parameters (water content, electrical conductivity, free acidity, pH, ash, water-insoluble matter, hydroxymethylfurfural (HMF), sugar content and composition, and diastase activity) were analyzed in 24 samples of sunflower honey collected from several localities in Vojvodina, Serbia. Crystallization indices were also calculated. Furthermore, the impact of eighteen months of room temperature storage ($22 \pm 2\,^{\circ}\mathrm{C}$) in a dark place on selected physicochemical parameters (water, HMF, diastase activity, pH value, and free acidity) was investigated. The results of the initial test indicated that the tested samples of sunflower honey from Vojvodina is of good quality because the parameters under examination revealed results that were within the legal bounds of both national and European legislations. Eighteen months of storage at room temperature reduced diastase activity by 2 times, increased HMF content by about 17 times, and decreased the pH value of honey from a mean value of 3.66 to 3.56. The water content was relatively stable at 17.01% before storage and 16.29% after storage. The storage of sunflower honey did not have an impact on the free acidity.

Keywords: sunflower honey; physicochemical properties; storage; shelf life

Citation: Živkov Baloš, M.; Popov, N.; Jakšić, S.; Mihaljev, Ž.; Pelić, M.; Ratajac, R.; Ljubojević Pelić, D. Sunflower Honey—Evaluation of Quality and Stability during Storage. *Foods* **2023**, *12*, 2585. https://doi.org/10.3390/foods12132585

Academic Editor: Paweł Kafarski

Received: 3 June 2023
Revised: 23 June 2023
Accepted: 26 June 2023
Published: 3 July 2023

1. Introduction

Honey is a natural food with special nutritional, sensorial, and potentially therapeutic properties [1,2]. These characteristics are connected to honey's chemical structure. Natural honey has a composition that consists of 80–85% glucose, fructose, and other carbohydrates; 15–17% moisture; 0.2% ash; 0.1–0.4% protein; and trace amounts of vitamins, enzymes, and other nutrients such as phenolic antioxidants [3]. The nutritional quality and chemical parameters of honey are mainly influenced by the species of bee, geographical region, and available floral source as well as processing temperature, packaging, storage, and climatic conditions [4–7].

Sunflower honey has remarkable medicinal and nutritional benefits. Since sunflower honey has a little amount of sucrose, it crystallizes quickly [8,9]. Only the southern regions, where there are plenty of sunlight and where the climate is suitable for cultivating this plant, are used for sunflower cultivation. Serbia has a well-developed honey industry because of its favorable temperature and position. One of the most common types of honey produced in Serbia is sunflower honey, but there is not enough information in the literature about the changes that can occur during the honey storage period. Honey can preserve its unique characteristics for several years if it is stored properly. Therefore, manufacturers need to determine an appropriate shelf life for their product while considering its properties.

Many of the components that provide honey with its distinctive aroma and some of its biological functions are thermolabile [10]. Appropriate storage is essential to maintain the quality of honey because the composition of honey could change during storage through oxidation and fermentation [11,12]. Therefore, physicochemical characteristics, microbiological features, and sensory qualities should be assessed in order to assure the authenticity and quality of honey. Since these factors affect the maturity and purity of fresh honey, it is crucial to monitor changes that might take place during storage. The change in properties and quality of honey can be induced by temperature, humidity, air, and light [13]. Sugar degradation, hydroxymethylfurfural (HMF) formation, a decreased diastase and invertase activity, an increased acidity, a lowered pH value, phenolic component degradation, and color changes are some of the alterations that may take place during storage [4,14–17]. Honey has an acidic pH value, which is connected to nectar, bee secretions, or organic acids (acetic, citric, tartaric, oxalic, etc.) [18]. The presence of organic acids in honey may influence fermentation processes, aroma, flavor, and the antibacterial characteristics of honey [19–21]. The formation of HMF and the decrease in honey enzyme activity can be a consequence of the aging or heating of honey when the dehydration of hexoses occurs. Hexoses break down into levulinic and formic acid, and consequently, the free acidity of honey increases [22].

The temperature plays a significant role in the long-term storage of honey. Some changes in honey composition could be catalyzed by higher ambient temperatures, so they are more expected in tropical regions. Honey may be processed using thermal treatment. Thermal processing eliminates spoilage microorganisms and reduces water content, prevents and delays crystallization, and reduces viscosity, which facilitates the processing and bottling of honey [10,23]. Liquid honey has a tendency to crystallize with time. Honey crystallization is an unfavorable process because it alters its textural characteristics, which makes it less appealing to consumers who prefer liquid and translucent honey [24]. The processing of honey during extraction, filtration, mixing, and bottling is affected by honey crystallization [25,26]. When honey has a higher glucose concentration and lower water content, the crystallization process occurs more quickly. Indicators of honey's crystallization potential include the fructose/glucose (F/G) ratio and the glucose/water (G/W) ratio. Most often, liquefying crystallized honey involves heating at 32–40 °C [27]. The amount of fructose and glucose in honey is reduced during overheating for better ripening, which also causes the formation of HMF [26,28]. Honey producers worldwide have practiced by using a variety of heating methods using temperatures between 30° and 140 °C from short periods of time to many hours [6]. Except for the temperature, factors such as heating time, storage conditions, the usage of metallic containers, and the physicochemical characteristics of honey may affect the development of HMF in honey [15,29]. Low temperatures slow down the crystallization process, prevent fermentation and other chemical processes, and reduce the viscosity [30]. Uncontrolled heating motivates the loss of thermolabile and aromatic substances and affects factors of quality like HMF and enzymatic activity. HMF has especially been used for detecting the intensity of changes during the thermal processing of food [10,27,31]. Relatively few microorganisms are capable to survive in honey due to its low moisture content. However, honey is a very hygroscopic substance, and its moisture content can change depending on the atmospheric humidity while being stored [32,33]. The probability that yeasts will ferment and affect the flavor of the honey during storage increases with the amount of moisture in the honey [7,21,34]. Microbiological contamination of honey can induce deterioration, i.e., affect its stability. The degree of the degradation of honey quality is affected by the honey type, manufacturing process, and storage conditions [13].

In this work, twenty-four Serbian sunflower honey samples from Vojvodina, Serbia, harvested in 2019, were evaluated regarding their physicochemical parameters. The samples of sunflower honey were then stored for 18 months at room temperature, and an investigation of stability was carried out to determine how storage affected selected physicochemical properties. The obtained results could significantly contribute to the determination of the accurate shelf life of honey, both for manufacturers and for customers to know how long they can store honey in their homes. The stability studies are very important for the industry, and it is very important to monitor product quality as a function of time as our study could contribute to the development of future stability study protocols and plans for shelf-life assessment.

2. Materials and Methods

2.1. Samples

Twenty-four samples of sunflower honey from various parts of Vojvodina Province, Republic of Serbia, were collected directly from beekeepers. All samples were delivered to the Scientific Veterinary Institute "Novi Sad" laboratory in their original packing for analysis. To determine the botanical origin of products, manufacturers used field observations. Only samples with verified botanical origin noted on the manufacturing specification label were used in our study. All samples were collected in sterile glass jars and kept in a dark place at room temperature (22 ± 2 °C). Honey analyses were carried out immediately after sampling and after storage for 18 months. All samples were examined in duplicate using the procedures outlined in the Harmonized methods of the International Honey Commission Methods [35].

2.2. Moisture Content

A conventional Abbetype refractometer was used to measure the refractive index (RI). The Chataway table was then used to determine the moisture content (%).

2.3. Sugar Composition Determination

An HPLC Dionex UltiMate 3000 Series system (Thermo Scientific, Germering, Germany) supplied with a refractive index detector RefractoMax521 (ERC Inc., Kawaguchi, Saitama, Japan) was used to measure the sugar content (fructose, glucose, and sucrose) at 35 °C. After dissolving in 25 mL of 25% methanol, the honey sample was filtered through a 0.22 m nylon filter and then injected (5 µL) into the HPLC. The HPLC column was a Hypersil GOLD Amino 150 × 3 mm (particle size 3 µm) (Thermo Scientific, Germany), equipped with a guard column Hypersil GOLD Amino 10 × 3 mm column with the same particle size. Acetonitrile and water (80:20, *v/v*) served as the mobile phase and were filtered using a 0.22 m membrane filter at a flow rate of 1 mL/min. At room temperature, all measurements were made. Thermo Scientific's Chromeleon®7 software (Version 7.1, Dionex, Sunnyvale, CA, USA) was used to control the system. Sugar concentrations in the samples were measured using external calibration curves created by standard solutions. By comparing honey sugars' retention periods and peak areas to those of standard sugar solutions, honey sugars were identified and quantified.

2.4. Electrical Conductivity

A conductometer Type Basic 30 (Crison, Spain) was used to measure the electrical conductivity of honey samples' solutions (20 g dry matter of honey in volume solution in 100 mL distilled water) at a temperature of 20 °C.

2.5. Free Acidity and pH Value

Ten grams of honey were dissolved in 75 mL of carbon dioxide-free water. Using a magnetic stirrer and a pH meter, the pH value was determined. The volumetric method was then used to determine the acidity of honey. With 0.1 mol/dm³ of NaOH, the sample solution was titrated to pH 8.30. Honey's acidity is expressed in milliequivalents per kilogram (mEq/kg).

2.6. Ash

To determine the ash, the residue was weighed after 5 g of honey samples were ashed in an electric furnace at 600 °C.

2.7. Hydroxymethylfurfural (HMF)

HPLC Dionex UltiMate 3000 Series (Thermo Scientific, Germering, Germany) equipped with UV detector was used to measure HMF. After dissolving 1 g of honey sample in 25 mL of water and filtered through a 0.45 μm nylon filter, 10 μL of sample was injected into the HPLC system. The HPLC column had a particle size of 3 μm and was a 150 × 3 mm Hypersil GOLD column (Thermo Scientific, Germany). Methanol and water (10:90, *v/v*) were used as the mobile phase at a flow rate of 1 mL/min. At 285 nm, HMF detection was carried out. Thermo Scientific's Chromeleon®7 software (Version 7.1, Dionex, Sunnyvale, CA, USA) was used to control the system. To calculate the amount of HMF in the samples, standard solutions' external calibration curves were employed.

2.8. Diastase Activity

The spectrophotometric method was used to determine the diastase activity (Megazyme International Ireland, Bray Business Park, Bray, Co. Wicklow, Ireland). The sample was dissolved in sodium maleate buffer and its volume was adjusted with water to the volumetric flask's mark. The buffer solution contained Amylazyme tablets (Megazyme International, Ireland). The substrate was hydrolyzed and colored compounds were produced when a α-amylase was present. Following the completion of the reaction, the absorbance of the filtrate was determined at 590 nm. The sample's diastase activity was inversely correlated with the absorbance. Diastase number (DN) was used to calculate diastase activity.

2.9. Water-Insoluble Matter

The gravimetric method was used to determine the amount of insoluble materials. By rinsing with warm water, the insoluble material was obtained on the filter with the defined pore size. Until a constant weight was reached, dried residues (135 °C) were weighed.

2.10. Statistical Analysis

The PAST software package, version 2.12 (Oslo, Norway) was used to conduct the statistical analysis. Univariate analysis (descriptive statistics) and analysis of variance (ANOVA) were used in the statistical data analysis.

3. Results and Discussion

The findings of Serbian sunflower honey's physicochemical analysis obtained immediately after sampling are shown in Tables 1 and 2.

Table 1. Physicochemical characteristics of sunflower honey samples before storage.

Sample	Moisture (%)	Electrical Conductivity (mS/cm)	Free Acidity (mEq/kg)	pH	Ash (%)	Insoluble Matter (%)	HMF (mg/kg)	Diastase Activity (DN)
1	16.2	0.40	35.4	3.83	0.17	<0.01	1.72	24.82
2	18.6	0.32	24.4	3.57	0.08	<0.01	3.05	15.50
3	18.6	0.22	21.4	3.49	0.03	<0.01	2.07	16.52
4	16.4	0.28	20.4	3.69	0.05	<0.01	2.66	19.64
5	18.4	0.30	20.8	3.52	0.05	<0.01	1.64	13.61
6	17.8	0.32	30.8	3.58	0.04	<0.01	4.41	25.57
7	16.4	0.40	31.0	3.50	0.24	0.03	1.75	19.43
8	17.8	0.30	26.4	3.63	0.09	0.19	2.17	17.73
9	17.6	0.32	32.2	3.70	0.09	<0.01	2.24	21.94
10	17.2	0.36	29.2	3.68	0.11	0.02	1.79	17.34
11	17.4	0.36	27.2	3.64	0.11	<0.01	1.23	16.24
12	16.4	0.54	30.0	3.63	0.21	<0.01	1.94	18.71
13	17.8	0.32	32.0	3.56	0.15	0.17	2.67	19.67
14	16.4	0.39	25.6	3.70	0.21	0.19	2.35	16.43
15	15.2	0.28	23.8	3.55	0.12	<0.01	0.82	18.71
16	16.4	0.42	36.4	3.61	0.13	0.07	2.51	16.03
17	16.0	0.22	26.8	3.46	0.09	0.05	4.30	10.88
18	15.4	0.26	24.5	3.59	0.09	<0.01	2.17	13.24
19	16.8	0.30	26.2	3.68	0.10	0.03	2.14	17.83
20	14.6	0.30	22.4	3.99	0.14	0.02	0.82	18.86
21	17.8	0.44	34.4	3.84	0.25	0.12	1.26	16.47
22	16.8	0.34	36.8	3.97	0.15	0.09	1.45	10.22
23	18.0	0.32	29.8	3.71	0.17	0.01	1.11	10.14
24	18.2	0.46	35.6	3.62	0.30	0.11	1.66	20.89

Table 1. *Cont.*

Sample	Moisture (%)	Electrical Conductivity (mS/cm)	Free Acidity (mEq/kg)	pH	Ash (%)	Insoluble Matter (%)	HMF (mg/kg)	Diastase Activity (DN)
Mean	17.01	0.34	28.48	3.66	0.13	0.08	2.08	17.35
SD	1.10	0.08	5.09	0.14	0.07	0.07	0.91	3.97
Min	14.60	0.22	20.40	3.46	0.03	0.01	0.82	10.14
Max	18.60	0.54	36.80	3.99	0.30	0.19	4.41	25.57
CV (%)	6.45	22.26	17.88	3.76	53.25	78.34	43.55	22.88

Table 2. Sugar profile and crystallization ratios of sunflower honey before storage.

Sample	Sucrose (%)	Glucose (%)	Fructose (%)	Reducing Sugars (%)	F/G *	G/W **
1	<0.250	37.66	40.94	78.60	1.09	2.32
2	0.299	37.04	39.46	76.50	1.07	1.99
3	0.253	36.22	39.78	76.00	1.10	1.95
4	0.568	37.33	41.53	78.86	1.11	2.28
5	0.267	38.47	40.85	79.32	1.06	2.09
6	<0.250	35.50	39.93	75.43	1.12	1.99
7	<0.250	38.25	40.64	78.89	1.06	2.33
8	<0.250	38.11	40.03	78.14	1.05	2.14
9	0.274	37.62	39.73	77.35	1.06	2.14
10	<0.250	38.58	40.14	78.72	1.04	2.24
11	<0.250	36.66	38.87	75.53	1.06	2.11
12	<0.250	36.35	38.87	75.22	1.07	2.22
13	<0.250	37.27	39.99	77.26	1.07	2.09
14	<0.250	37.77	40.84	78.61	1.08	2.30
15	<0.250	37.77	40.84	78.61	1.08	2.48
16	<0.250	37.17	40.06	77.23	1.08	2.27
17	<0.250	34.39	39.61	74.00	1.15	2.15
18	<0.250	37.93	41.36	79.29	1.09	2.46
19	<0.250	38.00	40.61	78.61	1.07	2.26
20	<0.250	37.14	39.31	76.45	1.06	2.54
21	<0.250	36.78	40.48	77.26	1.10	2.07
22	<0.250	37.38	40.93	78.31	1.09	2.23
23	<0.250	37.21	40.36	77.57	1.08	2.07
24	<0.250	33.02	40.70	73.72	1.23	1.81
Mean	0.332	37.07	40.24	77.31	1.09	2.19
SD	0.133	1.28	0.72	1.63	0.04	0.17
Min	0.253	33.02	38.87	73.72	1.04	1.81
Max	0.568	38.58	41.53	79.32	1.23	2.54
CV (%)	39.99	3.46	1.79	2.11	3.67	7.98

* F/G—the fructose/glucose ratio; ** G/W—the glucose/water ratio.

The results of the initial test indicated that Serbian sunflower honey is known for its good quality since the criteria that were closely correlated with the honey's quality were measured, and their levels complied with the limits established by international and national regulation. Analyses of water content, pH, free acidity, HMF, and diastase activity were performed to determine the effects of 18 months of room-temperature storage on the quality of honey. The findings of the selected physicochemical examination of Serbian sunflower honey, obtained after storage for 18 months, are shown in Table 3. The influence of the duration of storage on selected honey parameters is shown in Figures 1–5.

Table 3. Selected physicochemical characteristics of sunflower honey stored at room temperature for 18 months.

Sample	Moisture (%)	Free Acidity (mEq/kg)	pH	HMF (mg/kg)	Diastase Activity (DN)
1	14.0	37.0	3.61	36.49	6.60
2	18.4	22.8	3.56	27.38	7.08
3	18.4	20.4	3.39	35.90	7.94
4	16.6	21.0	3.54	38.45	9.78
5	18.0	22.0	3.66	26.98	6.50
6	14.8	28.4	3.63	42.61	12.92
7	15.8	32.0	3.75	36.45	8.56
8	16.8	25.2	3.51	38.02	7.94
9	15.4	32.0	3.35	31.14	9.40
10	15.2	28.6	3.60	35.14	6.88
11	16.2	27.8	3.56	27.70	6.50
12	15.4	28.4	3.83	29.22	8.40
13	17.2	34.0	3.39	33.06	7.42
14	16.6	30.8	3.66	38.91	6.48
15	15.0	23.6	3.51	32.14	9.76
16	16.8	31.4	3.64	30.25	9.24
17	15.4	25.6	3.57	44.65	5.32
18	15.4	25.6	3.57	42.13	6.86
19	16.2	25.2	3.39	33.92	8.74
20	18.0	23.2	3.54	30.00	11.34
21	15.8	34.4	3.60	27.10	9.96
22	14.6	31.0	3.45	37.94	6.08
23	17.6	26.0	3.53	33.57	5.30
24	16.2	37.2	3.65	42.88	11.30
Mean	16.29	28.07	3.56	34.67	8.18
SD	1.24	4.88	0.11	5.36	1.99
Min	14.00	20.40	3.35	26.98	5.30
Max	18.40	37.20	3.83	44.65	12.92
CV (%)	7.60	17.38	3.23	15.46	24.29

3.1. Moisture Content

All samples of sunflower honey that were investigated had moisture contents that were less than 20%, which is the maximum permitted amount for honey as specified by national legislation [36] (Table 1). The physical, microbiological, sensory, and economic value of honey are all influenced by the presence of water. Moreover, water content has a significant impact on the process of crystallization; thus, it is crucial to monitor and manage its amount in honey. The maturation of honey and harvest time is an important factor that could affect the percentage of water in honey [7]. The amount of water in sunflower honey after sampling corresponded with the regulations and indicated a timely harvest and good production practices. Similar values for moisture content in sunflower honey from Romania, Serbia, Greece, and Argentina were reported by other authors [37–40].

During storage, decrease in the mean content of water was found (Figure 1). All the sunflower honey samples that were examined had water content less than twenty percent. The mean water content was 16.29 ± 1.24% (Table 3). This is a slightly lower water content compared to the test results immediately after sampling. There were statistically significant variations in the mean values of water in honey between the time of sampling and 18 months afterward (p = 0.05). In general, all tested samples corresponded with the regulations even after storage, and water content remained relatively stable at 17.01% before storage and 16.29% after storage. These findings are in agreement with the findings presented by Soares et al. [11]. Seraglio et al. [41], however, did not find significant differences between moisture content before and after honey storage. Contrary to that, Da Silva et al. [4] established an increase in moisture content before and after honey storage, and the differences were statistically significant. Different results may be the consequence of the influence of numerous factors, like the environment, the time of harvest, and the level of honey maturation attained in the hive.

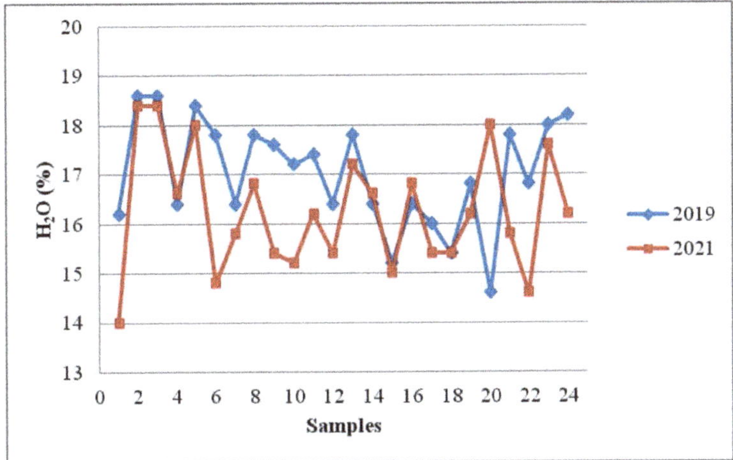

Figure 1. Effect of storage time on moisture content in sunflower honey.

3.2. Sugars

Our research showed that the levels of fructose and glucose ranged from 38.87 to 41.53 and from 33.02 to 38.58%, respectively, in all the sunflower honey samples we analyzed (Table 2). With a value of above 60 g/100 g for all samples of sunflower honey, the total amount of glucose and fructose complied with national and European criteria [36,42]. The amount of sucrose in each sample of honey that was examined was less than 5 g/100 g, which is the limit permitted by European legislation [42] and the national regulation for honey [36]. Sucrose level was below the detection limit of the used method in 19 out of the 24 total analyzed samples (79.2%) (Table 2). The method's detection limit is 0.25%. The F/G ratio ranged from 1.04 to 1.23, with a mean value of 1.09. The G/W ratio ranged from 1.81 to 2.54 with a mean value of 2.19 (Table 2).

Honey, a highly viscous mixture of sugars predominately contains glucose and fructose in amounts that are almost equal. As can be seen from the results (Table 2), fructose is the most prevalent reducing sugar in sunflower honey. Generally, the fructose content was higher than the glucose content, indicating that bee colonies were fed naturally. This supported the good quality of the different types of honey that were analyzed. If the beekeeper overfed the bees with sugar in the spring, the sucrose content might be used as a sign that artificial feeding was used. This sugar's high content also signals an early honey harvest [3]. Glucose and fructose levels in sunflower honey were in accordance with the literature data on sunflower honey [9,25,37,43]. The sucrose content in sunflower honey was lower than the results of the aforementioned authors.

The amount of sugar and water in honey, as well as their relative proportions, affect how quickly crystallization occurs. Parameters for the prediction of crystallization tendency are F/G and G/W proportions. During crystallization, due to its greater solubility, fructose remains in solution, while glucose crystallizes first. If the F/G ratio is greater than 1.33, the crystallization process proceeds slowly. In cases where the F/G ratio is less than 1.11, the honey crystallizes quickly [44]. According to our results (Table 2), sunflower honey crystallizes fast. The honey crystallizes faster when glucose content is higher and water content lower. Moreover, when the G/W ratio is less than 1.7, the crystallization process is either slower or absent; and when the ratio is larger than 2 [25] or 2.10 [45], the process is faster. According to this criterion, sunflower honey is a rapidly crystallizing honey.

3.3. Electrical Conductivity

The maximum electrical conductivity of sunflower honey in the Republic of Serbia is set at 0.8 mS/cm [36]. The values of electrical conductivity in the examined sunflower honey samples ranged from 0.22 to 0.54 mS/cm (Table 1). According to the standard, electrical conductivity is frequently used in the routine quality control of honey to differentiate between floral and honeydew honey. The concentrations of mineral salts, organic acids, and proteins are correlated with the electrical conductivity, total acidity (pH), and ash mass percentage [40]. Honey's electrical conductivity can also be affected by many factors, including flower source, the concentrations of organic acids and proteins, and storage period [20]. Data on electrical conductivity in sunflower honey that are similar to data from our research were provided by other researchers from Serbia, Romania, Greece, France, Argentina, and Spain [37–40,43,46,47].

3.4. Free Acidity and pH Value

According to the national regulation [36], the maximum allowable value of free acidity in any type of honey is set at 50 mEq/kg (with the exception of baker's honey). All examined samples of sunflower honey had free acidity less than 50 mEq/kg (Table 1). These findings showed that unfavorable fermentation was absent. The studied samples had an average acidity value of 28.48 mEq/kg. All examined samples of sunflower honey after storage had free acidity less than 50 mEq/kg.

The reference value for pH value is not prescribed by the regulations. According to our research, the tested samples of honey had pH values that ranged from 3.46 to 3.99 (on average, 3.66 ± 0.14) (Table 1). The mean acidity value and pH value in the examined samples after storage for 18 months (Table 3, Figures 2 and 3) were very similar to results obtained immediately after sampling (28.48 ± 5.09 mEq/kg versus 28.07 ± 4.88 mEq/kg and 3.66 ± 0.14 versus 3.56 ± 0.11). The differences between mean values for free acidity were not statistically significant ($p = 0.78$). However, differences between mean values for pH values in honey after sampling and 18 months later were statistically significant ($p = 0.01$).

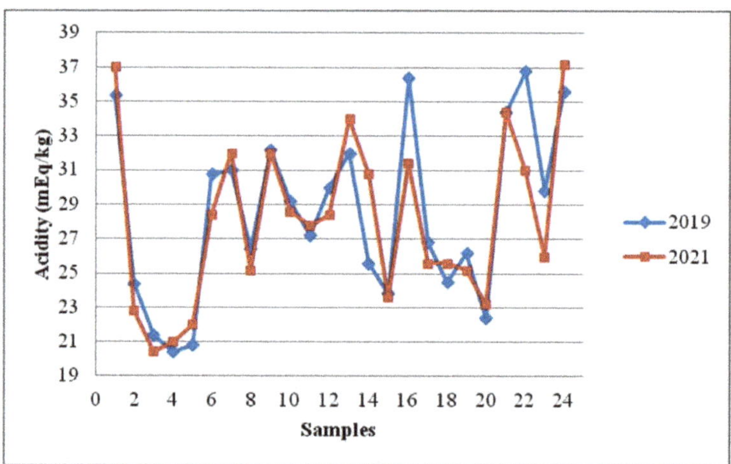

Figure 2. Effect of storage time on acidity in sunflower honey.

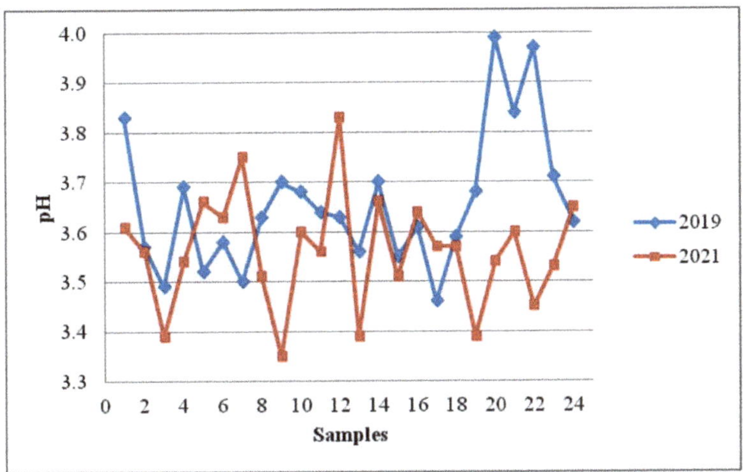

Figure 3. Effect of storage time on pH in sunflower honey.

Organic acids (tartaric, citric, oxalic, acetic, etc.), nectar, or bee secretions all contribute to the acidity of honey [48]. The pH value measurement (total acidity) or a titration with sodium hydroxide (free acidity) can be used to measure the acidity of honey. The acidity values (total and free acidity) obtained in our investigation of sunflower honey before storage were similar to the results published by Devillers et al. [43], Lazarević et al. [38], Sari and Ayyildiz [9], Prica et al. [21], and Đogo Mračević et al. [46]. The pH of sunflower honey decreased after 18 months of storage at room temperature (Figure 3). Differences in honey pH before and after storage were significant. However, changes in free acidity during storage for 18 months were not significant ($p > 0.05$). Although the honey's pH level changed during storage, it is important to emphasize that the acidity values (free acidity and pH) were in accordance with the values considered as normal for fresh honey. The normal pH value for fresh honey in the range of 3.2 to 4.5 (free acidity maximum 50 mEq/kg) inhibits most microorganisms and ensures that honey is safe for consumption [4]. Similar results that the pH value decreases during honey storage were reported by Seraglio et al. [41], Da Silva et al. [4], Soares et al. [11], and Evahelda et al. [13]. Czipa et al. [5] have reported that the pH value of honey did not change after a two-year storage. Seraglio et al. [41], Da Silva et al. [4], and Chou et al. [19] found that free acidity increased during storage time.

3.5. Ash Mass Fraction

The reference value for the ash mass fraction is not prescribed by the regulations. According to our research, the percentage of ash was in a range from 0.03 to 0.30% in all honey samples that were studied, while the average was $0.13 \pm 0.07\%$ (Table 1). The content of ash primarily depends on climate and soil properties [40]. The ashes in honey indicate environmental pollution and, consequently, its geographic origin. Data on ash in sunflower honey that are similar to data from our research were obtained by other researchers from Serbia, Romania, Greece, France, Argentina, and Spain [37–40,43,46,47].

3.6. HMF and Diastase Activity

Regularly used indicators to assess the freshness of honey and to provide details on processing and storage conditions include HMF content and diastase activity [11]. In general, all tested samples before storage complied with the provisions of the regulations concerning the level of HMF and diastase activity. The minimum permitted value of diastase activity in all kinds of honey (except from baker's honey) is 8 DN, as is stipulated by regulations [36,42]. According to the data presented in Table 1, it is possible to verify the initial freshness of honey, since HMF content and diastase activity were in compliance with

European and national legislations. The initial mean HMF level in examined sunflower honeys was 2.08 mg/kg, while it was in the range from 0.82 to 4.41 mg/kg (Table 1). The mean diastase activity was 17.35 DN, with the minimum value being 10.14 DN (Table 1).

Over the course of the 18 months of sunflower honey storage, there was a significant increase in HMF concentration and a decline in diastase activity (Table 3, Figures 4 and 5).

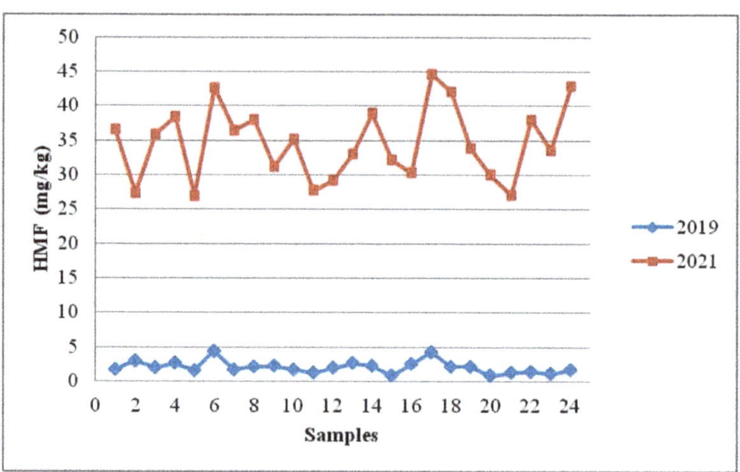

Figure 4. Effect of storage time on HMF content in sunflower honey.

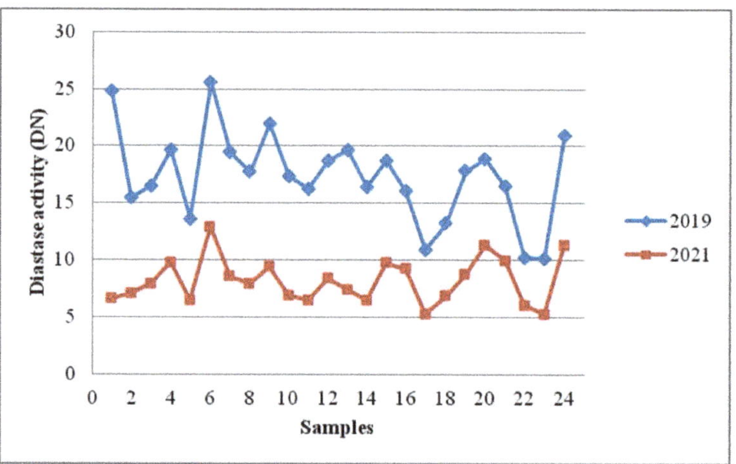

Figure 5. Effect of storage time on diastase activity in sunflower honey.

The most significant changes after storage for 18 months can be observed in the content of HMF ($p = 1.9 \times 10^{-31}$) and diastase activity ($p = 2.8 \times 10^{-13}$). Four samples (16.7%) of the 24 honey samples examined did not meet the national regulation for honey in terms of HMF level (maximum permitted is 40 mg/kg, mean HMF content 34.67 ± 5.36 mg/kg) (Table 3). HMF was present in concentrations ranging from 26.98 to 44.65 mg/kg. Diastase activity was lower than the minimum allowed (8 DN) in 13 out of 24 samples (54.2%). However, 18 months after being kept at room temperature, the HMF content increased by about 17 times (from a mean of 2.08 ± 0.91 to a mean of 34.67 ± 5.36 mg/kg). At the same time, diastase activity decreased by 2 times (from a mean of 17.35 ± 3.97 to a mean of 8.18 ± 1.99 DN). Nevertheless, 18 months after being kept at room temperature, nine samples (37.5%) of sunflower honey still presented acceptable values for HMF and diastase activity. Similar results of honey stability testing at room temperature during 18 months of storage were published by Seraglio et al. [41], Czipa et al. [5], Soares et al. [11], Korkmaz and Küplülü [49], Hasan [50], and Fallico et al. [51]. Da Silva et al. [4] and Chou et al. [19] reported that HMF and diastase activity did not change significantly 18 months after being kept at room temperature (20 ± 4 °C and 25 °C, respectively).

The potential for HMF formation is greater in honey that is more acidic than in those that have a higher pH value, such as darker honey [49,52]. Generally, the pH of honey is typically between 3.2 and 4.5 and depends on the content of organic acids. The mean pH value in the present paper was 3.66 ± 0.14 (ranging from 3.46 to 3.99). Based on this, it can be concluded that sunflower honey belongs to the group of honeys that are more acidic. The increase in HMF after storage is a possible consequence of the lower pH value of fresh honey. However, the small range of pH values was not suitable for confirming the linear dependence of pH and HMF. Our assumption that the increase in HMF concentration after storage for 18 months could be a consequence of the low initial pH value that was not confirmed by statistical analysis ($R^2 = 0.0289$).

3.7. Water-Insoluble Matter

Five samples (20.8%) of the twenty-four honey samples that were analyzed did not meet the requirements of national regulation for honey regarding the content of water-insoluble matter (maximum permitted is 0.1%) (Table 1). The insoluble matter remains after the extraction, centrifugation, and filtration of honey. This parameter provides data on the content of solids such as wax particles, parts of bee bodies, bee larvae, particles of plant origin, soil, and dust [53]. In 11 of the 24 honey samples, the content of insoluble matter was below 0.01%, and in 5 samples was between 0.01 and 0.03%. These data show that most beekeepers carry out the filtering operation more carefully. Albu et al. [53] found similar results in honey samples from Romania (20.59% of samples exceeded the 0.1% maximum permissible level). Taking into account that the examined samples were raw honey, i.e., honey that is not for sale as a commercial product, it is reasonable to assume that the honey is of high quality.

4. Conclusions

According to our findings, Serbian sunflower honey is of good quality. The closely connected quality-related criteria under examination revealed values that were consistent with the limitations imposed by national and European legislations. Those samples exceeding the insoluble matter content upper limit of 0.1% do not present a health risk for consumers. Taking into account the crystallization indices, we concluded that sunflower honey is a rapidly crystallizing honey.

The 18 months of storage at room temperature increased the concentration of HMF and decreased the activity of diastase, as well as the pH and the moisture content of sunflower honey. The acidity of sunflower honey was unaffected by storage.

In general, storage for 18 months at room temperature affects the quality of sunflower honey, thus, storing this type of honey at lower temperatures appears to be required. In order to predict the shelf life of sunflower honey and preserve the natural qualities of honey for as long as possible, additional research on the effects of storage time and storage temperature is required.

Author Contributions: Conceptualization, M.Ž.B. and D.L.P.; methodology, S.J., N.P. and Ž.M.; validation, S.J. and Ž.M.; formal analysis, N.P.; investigation, M.P.; resources, R.R.; data curation, M.P.; writing—original draft preparation, M.Ž.B.; writing—review and editing, D.L.P.; visualization, R.R.; supervision, D.L.P. and M.Ž.B. All authors have read and agreed to the published version of the manuscript.

Funding: This research was funded by the Ministry of Science, Technological Development and Innovation of Republic of Serbia by the Contract of implementation and funding of research work of NIV-NS in 2023, contract no.: 451-03-47/2023-01/200031.

Data Availability Statement: Data are contained within the article.

Conflicts of Interest: The authors declare no conflict of interest. The study's design, data collection, analysis, and interpretation, the drafting of the report, and the decision to publish the findings were all performed independently of the funder.

References

1. Przybyłowski, P.; Wilczyńska, A. Honey as an environmental marker. *Food Chem.* **2001**, *74*, 289–291. [CrossRef]
2. Seraglio, S.K.T.; Silva, B.; Bergamo, G.; Brugnerotto, P.; Gonzaga, L.V.; Fett, R.; Costa, A.C.O. An overview of physicochemical characteristics and health-promoting properties of honeydew honey. *Food Res. Int.* **2019**, *119*, 44–66. [CrossRef] [PubMed]
3. Živkov Baloš, M.; Popov, N.; Prodanov Radulović, J.; Stojanov, I.; Jakšić, S. Sugar profile of different floral origin honeys from Serbia. *J. Apic. Res.* **2020**, *59*, 398–405. [CrossRef]
4. da Silva, P.M.; Gonzaga, L.V.; Biluca, F.C.; Schulz, M.; Vitali, L.; Micke, G.A.; Costa, A.C.O. Stability of Brazilian *Apis mellifera* L. honey during prolonged storage: Physicochemical parameters and bioactive compounds. *LWT* **2020**, *129*, 109521. [CrossRef]
5. Czipa, N.; Philips, C.J.C.; Kovács, B. Composition of acacia honey following processing, storage, and adulteration. *J. Food Sci. Technol.* **2019**, *56*, 1245–1255. [CrossRef]
6. Eshete, Y.; Eshete, T.A. A Review of the effect of processing temperature and time duration on commercial honey quality. *Madr. J. Food Technol.* **2019**, *4*, 158–162. [CrossRef]
7. Živkov Baloš, M.; Jakšić, S.; Popov, N.; Mihaljev, Ž.; Ljubojević Pelić, D. Comparative study of water content in honey produced in different years. *Arch. Vet. Med.* **2019**, *12*, 43–53. [CrossRef]
8. Srinual, K.; Intipunya, P. Effects of crystallization and processing on sensory and physicochemical qualities of Thai sunflower honey. *Asian J. Food Agro-Ind.* **2009**, *2*, 749–754.
9. Sari, E.; Ayyildiz, N. Biological activities and some physicochemical properties of sunflower honeys collected from the Thrace region of Turkey. *Pak. J. Biol. Sci.* **2012**, *15*, 1102–1110. [CrossRef]
10. Tosi, E.; Martinet, R.; Ortega, M.; Lucero, H.; Ré, E. Honey diastase activity is modified by heating. *Food Chem.* **2008**, *106*, 883–887. [CrossRef]
11. Soares, S.; Pinto, D.; Rodigues, F.; Alves, R.C.; Oliveira, M.B.P.P. Portuguese honeys different geographical and botanical origins: A 4-year stability study regarding quality parameters and antioxidant activity. *Molecules* **2017**, *22*, 1338. [CrossRef] [PubMed]
12. Moreira, R.F.A.; De Maria, C.A.B.; Pietroluongo, M.; Trugo, L.C. Chemical changes in the volatile fractions of Brazilian honeys during storage under tropical conditions. *Food Chem.* **2010**, *104*, 1236–1241. [CrossRef]
13. Evahelda Pratama, F.; Malahayati, N.; Santoso, B. The changes in moisture content, pH, and total sugar content of honey originated from the flowers of the Bangka rubber tree during storage. *Int. J. Sci. Eng. Res.* **2017**, *5*, 33–36.
14. Kamboj, R.; Sandhu, R.S.; Kaler, R.S.S.; Bera, M.B.; Nanda, V. Optimization of process parameters on hydroxymethylfurfural content, diastase and invertase activity of coriander honey. *J. Food Sci. Technol.* **2019**, *56*, 3205–3214. [CrossRef] [PubMed]
15. Shapla, U.M.; Solayman, M.; Alam, N.; Khalil, I.; Hua Gan, S. 5-Hydroximethylfurfural (HMF) levels in honey and other food products: Effects on bees and human health. *Chem. Cent. J.* **2018**, *12*, 35. [CrossRef]
16. Mouhoubi-Tafinine, Z.; Ouchemoukh, S.; Bachir, M.; Louaileche, H.; Tamendjari, A. Effect of storage on hydroxymethylfurfural (HMF) and color of some Algerian honey. *Int. Food Res. J.* **2018**, *25*, 1044–1050.
17. Khan, Z.S.; Nanda, V.; Bhat, M.S.; Khan, A. Kinetic studies of HMF formation and diastase activity in two different honeys of Kashmir. *Int. J. Curr. Microbiol. Appl. Sci.* **2015**, *4*, 97–106.
18. Sousa, J.M.B.; Souza, E.L.; Marques, G.; Benassi, M.; Gullon, B.; Pintado, M.; Magnani, M. Sugar profile, physicochemical and sensory aspects of monofloral honeys produced by different stingless bee species in Brazilian semi-arid region. *LWT* **2016**, *65*, 645–651. [CrossRef]
19. Chou, W.; Liao, H.; Yang, Y.; Peng, C. Evaluation of Honey Quality with Stored Time and Temperatures. *J. Food Nutr. Res.* **2020**, *8*, 591–599. [CrossRef]
20. Živkov Baloš, M.; Popov, N.; Vidaković, S.; Ljubojević Pelić, D.; Pelić, M.; Mihaljev, Ž.; Jakšić, S. Electrical conductivity and acidity of honey. *Arch. Vet. Med.* **2018**, *11*, 91–101. [CrossRef]
21. Prica, N.; Živkov-Baloš, M.; Jakšić, S.; Mihaljev, Ž.; Kartalović, B.; Babić, J.; Savić, S. Moisture and acidity as indicators of the quality of honey originating from Vojvodina region. *Arch. Vet. Med.* **2014**, *7*, 99–109. [CrossRef]
22. Martínez, R.A.; Schvezov, N.; Brumovsky, L.A.; Pucciarelli Román, A.B. Influence of temperature and packaging type on quality parameters and antimicrobial properties during Yateí honey storage. *Food Sci. Technol.* **2018**, *38*, 196–202. [CrossRef]
23. Turhan, I.; Tetik, N.; Karhan, M.; Gurel, F.; Tavukcuoglu, H.R. Quality of honeys influenced by thermal treatment. *LWT* **2008**, *41*, 1396–1399. [CrossRef]

24. Kabbani, D.; Sepulcre, F.; Wedekind, J. Ultrasound-assisted liquefaction of rosemary honey: Influence on rheology and crystal content. *J. Food Eng.* **2011**, *107*, 173–178. [CrossRef]
25. Dobre, I.; Georgescu, I.A.; Alexe, P.; Escuerdo, O.; Seijo, M.C. Rheological behavior of different honey types from Romania. *Food Res. Int.* **2012**, *49*, 126–132. [CrossRef]
26. Laos, K.; Kirs, E.; Pall, R.; Martverk, K. The crystallization behavior of Estonian honeys. *Agron. Res.* **2011**, *9*, 427–432.
27. Al-Ghamdi, A.; Mohammed, S.E.A.; Ansari, M.J.; Adgaba, N. Comparison of physicochemical properties and effects of heating regimes on stored *Apis mellifera* and *Apis Florea* honey. *Saudi Biol. Sci.* **2019**, *26*, 845–848. [CrossRef] [PubMed]
28. Gao, H.F.; Wen, X.S.; Xian, C.J. Hydroxymethyl furfural in Chinese herbal medicines: Its formation, presence, metabolism, bioactivities and implications. *Afr. J. Tradit. Complem. Altern. Med.* **2015**, *12*, 43–54. [CrossRef]
29. Fallico, B.; Zappalá, M.; Arena, E.; Verzera, A. Effects of conditioning on HMF content in unifloral honeys. *Food Chem.* **2004**, *85*, 305–313. [CrossRef]
30. Kędzierska-Matysek, M.; Florek, M.; Wolanciuk, A.; Skałecki, P. Effect of freezing and room temperature storage for 18 months on quality of raw rapeseed honey (*Brassica napus*). *J. Food Sci. Technol.* **2016**, *53*, 3349–3355. [CrossRef]
31. Kowalski, S.; Lukasiewicz, M.; Duda-Chodak, A.; Zięć, G. 5-Hydroximethyl-2-Furfural (HMF)—Heat-Induced Formation, Occurrence in Food and Biotransformation—A Review. *Polish J. Food Nutr. Sci.* **2013**, *63*, 207–225. [CrossRef]
32. Bhattacharjee, D.; Patriwar, P.; Ravandale, V.; Bandekar, S.H.; Das, B.; Guduri, B.R.; Alam, T. Effect of lamitubes packaging on the shelf life of honey. *J. Package Technol. Res.* **2021**, *5*, 23–28. [CrossRef]
33. Olaitan, P.B.; Olufemi, E.A.; Ola, I.O. Honey: A reservoir for microorganisms and an inhibitory agent for microbes. *Afr. Health Sci.* **2007**, *7*, 159–165.
34. Fuad, A.M.A.; Anwar, N.Z.R.; Zakaria, A.J.; Shahidan, N.; Zakaria, Z. Physicochemical characteristics of Malaysian honeys influenced by storage time and temperature. *J. Fundam. Appl. Sci.* **2017**, *9*, 841–851. [CrossRef]
35. International Honey Commission. *World Network of Honey Science. Harmonized Methods of the International Honey Commission*; Swiss Bee Research Centre; FAM: Liebefeld, Switzerland, 2009.
36. Official Gazette RS. *Rulebook on Quality of Honey and Other Bee Products*; No. 101; Official Gazette RS: Belgrade, Serbia, 2015. Available online: https://www.pravno-informacioni-sistem.rs/SlGlasnikPortal/eli/rep/sgrs/ministarstva/pravilnik/2015/101/2 (accessed on 25 May 2023).
37. Juan-Borrás, M.; Domenech, E.; Hellebrandova, M.; Esriche, I. Effect of country origin on physicochemical, sugar and volatile composition of acacia, sunflower and tilia honeys. *Food Res. Int.* **2014**, *60*, 86–94. [CrossRef]
38. Lazarević, K.B.; Andrić, F.; Trifković, J.; Tešić, Ž.L.J.; Milojković-Opsenica, D.M. Characterisation of Serbian unifloral honeys according to their physicochemical parameters. *Food Chem.* **2012**, *132*, 2060–2064. [CrossRef]
39. Thrasyvoulou, A.; Manikis, J. Some physicochemical and microscopic characteristics of Greek unifloral honeys. *Apidologie* **1995**, *26*, 441–452. [CrossRef]
40. Acquarone, C.; Buera, P.; Elizalde, B. Pattern of pH and electric conductivity upon honey dilution as a complementary tool for discriminating geographical origin of honeys. *Food Chem.* **2007**, *101*, 695–703. [CrossRef]
41. Seraglio, S.K.T.; Bergamo, G.; Molognoni, L.; Daguer, H.; Silva, B.; Gonzaga, L.V.; Fett, R.; Costa, A.C.O. Quality changes during long-term storage of a peculiar Brazilian honeydew honey: "Bracatinga". *J. Food Compos. Anal.* **2021**, *97*, 103769. [CrossRef]
42. Council Directive of the European Union Council Directive 2001/110/EC of 20 December 2001 relating to honey. *Off J. Eur. Union.* **2002**, *10*, 47–52. Available online: https://eur-lex.europa.eu/legal-content/EN/ALL/?uri=celex%3A32001L0110 (accessed on 27 May 2023).
43. Devillers, J.; Morlot, M.; Pham-Delègue, M.H.; Doré, J.C. Classification of monofloral honeys on their quality control data. *Food Chem.* **2004**, *86*, 305–312. [CrossRef]
44. Escuredo, O.; Dobre, I.; Fernández-Gonzálezm, M.; Seijo, M.C. Contribution of botanical origin and sugar composition of honey on the crystallization phenomenon. *Food Chem.* **2014**, *149*, 84–90. [CrossRef]
45. Venir, E.; Spaziani, M.; Maltini, E. Crystallization in "Tarassaco" Italian honey studied by DSC. *Food Chem.* **2010**, *122*, 410–415. [CrossRef]
46. Đogo Mračević, S.; Krstić, M.; Lolić, A.; Ražić, S. Comparative study of the chemical composition and biological potential of honey from different regions of Serbia. *Microchem. J.* **2020**, *152*, 104420. [CrossRef]
47. Cimpoiu, C.; Hosu, A.; Miclaus, V.; Puscas, A. Determination of the floral origin of some Romanian honeys on the basis of physical and biochemical properties. *Spectrochim. Acta A Mol. Biomol. Spectrosc.* **2013**, *100*, 149–154. [CrossRef]
48. Yadata, D. Detection of the electrical conductivity and acidity of honey from different areas of Tepi. *Food Sci. Technol.* **2014**, *2*, 59–63. [CrossRef]
49. Korkmaz, S.D.; Küplülü, Ö. Effects of storage temperature on HMF and diastase activity of strained honeys. *Ank. Univ. Vet. Fak. Derg.* **2017**, *64*, 281–287.
50. Hasan, S.H. Effect of storage and processing temperature on honey quality. *JUBAS* **2013**, *6*, 2244–2253.
51. Fallico, B.; Arena, E.; Zappalá, M. Prediction of honey shelf life. *J. Food Qual.* **2009**, *32*, 352–368. [CrossRef]

52. Sajid, M.; Yasmin, T.; Asad, F.; Samina, Q. Changes in HMF content and diastase activity in honey after heating treatment. *Pure Appl. Bio* **2019**, *8*, 1668–1674. [CrossRef]
53. Albu, A.; Radu-Rusu, C.G.; Pop, I.M.; Frunza, G.; Nacu, G. Quality Assessment of Raw Honey Issued from Eastern Romania. *Agriculture* **2021**, *11*, 247. [CrossRef]

Communication

Metabolomics Reveals Distinctive Metabolic Profiles and Marker Compounds of Camellia (*Camellia sinensis* L.) Bee Pollen

Dandan Qi [1,2], Meiling Lu [3], Jianke Li [1] and Chuan Ma [1,*]

[1] State Key Laboratory of Resource Insects, Institute of Apicultural Research, Chinese Academy of Agricultural Sciences, No. 2 Yuanmingyuan West Road, Haidian District, Beijing 100193, China; qidandan07@126.com (D.Q.); apislijk@126.com (J.L.)

[2] Tea Research Institute, Shangdong Academy of Agricultural Sciences, Jinan 250000, China

[3] Agilent Technologies (China) Co., Ltd., Beijing 100102, China; mei-ling.lu@agilent.com

* Correspondence: machuan@caas.cn

Abstract: Camellia bee pollen (CBP) is a major kind of bee product which is collected by honeybees from tea tree (*Camellia sinensis* L.) flowers and agglutinated into pellets via oral secretion. Due to its special healthcare value, the authenticity of its botanical origin is of great interest. This study aimed at distinguishing CBP from other bee pollen, including rose, apricot, lotus, rape, and wuweizi bee pollen, based on a non-targeted metabolomics approach using ultra-high performance liquid chromatography–mass spectrometry. Among the bee pollen groups, 54 differential compounds were identified, including flavonol glycosides and flavone glycosides, catechins, amino acids, and organic acids. A clear separation between CBP and all other samples was observed in the score plots of the principal component analysis, indicating distinctive metabolic profiles of CBP. Notably, L-theanine (864.83–2204.26 mg/kg) and epicatechin gallate (94.08–401.82 mg/kg) were identified exclusively in all CBP and were proposed as marker compounds of CBP. Our study unravels the distinctive metabolic profiles of CBP and provides specific and quantified metabolite indicators for the assessment of authentic CBP.

Keywords: camellia bee pollen; L-theanine; epicatechin gallate; marker compound; botanical origin

Citation: Qi, D.; Lu, M.; Li, J.; Ma, C. Metabolomics Reveals Distinctive Metabolic Profiles and Marker Compounds of Camellia (*Camellia sinensis* L.) Bee Pollen. *Foods* **2023**, *12*, 2661. https://doi.org/10.3390/foods12142661

Academic Editor: Olga Escuredo

Received: 13 June 2023
Revised: 5 July 2023
Accepted: 6 July 2023
Published: 11 July 2023

1. Introduction

Bee pollen is collected from plant flowers and agglutinated into pellets by honeybees via oral secretion. As an indispensable nutrient source for honeybee development, bee pollen is rich in carbohydrates, proteins, amino acids, polyphenols, lipids, minerals, and vitamins [1]. Its chemical composition varies considerably according to its botanical origins [1,2]. The large amounts of bioactive constituents endow bee pollen with health beneficial properties, such as antioxidant, anti-inflammatory, anti-allergen, anti-aging, and anti-cancer effects [3–5]. Owing to its nutritional and therapeutic properties, bee pollen has gained increasing attention worldwide and is commercially consumed as a natural dietary supplement for human health promotion [6,7].

Camellia bee pollen (CBP) is among the most important bee pollen products that are extensively consumed in China. It is gathered by honeybees from the flowers of tea plants (*Camellia sinensis* L.), the leaves of which can be made into tea, a popular beverage worldwide with various health benefits. It has been reported that the chemical constituents of tea flowers are similar to those of tea leaves [8]. A share of some common bioactive constituents and functional properties between CBP and tea can thus be expected. Indeed, CBP has a special fragrance, similar to the aroma of tea. Moreover, it has been demonstrated that CBP possesses higher anti-inflammatory, antioxidant, and anti-tyrosinase activities relative to other types of bee pollen [9–11]. In recent years, the identification of bioactive constituents responsible for the observed functional properties has been attracting growing interest. Among them, caffeine, kaempferol, levulinic acid, and 5-hydroxymethyl furfural are reported to contribute partly to the anti-tyrosinase activities of CBP [12–14]. However, the metabolic basis for its functional properties is still far from being fully understood, thereby impeding the use of CBP in the cosmetics, food, and pharmaceutical industries.

The aforementioned superior functional properties promote increasing demand for CBP, which leads to fraudulent practices in the market [15]. To identify bee pollen of different botanical origins, sensory testing (e.g., color, aroma, and taste characteristics) and microscopic examination (e.g., size, form, and color of pollen grains) are widely used [16]. However, such subjective judgments based on sensory evaluation are easily biased by personal preference. Moreover, even with a microscope, it is still difficult to distinguish between different types of bee pollen with similar morphological and structural attributes [17,18]. The situation is even worse for CBP, which shows substantial morphological variation between tea cultivars [19]. The lack of accurate identification methods represents a loophole for the current adulteration chaos of bee pollen. A more sensitive method is, thus, urgently needed to ensure accurate identification for the long-term development of the bee product industry.

Non-targeted metabolomics based on high-resolution mass spectrometry provides a convenient method for the simultaneous analysis of hundreds or thousands of small molecules in food products, including various bee products [20–23]. This approach, combined with targeted metabolomics, plays a key role in screening and quantifying marker compounds for food authenticity [24–26]. For bee pollen authenticity, however, such research is currently limited [27].

Our study aimed to uncover distinctive metabolic components of CBP and to explore efficient metabolite indicators to identify authentic CBP. To this end, non-targeted metabolic profiling of CBP and other types of bee pollen was performed. We proposed epicatechin gallate (ECG) and L-theanine as marker compounds of CBP, and measured their content based on ultra-high performance liquid chromatography-quadruple-Exactive Orbitrap mass spectrometry (UHPLC-Q-Exactive Orbitrap-MS).

2. Materials and Methods

2.1. Reagents and Standards

Ultrapure water was produced using a Milli-Q water purification system (Millipore, St. Louis, MA, USA). Methanol of LC-MS grade was purchased from Merck (Darmstadt, Germany). Acetonitrile, formic acid, and ammonium formate of LC–MS grade were purchased from Thermo Fisher Scientific (Waltham, MA, USA). All of the authentic standards used for qualification are listed in Table S1.

2.2. Bee Pollen Sample Collection

Fifteen CBP samples were collected from Anhui, Fujian, Jiangsu, Sichuan, and Zhejiang Provinces in China ($n = 3$ for each), while fifteen non-CBP samples were obtained from five botanical plants ($n = 3$ for each; Table S1), i.e., rose (*Rosa rugosa* Thunb.), apricot (*Prunus armeniaca* L.), lotus (*Nelumbo nucifera* Gaertn.), rape (*Brassica campestris* L.), and wuweizi (*Frucus Schisandra chinensis*). To guarantee their authenticity, these samples were collected by professional beekeepers from their apiaries of *Apis mellifera* L. colonies using pollen traps, and were then identified using a scanning electron microscope (S-4800, Hitachi, Tokyo, Japan). Dead bee parts and other hive debris were removed manually. All samples were freeze-dried and stored at $-80\ ^{\circ}\text{C}$ until analysis.

2.3. Preparation of Bee Pollen Extracts

In brief, 25 mL of 80% methanol was added to accurately weighed 0.5 g samples of powdered bee pollen in a 50 mL vial. After supersonic extraction for 1.5 h at 4 °C, the mixture was kept still for 30 min, followed by 0.22 μm membrane filtration (Shimadzu,

Shanghai, China). Tolbutamide and sulfacetamide, at final concentrations of 2 μg/mL and 4 μg/mL, respectively, were added as internal standards for retention time correction.

2.4. UHPLC-QTOF/MS-Based Non-Targeted Metabolomics Analysis

Non-targeted metabolomics analysis was performed on an Infinity 1290 UHPLC system (Agilent Technologies, Santa Clara, CA, USA) coupled to an Agilent 6545 QTOF mass spectrometer (Agilent Technologies, Santa Clara, CA, USA). Chromatographic separation was carried out on a Zorbax Eclipse Plus C18 column (3.0 × 150 mm, 1.8 μm, Agilent Technologies, Santa Clara, CA, USA) at 40 °C. Water with 5 mmol/L ammonium acetate and methanol with 5 mmol/L ammonium acetate were used for mobile phases A and B, respectively, which were kept at a flow rate of 0.40 mL/min with a gradient elution profile. The proportion of solvent B was linearly applied as follows: 0–5 min, 0–12%; 5–15 min, 12–35%; 15–18 min, 35–45%; 18–26 min, 45–75%; 26–33 min, 75–95%; and 33–35 min, 95–5%. The injection volume was 3 μL. The post-time between each two consecutive injections was 3 min. Dual jet stream electrospray ionization (ESI) was performed under negative ionization mode. The parameters were as follows: nebulizer pressure: 35 psi; capillary voltage: 3.5 kV; gas flow rate: 8 L/min; fragmentator voltage: 130 V; sheath gas temperature: 350 °C; and sheath gas flow rate: 8 L/min. The TOF scan was set at an m/z of 100–1100 with an acquisition rate of 2 spectra per second. The auto MS/MS model was applied for compound identification with fixed collision energies (10 V, 20 V and 40 V). Reference ions with m/z 112.9856 and 1033.9881 were utilized for real-time mass calibration during both the TOF scan and the auto MS/MS scan.

The obtained raw data were imported into Masshunter Qualitative Analysis software (B.07.00 SP1, Agilent Technologies, Santa Clara, CA, USA) to extract all feature ions, then exported as .cef documents. These .cef documents were imported into MPP (Mass Profiler Professional software package, version B.14.5, Agilent Technologies, Santa Clara, CA, USA) for retention time correction using internal standards and subsequent peak alignment within the specified retention time window (±2.5%). The entities with an occurrence frequency >60% and a coefficient of variability (CV) < 25% were retained. After Pareto scaling and logarithmic transformation of the quantitative data, principal component analysis (PCA) was performed using SIMCA 14.1 (Umetrics AB, Umeå, Sweden) to provide an intuitionistic demonstration of an overall clustering pattern of the bee pollen samples.

Differential entities ($p < 0.05$ in analysis of variance, ANOVA) among the bee pollen samples from different botanical origins were identified using SPSS 20.0 (Chicago, IL, USA) for further analysis. Metabolite identification was performed by searching for exact mass and MS/MS spectra in the Metlin database (http://metlin.scripps.edu, accessed on 10 January 2021) and Human Metabolome Database (HMDB, https://hmdb.ca/, accessed on 10 January, 2021). The retention time and MS/MS spectra of putatively identified compounds were validated by authentic standards analyzed under the same conditions. To show the abundance differences in identified compounds among these bee pollen samples, heatmap visualization was carried out using MetaboAnalyst 4.0 [28] with Pareto scaling and logarithmic transformation. To improve the classification of CBP and non-CBP samples, orthogonal projections to latent structures discriminant analysis (OPLS-DA) was conducted in SIMCA 14.1. The OPLS-DA model was cross-validated by permutation tests with 200 iterations. The values of variable importance in projection (VIP) were used to rank the overall contribution of each compound to the OPLS-DA model. Compounds with VIP > 1.0, $p < 0.05$ according to Student's t-test and fold change (FC) > 1.5 were regarded as discriminating compounds driving the observed group separation.

2.5. Targeted Quantification of ECG and L-Theanine in CBP

The contents of ECG and L-theanine in CBP were determined using the UHPLC system (Dionex Ultimate 3000, Thermo Scientific, Waltham, CA, USA) coupled with Q-Exactive Orbitrap-MS (Thermo Scientific, Waltham, CA, USA) in parallel reaction monitoring (PRM) mode. Compound separation was performed using hydrophilic interaction liquid chromatography (HILIC) with an ACQUITY BEH Amide column (150 mm × 2.1 mm, 1.7 μm, Waters, Milford, MA, USA) at 50 °C. Mobile phase A and B were 30% and 95% acetonitrile, respectively, both containing 10 mmol/L ammonium formate and 0.1% formic acid. A gradient elution profile at a flow rate of 0.3 mL/min was set as follows: 1 min, 100% B; 11 min, 30% B; 11.5 min, 100% B; and 15 min, 100% B. The injection volume was 2 μL. The Q-Exactive Orbitrap-MS with a heated HESI source was operated in negative mode with stepped normalized collision energies (NCE 10, 20, and 30). The mass resolution was 70,000 full width at half maximum (FWHM) for the full MS mode and 17,500 FWHM for the MS/MS scan. Calibration curves were established with standard solutions (0.5, 1.0, 2.0, 4.0, 6.0, 8.0, and 10.0 μg/mL for ECG and 5, 10, 20, 40, 60, 80, and 100 μg/mL for L-theanine) and used to calculate the ECG and L-theanine content in Xcalibur (v4.0.27.19, Thermo Fisher Scientific, Waltham, MA, USA).

3. Results

3.1. Overall Metabolic Profiles of the Bee Pollen

To obtain an overview of grouping patterns of the bee pollen samples of different botanical origins, 670 valid entities were submitted with which to perform unsupervised PCA. The first two principal components of PCA explained 74.46% of the total variance (PC1 = 65.70% and PC2 = 8.76%). In the PCA score plots (Figure 1A), a tight clustering of the pollen samples from the same botanical origins was observed, and their distribution patterns were found to be affected by their botanical origins. Remarkably, a clear separation between the CBP and non-CBP samples was observed. Specifically, the CBP samples were all distributed in the negative part of the PC1 axis, while the non-CBP samples were located in the positive part of the PC1 axis and separated along the PC2 axis.

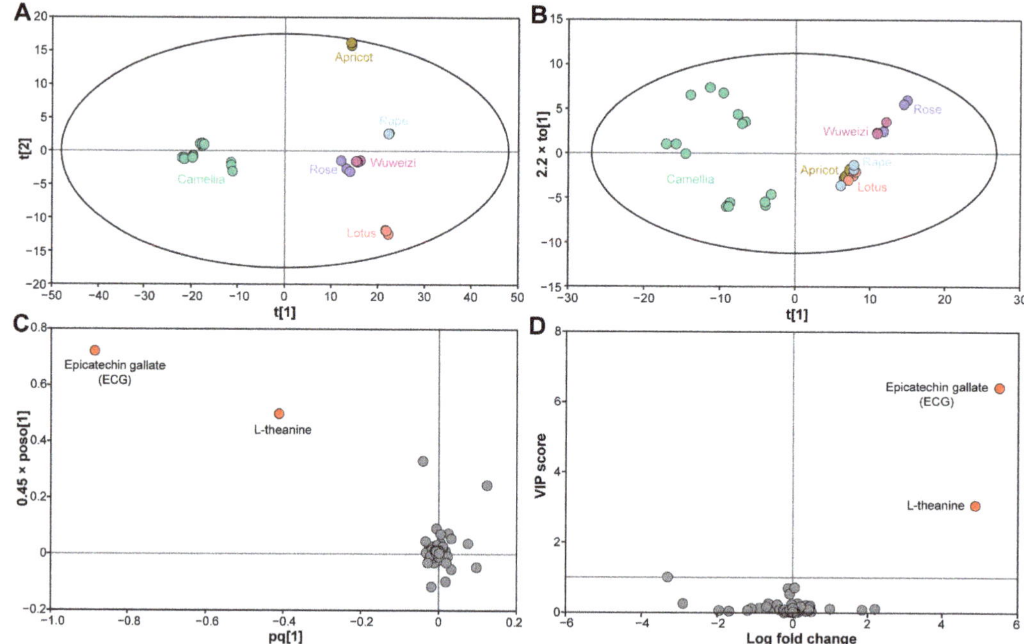

Figure 1. Multivariate and univariate analysis of CBP and non-CBP (rose, apricot, lotus, rape, and wuweizi bee pollen) metabolomics data. PCA score plots were generated from all valid entities (**A**); score plots (**B**) and loading plots (**C**) of OPLS-DA for the 54 identified compounds. The ellipses indicate 95% confidence limits according to Hotelling's T2 statistics. VIP score and log fold change (CBP/non-CBP) of the identified compounds are shown (**D**).

3.2. Metabolite Identification

To screen the compounds explaining the overall grouping patterns observed in the PCA score plots, differential entities among the bee pollen samples ($p < 0.05$ in ANOVA) were subjected to compound identification. Finally, 54 compounds were identified, including 15 flavonol glycosides and flavone glycosides, 3 catechins, 11 amino acids, 8 organic acids, 4 fatty acids, 4 nucleotides and their derivatives, 2 aldehydes, and 7 other compounds (Table S1). Among them, four compounds were detected exclusively in all CBP samples, including ECG, L-theanine, gallic acid (GA), and kaempferol. The CBP samples were all clustered into a single clade, whereas the non-CBP samples formed a different clade in the clustering heatmap of the identified compounds (Figure 2).

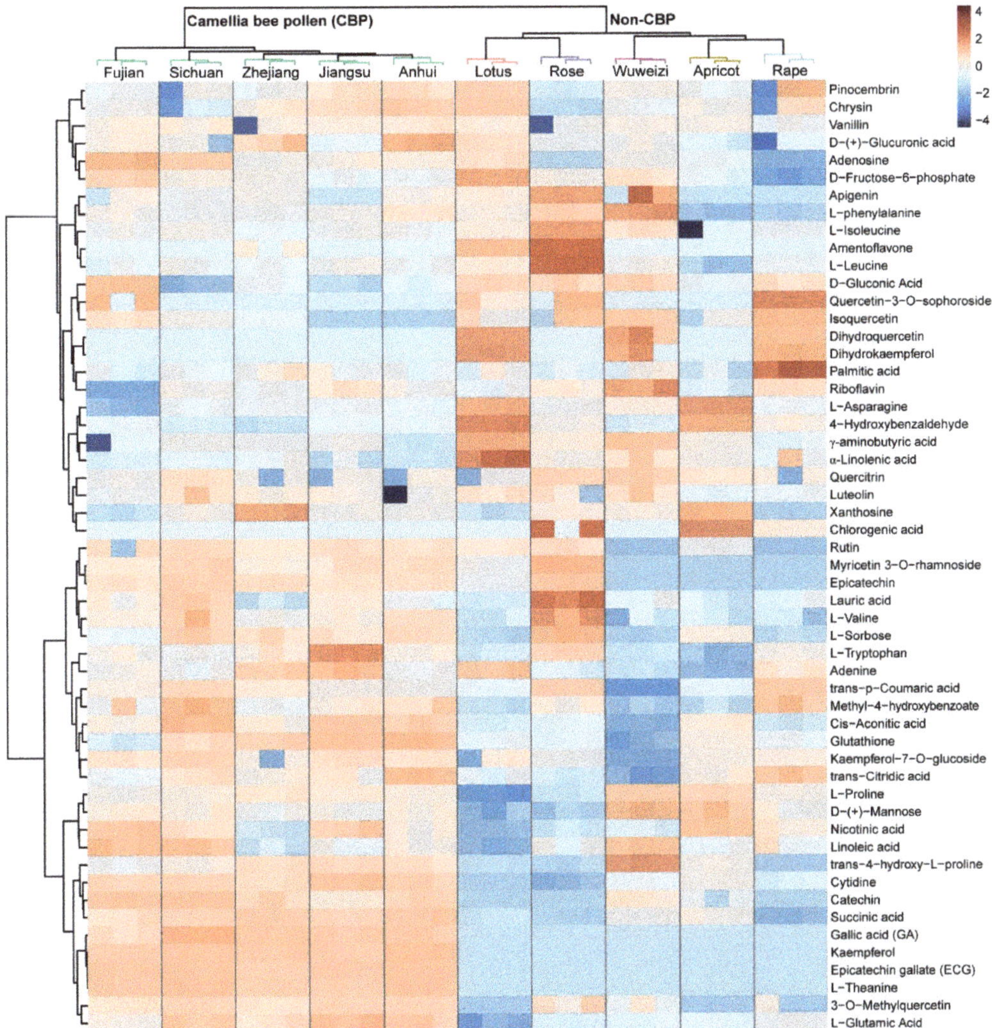

Figure 2. Clustering heatmap of the 54 identified compounds in the 30 bee pollen samples.

3.3. Marker Compound Selection of CBP

To pick out the most discriminating compounds between the CBP and non-CBP samples, univariate and multivariate analyses were conducted based on the relative abundance levels of the 54 identified compounds (Table S2). In total, 16 compounds with FC > 1.5 showed a significant difference ($p < 0.05$) between the CBP and non-CBP samples (Table S1). A reliable OPLS-DA model was established ($R^2Y = 0.873$, $Q^2 = 0.845$, R^2 intercepts = 0.0902, and Q^2 intercepts = -0.4112 in a 200-time permutation test), and the resulting score plots supported a clear separation between the CBP and non-CBP samples (Figure 1B), as is consistent with the grouping patterns in the PCA score plots (Figure 1A). Further filtering with VIP values > 1.0 in the OPLS-DA resulted in a final selection of two compounds, i.e., ECG and L-theanine, which had the highest VIP values (Figure 1D). Moreover, ECG and L-theanine exhibited the greatest distance from the origin in the loading plots (Figure 1C) and, hence, had the highest discriminatory power. Taken together, ECG and L-theanine could be used as marker compounds to distinguish CBP from non-CBP samples.

3.4. ECG and L-Theanine Quantification

A targeted quantification method based on a PRM assay was carried out to measure the ECG and L-theanine content in the CBP samples. The established calibration curve showed good linearity for ECG ($r^2 = 0.9965$) and L-theanine ($r^2 = 0.9932$). A significant difference was observed in the ECG (94.08–401.82 mg/kg) and L-theanine (864.83–2204.26 mg/kg) content among the CBP samples (Figure 3). The highest ECG content was found in the CBP samples from Anhui Province (387.04–430.70 mg/kg), followed by Sichuan Province (272.52–280.56 mg/kg), and the lowest content was found in the samples from Zhejiang Province (90.95–98.33 mg/kg). The CBP samples from Sichuan Province had the highest L-theanine content (2138.56–2314.01 mg/kg), while those from Jiangsu and Zhejiang Provinces had the lowest content (819.45–964.43 mg/kg).

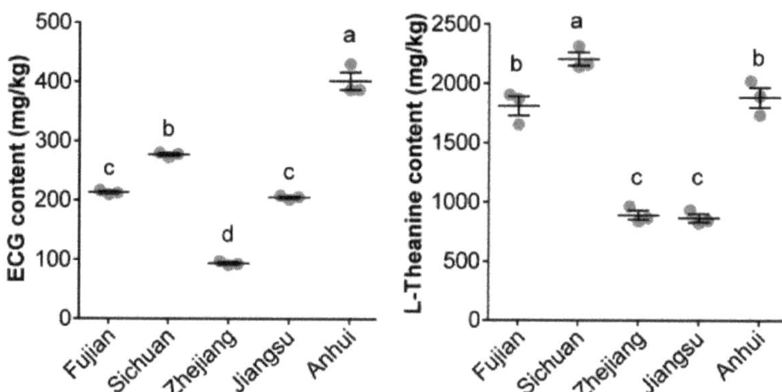

Figure 3. ECG and L-theanine content (mean ± standard error) in the CBP samples from different locations. Different letters indicate statistically significant differences ($p < 0.05$ in ANOVA).

4. Discussion

4.1. Distinctive Metabolic Profiles of CBP

As a mixture of flower pollen and honeybee saliva, bee pollen from different botanical origins shows differing chemical composition and functional properties [2,29]. Our metabolomics analysis confirmed the presence of different metabolic profiles of bee pollen samples according to botanical origins. Remarkably, one of our key findings was the distinctive metabolic profile of CBP samples compared with others. It has been reported that tea plants synthesize unique metabolites and transport quality-related components to their organs, including tea flowers [30], which could account for our observed special metabolic profile of CBP. Indeed, our study revealed four characteristic compounds (L-theanine, ECG, kaempferol, and GA) which were found to be specific to the CBP samples and 12 other differential compounds between CBP and non-CBP samples.

A wide diversity of bioactive functions has been reported for our identified differential compounds. Among them, L-theanine, a non-protein amino acid, is reported to naturally occur mainly in tea plants and shows a wide range of beneficial effects, such as antioxidant, anti-cancer, and immune-modulating activities [31,32]. ECG, a highly abundant catechin in green tea, has been demonstrated to possess antioxidant, anti-inflammatory, and anti-tumor effects [33]. Other polyphenolic compounds, such as kaempferol, GA, and rutin, show similar biological activities [34,35]. Collectively, these bioactive compounds account for at least some of the superior functional activities of CBP which have been reported in previous studies [9–11].

4.2. ECG and L-Theanine as Maker Compounds of CBP

Metabolomics approaches have been widely adopted for the global evaluation of marker compounds for food authenticity [24]. If some compounds are detected exclusively in certain bee products, or are significantly more abundant or profile-defining, they could be considered as markers of these products [36]. Typically, compounds with VIP scores greater than 1.0 in the OPLS-DA are generally considered to have the highest discrimination potential [37]. With these methods, marker compounds of bee products of different origins, such as honey [38] and propolis [23], have been proposed. In our study, ECG and L-theanine satisfied the conditions mentioned above (FC > 1.5, $p < 0.05$, and VIP score > 1.0), and could, thus, be regarded as the best potential candidates for CBP marker compounds. In addition, flavonoid glycosides, which are present in lower quantities in CBP, have been proposed to distinguish CBP from several kinds of bee pollen [17]. Unlike these flavonoid glycosides, ECG and L-theanine were found exclusively in CBP in our study. Based on the measured content in our study, a minimum content of 90.95 mg/kg for ECG and 819.45 mg/kg for L-theanine are required for the authentication of CBP. The combination of the two special components specific to CBP could, thus, assist in distinguishing CBP from adulterated CBP or other bee pollen.

It should be noted that significant variation in the content of both ECG (90.95–430.70 mg/kg) and L-theanine (819.45–2314.01 mg/kg) was observed in our CBP samples from different geographical locations. This finding could be explained by diverse tea germplasm resources and environmental conditions, which have been reported to affect L-theanine content in the young shoots of tea trees [39,40]. It is, thus, likely that the geographical origins of CBP could be predicted by means of ECG and L-theanine content after extensive sampling of CBP in future studies.

5. Conclusions

Our comparative metabolomics analysis revealed distinctive metabolic profiles of the CBP relative to other bee pollen, including rose, apricot, lotus, rape, and wuweizi bee pollen. Among the differential compounds, L-Theanine and ECG were detected exclusively in all the CBP samples, and showed the highest discriminatory power. Further quantification based on targeted metabolomics demonstrated the content of L-theanine (819.45–2314.01 mg/kg) and ECG (90.95–430.70 mg/kg) in the CBP samples. The feasibility of easy detection and quantification of ECG and L-theanine in bee pollen demonstrates their possible practical application as marker compounds for CBP authentication.

Supplementary Materials: The following supporting information can be downloaded at: https://www.mdpi.com/article/10.3390/foods12142661/s1, Table S1: Compounds identified in bee pollen samples. Table S2: Relative abundance of the identified compounds in bee pollen samples.

Author Contributions: D.Q.: conceptualization, formal analysis, writing—original draft. M.L.: formal analysis, writing—review and editing. J.L.: resources, supervision, validation, writing—review and editing. C.M.: project administration, formal analysis, writing—original draft. All authors have read and agreed to the published version of the manuscript.

Funding: The study was funded by the Agricultural Science and Technology Innovation Program in China (CAAS-ASTIP-2021-IAR) and the Earmarked Fund for Modern Agro-Industry Technology Research System in China (CARS-44).

Data Availability Statement: Data is contained within the article.

Conflicts of Interest: Author M.L. was employed by the company Agilent Technologies (China) Co. Ltd. The remaining authors declare that the research was conducted in the absence of any commercial or financial relationships that could be construed as a potential conflict of interest.

References

1. Thakur, M.; Nanda, V. Composition and functionality of bee pollen: A review. *Trends Food Sci. Technol.* **2020**, *98*, 82–106. [CrossRef]
2. Barbieri, D.; Gabriele, M.; Summa, M.; Colosimo, R.; Leonardi, D.; Domenici, V.; Pucci, L. Antioxidant, nutraceutical properties, and fluorescence spectral profiles of bee pollen samples from different botanical origins. *Antioxidants* **2020**, *9*, 1001. [CrossRef] [PubMed]
3. Algethami, J.S.; Abd El-Wahed, A.A.; Elashal, M.H.; Ahmed, H.R.; Elshafiey, E.H.; Omar, E.M.; Al Naggar, Y.; Algethami, A.F.; Shou, Q.Y.; Alsharif, S.M.; et al. Bee pollen: Clinical trials and patent applications. *Nutrients* **2022**, *14*, 2858. [CrossRef] [PubMed]
4. Khalifa, S.A.M.; Elashal, M.H.; Yosri, N.; Du, M.; Musharraf, S.G.; Nahar, L.; Sarker, S.D.; Guo, Z.M.; Cao, W.; Zou, X.B.; et al. Bee pollen: Current status and therapeutic potential. *Nutrients* **2021**, *13*, 1876. [CrossRef]
5. Margaoan, R.; Strant, M.; Varadi, A.; Topal, E.; Yucel, B.; Cornea-Cipcigan, M.; Campos, M.G.; Vodnar, D.C. Bee collected pollen and bee bread: Bioactive constituents and health benefits. *Antioxidants* **2019**, *8*, 568. [CrossRef]
6. Kostic, A.Z.; Milincic, D.D.; Barac, M.B.; Ali Shariati, M.; Tesic, Z.L.; Pesic, M.B. The application of pollen as a functional food and feed ingredient-The present and perspectives. *Biomolecules* **2020**, *10*, 84. [CrossRef]
7. Baky, M.H.; Abouelela, M.B.; Wang, K.; Farag, M.A. Bee pollen and bread as a super-food: A comparative review of their metabolome composition and quality assessment in the context of best recovery conditions. *Molecules* **2023**, *28*, 715. [CrossRef]
8. Chen, D.; Chen, G.J.; Sun, Y.; Zeng, X.X.; Ye, H. Physiological genetics, chemical composition, health benefits and toxicology of tea (*Camellia sinensis* L.) flower: A review. *Food Res. Int.* **2020**, *137*, e109584. [CrossRef]
9. Li, Q.Q.; Sun, M.H.; Wan, Z.R.; Liang, J.S.; Betti, M.; Hrynets, Y.; Xue, X.F.; Wu, L.M.; Wang, K. Bee pollen extracts modulate serum metabolism in lipopolysaccharide-induced acute lung injury mice with anti-inflammatory effects. *J. Agric. Food Chem.* **2019**, *67*, 7855–7868. [CrossRef]
10. Su, J.; Yang, X.Y.; Lu, Q.; Liu, R. Antioxidant and anti-tyrosinase activities of bee pollen and identification of active components. *J. Apic. Res.* **2021**, *60*, 297–307. [CrossRef]
11. Zhang, H.C.; Wang, X.; Wang, K.; Li, C.Y. Antioxidant and tyrosinase inhibitory properties of aqueous ethanol extracts from monofloral bee pollen. *J. Apic. Sci.* **2015**, *59*, 119–128. [CrossRef]
12. Yang, Y.F.; Sun, X.; Ni, H.; Du, X.P.; Chen, F.; Jiang, Z.D.; Li, Q.B. Identification and characterization of the tyrosinase inhibitory activity of caffeine from Camellia pollen. *J. Agric. Food Chem.* **2019**, *67*, 12741–12751. [CrossRef]
13. Yang, Y.F.; Lai, X.Y.; Huang, G.L.; Chen, Y.H.; Du, X.P.; Jiang, Z.D.; Chen, F.; Ni, H. Isolation and purification of two tyrosinase inhibitors from camellia pollen by high-speed counter-current chromatography. *Acta Chromatogr.* **2017**, *29*, 477–483. [CrossRef]
14. Yang, Y.F.; Lai, X.Y.; Lai, G.Y.; Jiang, Z.D.; Ni, H.; Chen, F. Purification and characterization of a tyrosinase inhibitor from Camellia pollen. *J. Funct. Food.* **2016**, *27*, 140–149. [CrossRef]
15. Wang, Z.Y.; Ren, P.P.; Wu, Y.N.; He, Q.H. Recent advances in analytical techniques for the detection of adulteration and authenticity of bee products—A review. *Food Addit. Contam. Part A-Chem.* **2021**, *38*, 533–549. [CrossRef] [PubMed]
16. Barth, O.M.; Freitas, A.S.; Oliveira, E.S.; Silva, R.A.; Maester, F.M.; Andrella, R.R.S.; Cardozo, G.M.B.Q. Evaluation of the botanical origin of commercial dry bee pollen load batches using pollen analysis: A proposal for technical standardization. *An. Acad. Bras. Cienc.* **2010**, *82*, 893–902. [CrossRef] [PubMed]
17. Zhou, J.H.; Qi, Y.T.; Ritho, J.; Zhang, Y.X.; Zheng, X.W.; Wu, L.M.; Li, Y.; Sun, L.P. Flavonoid glycosides as floral origin markers to discriminate of unifloral bee pollen by LC-MS/MS. *Food Control.* **2015**, *57*, 54–61. [CrossRef]
18. Khansari, E.; Zarre, S.; Alizadeh, K.; Attar, F.; Aghabeigi, F.; Salmaki, Y. Pollen morphology of *Campanula* (Campanulaceae) and allied genera in Iran with special focus on its systematic implication. *Flora* **2012**, *207*, 203–211. [CrossRef]
19. Xie, W.; Yu, W.; Yang, G.; Chen, J.; Pan, Y.; Ye, N. Micromorphological observation on pollen of 14 cultivars of tea tree (*Camellia sinensis*). *J. South. Agric.* **2018**, *49*, 1698–1704.
20. Zhang, H.F.; Zhu, X.L.; Huang, Q.; Zhang, L.; Liu, X.H.; Liu, R.; Lu, Q. Antioxidant and anti-inflammatory activities of rape bee pollen after fermentation and their correlation with chemical components by ultra-performance liquid chromatography-quadrupole time of flight mass spectrometry-based untargeted metabolomics. *Food Chem.* **2023**, *409*, e135342. [CrossRef]
21. Ma, C.; Ma, B.; Li, J.; Fang, Y. Changes in chemical composition and antioxidant activity of royal jelly produced at different floral periods during migratory beekeeping. *Food Res. Int.* **2022**, *155*, e111091. [CrossRef]
22. Liu, T.; Qiao, N.; Ning, F.J.; Huang, X.Y.; Luo, L.P. Identification and characterization of plant-derived biomarkers and physico-chemical variations in the maturation process of *Triadica cochinchinensis* honey based on UPLC-QTOF-MS metabolomics analysis. *Food Chem.* **2023**, *408*, e135197. [CrossRef]

23. Stavropoulou, M.I.; Termentzi, A.; Kasiotis, K.M.; Cheilari, A.; Stathopoulou, K.; Machera, K.; Aligiannis, N. Untargeted ultrahigh-performance liquid chromatography-hybrid quadrupole-orbitrap mass spectrometry (UHPLC-HRMS) metabolomics reveals propolis markers of Greek and Chinese origin. *Molecules* **2021**, *26*, 456. [CrossRef]

24. Selamat, J.; Rozani, N.A.A.; Murugesu, S. Application of the metabolomics approach in food authentication. *Molecules* **2021**, *26*, 7565. [CrossRef]

25. Li, S.B.; Tian, Y.F.; Jiang, P.Y.Z.; Lin, Y.; Liu, X.L.; Yang, H.S. Recent advances in the application of metabolomics for food safety control and food quality analyses. *Crit. Rev. Food Sci. Nutr.* **2021**, *61*, 1448–1469. [CrossRef]

26. Yan, S.; Wang, X.; Wu, Y.C.; Wang, K.; Shan, J.H.; Xue, X.F. A metabolomics approach revealed an Amadori compound distinguishes artificially heated and naturally matured acacia honey. *Food Chem.* **2022**, *385*, e132631. [CrossRef]

27. Rocchetti, G.; Castiglioni, S.; Maldarizzi, G.; Carloni, P.; Lucini, L. UHPLC-ESI-QTOF-MS phenolic profiling and antioxidant capacity of bee pollen from different botanical origin. *Int. J. Food Sci. Technol.* **2019**, *54*, 335–346. [CrossRef]

28. Chong, J.; Wishart, D.S.; Xia, J. Using MetaboAnalyst 4.0 for comprehensive and integrative metabolomics data analysis. *Curr. Protoc. Bioinform.* **2019**, *68*, e86. [CrossRef]

29. Zhang, X.X.; Yu, M.H.; Zhu, X.L.; Liu, R.; Lu, Q. Metabolomics reveals that phenolamides are the main chemical components contributing to the anti-tyrosinase activity of bee pollen. *Food Chem.* **2022**, *389*, e133071. [CrossRef]

30. Chen, Y.Y.; Zhou, Y.; Zeng, L.T.; Dong, F.; Tu, Y.Y.; Yang, Z.Y. Occurrence of functional molecules in the flowers of tea (*Camellia sinensis*) plants: Evidence for a second resource. *Molecules* **2018**, *23*, 790. [CrossRef]

31. Sharma, E.; Joshi, R.; Gulati, A. L-Theanine: An astounding sui generis integrant in tea. *Food Chem.* **2018**, *242*, 601–610. [CrossRef] [PubMed]

32. Li, M.Y.; Liu, H.Y.; Wu, D.T.; Kenaan, A.; Geng, F.; Li, H.B.; Gunaratne, A.; Li, H.; Gan, R.Y. L-theanine: A unique functional amino acid in tea (*Camellia sinensis* L.) with multiple health benefits and food applications. *Front. Nutr.* **2022**, *9*, e853846. [CrossRef] [PubMed]

33. Yu, J.J.; Li, W.F.; Xiao, X.; Huang, Q.X.; Yu, J.B.; Yang, Y.J.; Han, T.F.; Zhang, D.Z.; Niu, X.F. (-)-Epicatechin gallate blocks the development of atherosclerosis by regulating oxidative stress in vivo and in vitro. *Food Funct.* **2021**, *12*, 8715–8727. [CrossRef] [PubMed]

34. Gullon, B.; Lu-Chau, T.A.; Moreira, M.T.; Lema, J.M.; Eibes, G. Rutin: A review on extraction, identification and purification methods, biological activities and approaches to enhance its bioavailability. *Trends Food Sci. Technol.* **2017**, *67*, 220–235. [CrossRef]

35. Kahkeshani, N.; Farzaei, F.; Fotouhi, M.; Alavi, S.S.; Bahramsoltani, R.; Naseri, R.; Momtaz, S.; Abbasabadi, Z.; Rahimi, R.; Farzaei, M.H.; et al. Pharmacological effects of gallic acid in health and diseases: A mechanistic review. *Iran. J. Basic Med. Sci.* **2019**, *22*, 225–237. [PubMed]

36. Oelschlaegel, S.; Gruner, M.; Wang, P.N.; Boettcher, A.; Koelling-Speer, I.; Speer, K. Classification and characterization of manuka honeys based on phenolic compounds and methylglyoxal. *J. Agric. Food Chem.* **2012**, *60*, 7229–7237. [CrossRef]

37. Ivanisevic, J.; Want, E.J. From samples to insights into metabolism: Uncovering biologically relevant information in LC-HRMS metabolomics data. *Metabolites* **2019**, *9*, 308. [CrossRef]

38. Wang, X.R.; Li, Y.; Chen, L.Z.; Zhou, J.H. Analytical strategies for LC-MS-based untargeted and targeted metabolomics approaches reveal the entomological origins of honey. *J. Agric. Food Chem.* **2022**, *70*, 1358–1366. [CrossRef]

39. Fang, K.X.; Xia, Z.Q.; Li, H.J.; Jiang, X.H.; Qin, D.D.; Wang, Q.S.; Wang, Q.; Pan, C.D.; Li, B.; Wu, H.L. Genome-wide association analysis identified molecular markers associated with important tea flavor-related metabolites. *Hortic. Res.* **2021**, *8*, e42. [CrossRef]

40. Gong, A.D.; Lian, S.B.; Wu, N.N.; Zhou, Y.J.; Zhao, S.Q.; Zhang, L.M.; Cheng, L.; Yuan, H.Y. Integrated transcriptomics and metabolomics analysis of catechins, caffeine and theanine biosynthesis in tea plant (*Camellia sinensis*) over the course of seasons. *BMC Plant Biol.* **2020**, *20*, e294. [CrossRef]

Article

Quality of Commercially Available Manuka Honey Expressed by Pollen Composition, Diastase Activity, and Hydroxymethylfurfural Content

Alicja Sęk *, Aneta Porębska and Teresa Szczęsna

The National Institute of Horticultural Research, Konstytucji 3 Maja 1/3, 96-100 Skierniewice, Poland;
aneta.porebska@inhort.pl (A.P.); teresa.szczesna@inhort.pl (T.S.)
* Correspondence: alicja.sek@inhort.pl

Abstract: Manuka honey plays a significant role in modern medical applications as an antibacterial, antiviral, and antibiotic agent. However, although the importance of manuka honey is well documented in the literature, information regarding its physicochemical characteristics remains limited. Moreover, so far, only a few papers address this issue in conjunction with the examination of the pollen composition of manuka honey samples. Therefore, in this study, two parameters crucial for honey quality control—the diastase number (DN) and the hydroxymethylfurfural (HMF) content—as well as the melissopalynological analysis of manuka honey, were examined. The research found a large variation in the percentage of *Leptospermum scoparium* pollen in honeys labeled and sold as manuka honeys. Furthermore, a significant proportion of these honeys was characterized by a low DN. However, since low diastase activity was not associated with low HMF content, manuka honey should not be considered as a honey with naturally low enzymatic activity. Overall, the DN and HMF content results indicate that the quality of commercially available manuka honey is questionable.

Keywords: manuka honey; *Leptospermum scoparium*; melissopalynology; diastase number (DN); hydroxymethylfurfural (HMF)

Citation: Sęk, A.; Porębska, A.; Szczęsna, T. Quality of Commercially Available Manuka Honey Expressed by Pollen Composition, Diastase Activity, and Hydroxymethylfurfural Content. *Foods* **2023**, *12*, 2930. https://doi.org/10.3390/foods12152930

Academic Editors: Liming Wu and Qiangqiang Li

Received: 30 June 2023
Revised: 26 July 2023
Accepted: 30 July 2023
Published: 2 August 2023

1. Introduction

Manuka honey is a dark monofloral honey derived from the New Zealand manuka tree (*Leptospermum scoparium*, *L. scoparium*); known for its distinctive taste and aroma, it plays a significant role in modern medical applications, especially as an antibacterial, antiviral, and antibiotic agent [1,2]. Manuka honey showed, for instance, bactericidal potency against *Pseudomonas aeruginosa* and *Staphylococcus aureus* (strains responsible for the inflammation of the mucous lining of the paranasal sinuses) and antiviral ability against Varicella zoster (the virus that causes chickenpox and shingles) [3,4]. In turn, a study of the combination of five novel antibiotics and manuka honey revealed the improved activity of these antibiotics against wound pathogens [5]. In addition, the honey from New Zealand promotes the growth of certain types of probiotics such as *Lactobacillus reuteri* and *Lactobacillus rhamnosus* [6]. In addition, manuka honey exhibits strong antioxidant and anti-inflammatory properties due to the presence of various bioactive compounds such as phenols, flavonoids, and enzymes, which makes it effective in the treatment of chronic ulcer and topical clinical inflammation [7–9]. Furthermore, numerous data highlight the anticancer effect of manuka honey [10,11]. A cytotoxic influence on human lung, breast, colon, and metastatic cancer cell lines, as well as on murine melanoma, colorectal carcinoma, and human hepatocarcinoma was observed [12–14].

Although the importance of manuka honey for medicinal purposes is well documented in the literature, information regarding its physicochemical characteristics remains limited. The European Union has established regulations governing the physicochemical parameters of honey in order to ensure its quality, authenticity, and safety [15]. These properties includes moisture content, apparent reducing sugars content, apparent sucrose content, water-insoluble solids content, ash content, acidity, diastase activity, hydroxymethylfurfural (HMF) content, and electrical conductivity [15]. Specifying the above-mentioned characteristics of honey is also intended to protect consumers against deceptive practices of sellers of this rarely cheap product. Two physicochemical parameters of honey, which deserve particular attention as they are crucial for honey quality control, are not clearly defined for manuka honey: the diastase activity (diastase number, DN) and the HMF content. Diastase (α-amylase) is one of the main enzymes found in honey; its activity level is considered an indicator of honey freshness and its proper processing [16]. Similarly, the content of HMF, a furanic compound formed when sugar-containing products are acidified or heated, informs about the quality of honey. Since a high level of HMF can be a result of adulteration with sugar additives, the HMF amount is also useful for assessing the authenticity of honey samples [17].

Most of the studies on the DN and HMF content in manuka honey carried out so far have not linked these experiments with melissopalynological analysis [17–24]. Only a few papers discuss this issue using these three methods [25–27]. However, in these studies, the results are presented for individual samples of manuka honeys without using the relevant statistics. It was found that the manuka honey with 75.8% of *L. scoparium* pollen grains contains a DN and HMF content equal to 19.03 (Schade units) and 16.42 mg·kg^{-1}, respectively [25]. In turn, the honey samples with an *L. scoparium* pollen content of about 31 and 51% are characterized by a diastase activity of about 7.2 and 9.1 Schade, respectively, and by an HMF amount of 27.9 and 33.5 mg·kg^{-1}, respectively [26]. Another study indicates that the manuka honey DN and HMF content was equal to 8.0 ± 1.0 Schade and 18.1 ± 3.2 mg·kg^{-1}, respectively, but declared that the pollen analysis was performed; however, the percentage composition of pollen grains in the tested honey was not given [27]. Importantly, since the manuka pollen was considered over-representative and the manuka tree shares similar pollens with the kanuka bush, Moar stated that to be called manuka honey, honey must contain at least 70% of manuka pollen grains [28]. Thus, it is likely that some of the honeys mentioned should not be called manuka honey.

The problem of manuka honey falsification has been recognized earlier; in response to this issue, the New Zealand Ministry for Primary Industries reported the combination of five attributes required to distinguish manuka honey from other honey types, as well as to identify monofloral and multifloral manuka honeys [29,30]. Four of the attributes refer to the four specific compounds found in manuka honey by liquid chromatography–tandem mass spectrometry, i.e., 2′-methoxyacetophenone, 2-methoxybenzoic acid, 3-phenyllactic acid, and 4-hydroxyphenyllactic acid; the other relates to the DNA markers of manuka pollen determined by the multiplex qPCR assay [29,30]. Other studies have described the potential of compact atmospheric solids analysis probe mass spectrometry, liquid chromatography–high resolution mass spectrometry, fluorescence spectroscopy, and ^{1}H NMR spectroscopy for the evaluation of manuka honey authenticity [31–34]. Another unique feature of manuka honey is its high methylglyoxal (MGO) content [35]. This compound can be formed in foods through five pathways. The first pathway is the Maillard reaction, i.e., a reaction between amino acids and reducing sugars that occurs during food storage at room temperature and during thermal processing [36]. The other pathways include the autoxidation of hexoses, the oxidation of unsaturated fatty acids in lipids, the dehydration of dihydroxyacetone (DHA), and the microbial metabolism of dihydroxyacetone phosphate [36]. In manuka honey, MGO is naturally formed as a result of the dehydration of dihydroxyacetone, a compound present in the nectar of manuka tree flowers [36]. The chemical transformation of DHA to MGO occurs in moderate heat, and the period of storage, which means that the MGO in manuka honey has already been produced in the beehive, as well as being produced in storage drums after humans collect the honey [37]. It is worth mentioning that manuka honey is frequently stored for several years without any temperature regulation [37]. In some cases, this is to intentionally increase the amount of MGO formed, especially since the MGO is thought to be responsible for the antibacterial activity of manuka honey [37,38]. It is easy to determine that advanced and often sophisticated equipment is needed to apply above described techniques used for verifying the authenticity of manuka honey; therefore, pollen analysis appears to be the most suitable first-choice method for manuka honey recognition. Thus, in this paper, using the example of the Polish market, we demonstrate the variation in the percentage of *L. scoparium* pollen in honeys labeled and sold as manuka honeys. Then, we present the DN value and HMF content based on relevant statistics. The pollen analysis results can be regarded as an alert both for consumers of manuka honey and for sellers purchasing the product from unproven suppliers, and above all for scientists who conduct research on manuka honeys without confirmation of their authenticity. In turn, the DN and HMF content results point to the questionable quality of the imported manuka honey.

2. Materials and Methods

2.1. Chemicals

Glycerol (pure per analysis grade, ppa grade), gelatine (ppa grade), sodium hydroxide 0.1 M analytical weighed amount, sodium acetate trihydrate (ppa grade), potassium hexacyanoferrate (II) trihydrate (ppa grade), and zinc acetate dihydrate (ppa grade) were purchased from Chempur (Piekary Śląskie, Poland). The Phadebas Honey Diastase Test tablets were bought from Magle Life Sciences (Malmö, Sweden) and glacial acetic acid (ppa grade) from Pol-Aura (Zawroty, Poland). Methanol and hydroxymethylfurfural were obtained from J.T. Baker (Gliwice, Poland) and Merck (Darmstadt, Germany), respectively, and were of HPLC grade. For the analyses, distilled or ultrapure water from the Milli-Q system (Merck, Darmstadt, Germany, resistivity 18.3 MΩ cm) was used.

2.2. Samples

The study included thirty honey samples labeled as manuka honey available on the Polish market. All honeys had current best before dates, and the analyses were carried out immediately after obtaining the samples for testing.

2.3. Melissopalynological Analysis

The pollen analysis of honey was performed according to the previously described procedure with slight modifications [39]. Briefly, ten grams of honey was dissolved in 50 mL distilled water at about 50 °C. The solution was then centrifuged for 8 min (3000 rpm) and, after removing the supernatant, 50 mL of distilled water was added to the sample again and the centrifugation was repeated. After the second centrifugation, the supernatant was not completely removed, leaving a tiny amount of solution above the sediment. The residue was mixed and placed on a microscope slide. Next, the slide was dried and secured with glycerol/gelatine and a coverslip. The collected material was analyzed using an Olympus BX41 microscope (Olympus America, PA, USA)) at 400× magnification. At least 300 consecutive pollen grains of nectar-producing plants were determined on each microscopic slide, and then the percentage content of *L. scoparium* pollen was calculated. Two repetitions were made for each honey sample and the final result is presented as an average of two results not differing between individual determinations by more than 5%, and rounded to the whole unit (%).

2.4. Diastase Analysis

The diastatic activity was determined by the Phadebas method in which the α-amylase activity is expressed as the diastase number, and is reported in Schade units [40]. One Schade unit corresponds to the enzyme activity contained in 1 g of honey, which can hydrolyze 0.01 g of starch in 1 hour at 40 °C. The procedure was as follows: first, 1 g of the analyzed honey was weighed, transferred to a 100 mL volumetric flask, and filled up to its volume with 0.1 M acetate buffer at pH = 5.2. Then, 5 milliliters of the sample was transferred to the test tube, placed in a water bath at 40 °C, and after 15 min a Phadebas tablet was added to the solution. The solution was mixed and heated again in a water bath for 30 min. After this time, 1 mL of 0.5 M sodium hydroxide solution was added to complete the enzymatic reaction. Next, the solution was filtered through a filter paper (φ = 70 mm) and the absorbance at 620 nm was measured using a Specord 200 spectrophotometer (Analytic Jena, Jena, Germany) as it is in proportion to the enzyme activity in the analyzed honey sample. The measurements were collected in triplicate and the final result is their average.

2.5. Hydroxymethylfurfural Analysis

The HMF content in the analyzed honey samples was determined chromatographically with the application of a Knauer HPLC system (Knauer GmbH, Berlin, Germany) equipped with an UV K-2501 detector and with a reversed-phase C-18 Vertex Plus Eurospher column (BGB Analytik Vertrieb GmbH, Lörrach, Germany), as described previously [41]. The internal diameter of column was 4 mm, length 250 mm, and particle size 5 μm. The column and detector were placed on a thermostat at 30 °C. As a mobile phase, a mixture of methanol and deionized water (10:90, v/v) was used, and the flow rate was 1 mL·min^{-1}. For the quantitative analysis of HMF, the external standard method was applied. The determinations were carried out in triplicate. Honey samples were prepared based on the European Honey Commission procedure [40].

3. Results and Discussion

The first stage of the research included the melissopalynological analysis of the examined honey samples. For each honey, at least 300 consecutive pollen grains of nectar-producing plants, including *L. scoparium* pollen, were identified and counted, and then the percentage of *L. scoparium* pollen content was calculated on this basis. The microscopic image of *L. scoparium* pollen found in the analyzed honeys is presented in Figure 1. As can be seen, the pollen grain is small (~15–20 μm), triangular in polar view, isopolar, and tricolporate, which is consistent with the literature data [42,43]. It is also worth noting that the sides of the pollen sometimes appear concave, while the angles appear extended [43]. Most of the studies to date report a significant resemblance between manuka and kanuka pollen; consequently, melissopalynological studies tend to combine them [28,43,44]. Despite its imperfections, this method is still useful as it allows for the simple verification of the presence and quantity of manuka-like pollen in a honey sample.

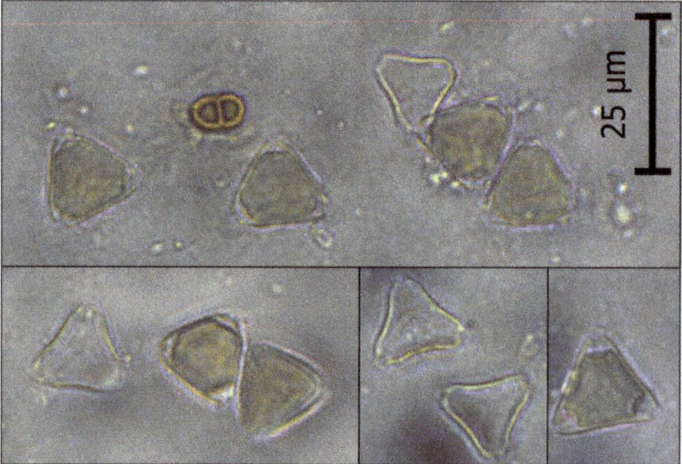

Figure 1. Microscopic image of *L. scoparium* pollen under 400× magnification.

The results obtained from the pollen analysis of thirty honey samples labeled as manuka honeys are listed in Table 1. As shown, the percentage of *L. scoparium* pollen grains in the analyzed honeys varied from 45 to 90%. Importantly, as much as 47% of the tested honeys indicate a low content of manuka tree pollen, lower than the Moar's limit (70%) for honey to be called manuka honey [28]. The obtained results point to the wide diversity in the manuka pollen composition of commercially available manuka honeys, and simultaneously prove the importance of the melissopalynological analysis of purchased honeys, especially in order to verify the authenticity of honey without the use of advanced methods.

Table 1. The percentage of *L. scoparium* (*LS*) pollen grains in the analyzed honeys.

Sample Number	LS Pollen Grains
1.	45%
2.	51%
3.	52%
4.	52%
5.	54%
6.	55%
7.	61%
8.	64%
9.	65%
10.	65%
11.	66%
12.	67%
13.	68%
14.	68%
15.	70%
16.	71%
17.	71%
18.	73%
19.	74%
20.	75%
21.	76%
22.	76%
23.	78%
24.	78%
25.	78%
26.	80%
27.	83%
28.	85%
29.	85%
30.	90%

In the next step, the activity level of α-amylase was determined and plotted in Figure 2. While the DN was estimated for all analyzed honey samples, we focus our discussions on the results obtained for the melissopalynologically classified manuka honey samples; that is, honeys that contain at least 70% of manuka pollen (shaded area of the graph). As illustrated, the diastase level ranged from 1.8 to 15.2 Schade units, but was mostly lower than 8 Schade units. According to the European Directive that governs the standards for honey sold in the European market, a minimum diastase level of 8 Schade units is necessary for honey to be deemed of good quality and acceptable; thus, the vast majority of the analyzed manuka honeys do not meet this requirement [15]. In fact, merely five manuka honeys (31%) had a diastase activity higher than 8 Schade units. Nevertheless, honeys with naturally low enzymatic activity are known, such as *Acacia, Becium grandiflorum, Croton macrostachyus, Eucalyptus globulus, Hypoestes, Leucas abyssinica, Schefflera abyssinica, Syzygium guineense,* and *Vernonia amygdalina* monofloral honeys [15,45]. Such honeys are characterized by a DN value between 3 and 8 Schade units and an HMF content lower than 15 mg·kg^{-1} [15,46].

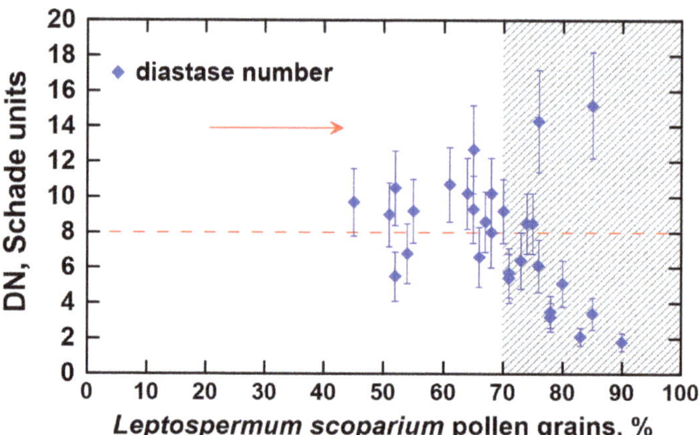

Figure 2. The diastase number (DN) in the analyzed honeys.

Figure 3 provides the HMF content in the manuka honey samples evaluated in this study. Only two melissopalynologically classified manuka honeys showed the HMF content below 15 mg·kg^{-1}, more precisely 13.5 ± 1.8 mg·kg^{-1} (when the percentage of *L. scoparium* pollen grains was 76%) and 5.1 ± 1.5 mg·kg^{-1} (when the percentage of *L. scoparium* pollen grains was 85%). However, the DN for these samples was 14.3 ± 2.9 Schade units and 15.2 ± 3.0 Schade units, respectively. This denotes that the low level of HMF present in this honey does not coincide with a low level of diastase activity, which would typically be expected in honeys with naturally low enzymatic activity. Thus, even though manuka honey is most often characterized by a low level of α-amylase activity, it should not be considered a honey with intrinsically low enzyme activity as it does not fulfill the second condition for such honeys [15]. The low DN may result from thermal treatments or long-term storage, as well as from storage under inappropriate conditions [47].

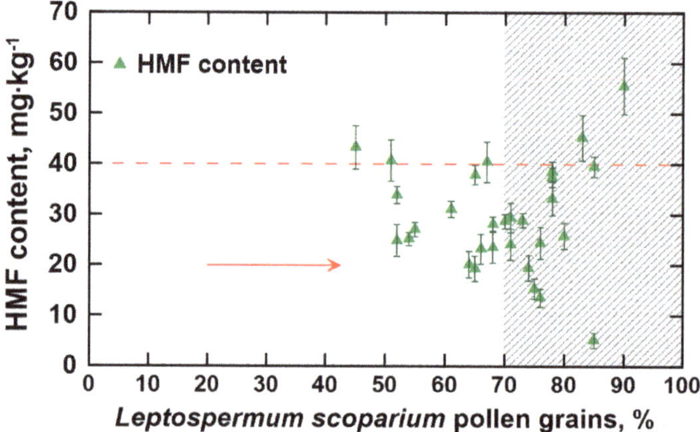

Figure 3. The hydroxymethylfurfural (HMF) content in the analyzed honeys.

Nearly all of the manuka honey samples with a high content of manuka pollen (shaded area in Figure 3) satisfy the maximum limit of HMF content allowed by the European Directive, which is 40 mg·kg^{-1} [15]. Specifically, the HMF content ranged from 5.1 to 55.5 mg·kg^{-1} and two samples (13%) exceeded the permissible limit. These two samples had HMF values of 45.2 ± 4.5 mg·kg^{-1} and 55.5 ± 5.6 mg·kg^{-1}, and were characterized by a DN equal to 2.1 ± 0.5 Schade units and 1.8 ± 0.5 Schade units, respectively. Such a combination of HMF content and DN clearly indicates that the honeys were of poor quality, presumably due to overheating or improper/prolonged storage [48]. According to European regulations, these honeys should be withdrawn from sale, especially since the HMF impact on human health is ambiguous [15,41,49,50]. Other melissopalynologically classified manuka honeys, based on the HMF content, may have been released for consumption.

To gain further insight into the characteristic of commercially available manuka honeys, the correlation between the DN value and HMF content was analyzed. As can be seen in Figure 4, showing a comparison of the regression graph between the diastatic activity and HMF content obtained for thirty honey samples labeled and sold as manuka honeys (Figure 4a) and for the manuka honeys with a high content of manuka pollen, that is, honeys that contain at least 70% of manuka pollen (Figure 4b); in both cases, the DN presents a substantial negative relationship with HMF. Regression equations for diastase activity to HMF content for all analyzed honey samples and for melissopalynologically classified manuka honeys were y = −0.2086x + 13.759 and y = −0.2762x + 14.361, respectively. This shows that the negative regression is even more significant when the manuka honeys with a high content of manuka pollen are considered. Notably, the obtained dependencies are in good agreement with the knowledge that a high HMF content is a marker of excessive heating or improper storage of honey, which in turn causes a decrease in the enzymatic activity. Similar behavior was reported in the study of the Ethiopian monofloral honeys [45]. As a result of the regression analysis of these honeys, the following regression equation for DN to HMF was obtained: y = −0.1389x + 6.3701 [45]. In addition, the study of honeys from Bosnia and Herzegovina and Algeria also revealed a negative correlation between the diastase activity and HMF content [51,52].

A concise summary of the results obtained for the melissopalynologically classified manuka honeys is presented in Table 2. As depicted, the average percentage of *L. scoparium* pollen grains in the analyzed honeys was 77.7 ± 5.7%, while the mean DN and the mean HMF content were 6.4 ± 4.0 Schade units and 29.0 ± 12.7 mg·kg^{-1}, respectively. These results indicate there is a wide variety of manuka honeys available on the Polish market in terms of enzymatic activity and HMF content. In addition, a significant part of these honeys do not meet the requirements of the European Directive authorizing honey for use, mainly due to the low DN. The obtained results suggest that the analyzed manuka honeys were stored under inappropriate conditions for long periods of time or were intentionally heated, for example in order to achieve a higher content of methylglyoxal, a compound believed to be the predominant antibacterial constituent of manuka honey [53]. Thus, the physicochemical quality of these imported honeys is questionable.

Table 2. Summary of the characteristics of the melissopalynologically classified manuka honeys.

Parameter	Unit	Min	Max	Mean	SD
L. scoparium pollen	%	70.0	90.0	77.7	5.7
Diastase number	Schade	1.8	15.2	6.4	4.0
HMF content	mg·kg^{-1}	5.1	55.5	29.0	12.7

SD—Standard Deviation.

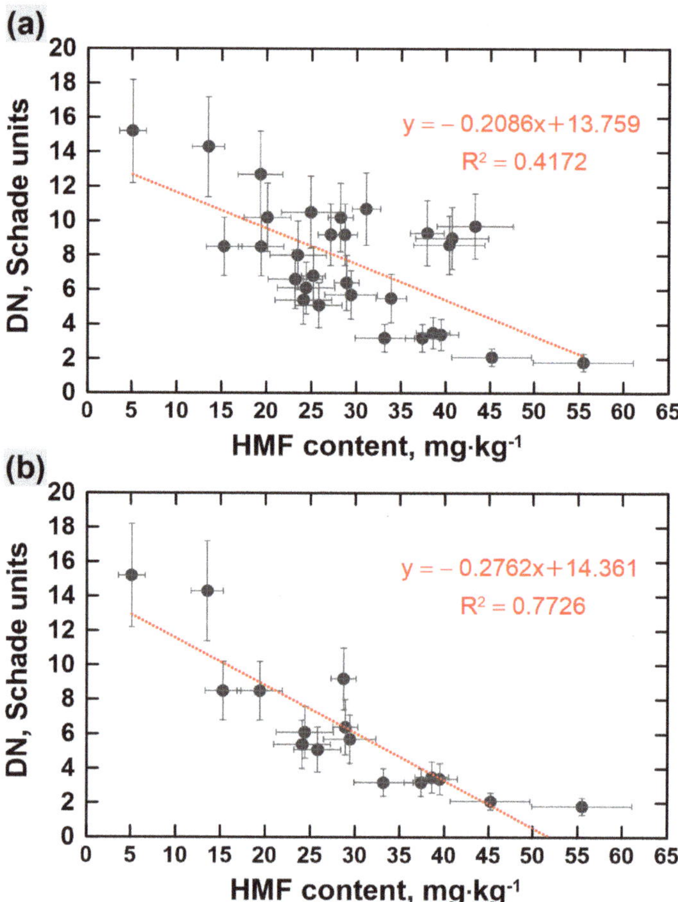

Figure 4. The correlation between the diastase number (DN) and hydroxymethylfurfural (HMF) content obtained for: (**a**) thirty honey samples labeled and sold as manuka honeys; (**b**) the part of the examined manuka honeys that contains at least 70% of manuka pollen.

4. Conclusions

In the paper, thirty honey samples labeled and sold as manuka honey were tested for their pollen composition, diastase number, and HMF content. The obtained results indicate the large diversity of manuka honey available on the Polish market both in terms of the percentage of *L. scoparium* pollen grains and the enzymatic activity or HMF content. The most striking observation from the experiments is that, according to Moar's statement and based on the pollen analysis, almost half of the analyzed honeys should not be classified as manuka honeys since their *L. scoparium* pollen content did not meet the required minimum. Presumably, the Polish market is not the only market facing such a problem. Thus, the presented results can be regarded as a general warning for consumers of manuka honey and for sellers purchasing the product from unverified suppliers, and above all for scientists conducting research on manuka honey without confirmation of its authenticity. It is also worth emphasizing that a significant proportion of the analyzed honeys had a low DN, lower than the European Directive limit for the honey to be approved for use. The low diastase activity was not associated with the low HMF content, and therefore it would be incorrect to assume that manuka honey naturally has low enzymatic activity. The low value of DN and relatively high HMF content indicate, in turn, the flawed physicochemical quality of these commercially available honeys. The present research encourages further study of natural (without any additional processing) and commercially available manuka honeys. It is reasonable, for example, to extend the research by using other hallmarks of manuka honeys in order to establish their authenticity or adulteration on the basis of further scientific evidence.

Author Contributions: Conceptualization, A.S. and T.S.; formal analysis, A.S. and T.S.; investigation, A.P.; data curation, A.S.; writing—original draft preparation, A.S.; writing—review and editing, A.S. and T.S.; visualization, A.S.; supervision, T.S. All authors have read and agreed to the published version of the manuscript.

Funding: This research did not receive any specific grant from funding agencies in the public, commercial, or not-for-profit sectors.

Data Availability Statement: All data generated or analyzed during this study are available from the corresponding author on reasonable request.

Conflicts of Interest: The authors declare no conflict of interest.

References

1. Patel, S.; Cichello, S. Manuka Honey: An Emerging Natural Food with Medicinal Use. *Nat. Prod. Bioprospect.* **2013**, *3*, 121–128. [CrossRef]
2. El-Senduny, F.F.; Hegazi, N.M.; Abd Elghani, G.E.; Farag, M.A. Manuka Honey, a Unique Mono-Floral Honey. A Comprehensive Review of Its Bioactives, Metabolism, Action Mechanisms, and Therapeutic Merits. *Food Biosci.* **2021**, *42*, 101038. [CrossRef]
3. Kilty, S.J.; Duval, M.; Chan, F.T.; Ferris, W.; Slinger, R. Methylglyoxal: (Active Agent of Manuka Honey) in Vitro Activity against Bacterial Biofilms. *Int. Forum Allergy Rhinol.* **2011**, *1*, 348–350. [CrossRef]
4. Shahzad, A.; Cohrs, R.J. In Vitro Antiviral Activity of Honey against Varicella Zoster Virus (VZV): A Translational Medicine Study for Potential Remedy for Shingles. *Transl. Biomed.* **2012**, *3*, 2. [CrossRef] [PubMed]
5. Jenkins, R.; Cooper, R. Improving Antibiotic Activity against Wound Pathogens with Manuka Honey In Vitro. *PLoS ONE* **2012**, *7*, e45600. [CrossRef] [PubMed]
6. Rosendale, D.I.; Maddox, I.S.; Miles, M.C.; Rodier, M.; Skinner, M.; Sutherland, J. High-Throughput Microbial Bioassays to Screen Potential New Zealand Functional Food Ingredients Intended to Manage the Growth of Probiotic and Pathogenic Gut Bacteria. *Int. J. Food Sci. Technol.* **2008**, *43*, 2257–2267. [CrossRef]
7. Almasaudi, S.B.; Abbas, A.T.; Al-Hindi, R.R.; El-Shitany, N.A.; Abdel-Dayem, U.A.; Ali, S.S.; Saleh, R.M.; Al Jaouni, S.K.; Kamal, M.A.; Harakeh, S.M. Manuka Honey Exerts Antioxidant and Anti-Inflammatory Activities That Promote Healing of Acetic Acid-Induced Gastric Ulcer in Rats. *Evid.-Based Complement. Altern. Med.* **2017**, *2017*, 5413917. [CrossRef]
8. Leong, A.G.; Herst, P.M.; Harper, J.L. Indigenous New Zealand Honeys Exhibit Multiple Anti-Inflammatory Activities. *Innate Immun.* **2012**, *18*, 459–466. [CrossRef]
9. Alvarez-Suarez, J.M.; Gasparrini, M.; Forbes-Hernández, T.Y.; Mazzoni, L.; Giampieri, F. The Composition and Biological Activity of Honey: A Focus on Manuka Honey. *Foods* **2014**, *3*, 420–432. [CrossRef]
10. Martinotti, S.; Pellavio, G.; Patrone, M.; Laforenza, U.; Ranzato, E. Manuka Honey Induces Apoptosis of Epithelial Cancer Cells through Aquaporin-3 and Calcium Signaling. *Life* **2020**, *10*, 256. [CrossRef]
11. Al Refaey, H.R.; Newairy, A.-S.A.; Wahby, M.M.; Albanese, C.; Elkewedi, M.; Choudhry, M.U.; Sultan, A.S. Manuka Honey Enhanced Sensitivity of HepG2, Hepatocellular Carcinoma Cells, for Doxorubicin and Induced Apoptosis through Inhibition of Wnt/β-Catenin and ERK1/2. *Biol. Res.* **2021**, *54*, 16. [CrossRef]
12. Halawani, E.M. Potential Effects of Saudi Shaoka (*Fagonia bruguieri*) Honey against Multi-Drug-Resistant Bacteria and Cancer Cells in Comparison to Manuka Honey. *Saudi J. Biol. Sci.* **2021**, *28*, 7379–7389. [CrossRef]
13. Aryappalli, P.; Shabbiri, K.; Masad, R.J.; Al-Marri, R.H.; Haneefa, S.M.; Mohamed, Y.A.; Arafat, K.; Attoub, S.; Cabral-Marques, O.; Ramadi, K.B.; et al. Inhibition of Tyrosine-Phosphorylated STAT3 in Human Breast and Lung Cancer Cells by Manuka Honey Is Mediated by Selective Antagonism of the IL-6 Receptor. *Int. J. Mol. Sci.* **2019**, *20*, 4340. [CrossRef] [PubMed]
14. Afrin, S.; Forbes-Hernandez, T.Y.; Gasparrini, M.; Bompadre, S.; Quiles, J.L.; Sanna, G.; Spano, N.; Giampieri, F.; Battino, M. Strawberry-Tree Honey Induces Growth Inhibition of Human Colon Cancer Cells and Increases ROS Generation: A Comparison with Manuka Honey. *Int. J. Mol. Sci.* **2017**, *18*, 613. [CrossRef]
15. Bogdanov, S.; Lullmann, C.; Mossel, B.; D'Arcy, B.; Russmann, H.; Vorwohl, G.; Oddo, L.; Sabatini, A.; Marcazzan, G.; Piro, R.; et al. Honey Quality, Methods of Analysis and International Regulatory Standards: Review of the Work of the International Honey Commission. *Mitt. Lebensm. Hyg.* **1999**, *90*, 108–125.
16. Pasias, I.N.; Kiriakou, I.K.; Proestos, C. HMF and Diastase Activity in Honeys: A Fully Validated Approach and a Chemometric Analysis for Identification of Honey Freshness and Adulteration. *Food Chem.* **2017**, *229*, 425–431. [CrossRef]

17. Fauzi, N.A.; Farid, M.M. High-Pressure Processing of Manuka Honey: Brown Pigment Formation, Improvement of Antibacterial Activity and Hydroxymethylfurfural Content. *Int. J. Food Sci. Technol.* **2015**, *50*, 178–185. [CrossRef]

18. Stephens, J.M.; Schlothauer, R.C.; Morris, B.D.; Yang, D.; Fearnley, L.; Greenwood, D.R.; Loomes, K.M. Phenolic Compounds and Methylglyoxal in Some New Zealand Manuka and Kanuka Honey. *Food Chem.* **2010**, *120*, 78–86. [CrossRef]

19. Al-Habsi, N.A.; Niranjan, K. Effect of High Hydrostatic Pressure on Antimicrobial Activity and Quality of Manuka Honey. *Food Chem.* **2012**, *135*, 1448–1454. [CrossRef]

20. Moniruzzaman, M.; Sulaiman, S.A.; Khalil, M.I.; Gan, S.H. Evaluation of Physicochemical and Antioxidant Properties of Sourwood and Other Malaysian Honeys: A Comparison with Manuka Honey. *Chem. Central J.* **2013**, *7*, 138. [CrossRef]

21. Alqarni, A.S.; Owayss, A.A.; Mahmoud, A.A. Physicochemical Characteristics, Total Phenols and Pigments of National and International Honeys in Saudi Arabia. *Arab. J. Chem.* **2016**, *9*, 114–120. [CrossRef]

22. Grainger, M.N.C.; Owens, A.; Manley-Harris, M.; Lane, J.R.; Field, R.J. Kinetics of Conversion of Dihydroxyacetone to Methylglyoxal in New Zealand Mānuka Honey: Part IV—Formation of HMF. *Food Chem.* **2017**, *232*, 648–655. [CrossRef]

23. Chernyshev, A.; Braggins, T. Investigation of Temporal Apparent C4 Sugar Change in Manuka Honey. *J. Agric. Food Chem.* **2020**, *68*, 4261–4267. [CrossRef]

24. Septiani, A.; Suryati, T.; Apriantini, A.; Endrawati, Y.C. Characteristics of Forest and Manuka Honey As Well As Their Application as Herbal Honey Drinks with The Addition of Qusthul Hindi and Turmeric. *J. Ilmu Dan Teknol. Has. Ternak (JITEK)* **2022**, *17*, 183–196. [CrossRef]

25. Gkoutzouvelidou, M.; Panos, G.; Xanthou, M.N.; Papachristoforou, A.; Giaouris, E. Comparing the Antimicrobial Actions of Greek Honeys from the Island of Lemnos and Manuka Honey from New Zealand against Clinically Important Bacteria. *Foods* **2021**, *10*, 1402. [CrossRef] [PubMed]

26. Zhang, Y.-Z.; Si, J.-J.; Li, S.-S.; Zhang, G.-Z.; Wang, S.; Zheng, H.-Q.; Hu, F.-L. Chemical Analyses and Antimicrobial Activity of Nine Kinds of Unifloral Chinese Honeys Compared to Manuka Honey (12+ and 20+). *Molecules* **2021**, *26*, 2778. [CrossRef] [PubMed]

27. Pasias, I.N.; Kiriakou, I.K.; Kaitatzis, A.; Koutelidakis, A.E.; Proestos, C. Effect of Late Harvest and Floral Origin on Honey Antibacterial Properties and Quality Parameters. *Food Chem.* **2018**, *242*, 513–518. [CrossRef]

28. Moar, N.T. Pollen Analysis of New Zealand Honey. *N. Z. J. Agric. Res.* **1985**, *28*, 39–70. [CrossRef]

29. Ministry for Primary Industries. *Determination of Four Chemical Characterisation Compounds in Honey by Liquid Chromatography Tandem Mass Spectrometry (LC-MS/MS)*; MPI Technical Paper No: 2017/30; New Zealand Government: Wellington, New Zealand, 2017.

30. Ministry for Primary Industries. *Multiplex QPCR for Detection of Leptospermum Scoparium DNA from Pollen in Honey*; MPI Technical Paper No 2017/31; New Zealand Government: Wellington, New Zealand, 2017.

31. Loh, L.X.; Lee, H.H.; Stead, S.; Ng, D.H.J. Manuka Honey Authentication by a Compact Atmospheric Solids Analysis Probe Mass Spectrometer. *J. Food Compos. Anal.* **2022**, *105*, 104254. [CrossRef]

32. Yan, L.; Xu, D.; Xue, X.; Lin, L.; Lai, K.; Wang, J.; Zhang, Z. Authenticity identification of manuka honey using liquid chromatography-high resolution mass spectrometry based metabolomic technique. *Chin. J. Chromatogr.* **2019**, *37*, 589–596. [CrossRef]

33. Bong, J.; Prijic, G.; Braggins, T.J.; Schlothauer, R.C.; Stephens, J.M.; Loomes, K.M. Leptosperin Is a Distinct and Detectable Fluorophore in Leptospermum Honeys. *Food Chem.* **2017**, *214*, 102–109. [CrossRef] [PubMed]

34. Spiteri, M.; Rogers, K.M.; Jamin, E.; Thomas, F.; Guyader, S.; Lees, M.; Rutledge, D.N. Combination of 1H NMR and Chemometrics to Discriminate Manuka Honey from Other Floral Honey Types from Oceania. *Food Chem.* **2017**, *217*, 766–772. [CrossRef]

35. Majtan, J. Methylglyoxal-a Potential Risk Factor of Manuka Honey in Healing of Diabetic Ulcers. *Evid.-Based Complement. Altern. Med.* **2011**, *2011*, 295494. [CrossRef]

36. Zheng, J.; Guo, H.; Ou, J.; Liu, P.; Huang, C.; Wang, M.; Simal-Gandara, J.; Battino, M.; Jafari, S.M.; Zou, L.; et al. Benefits, Deleterious Effects and Mitigation of Methylglyoxal in Foods: A Critical Review. *Trends Food Sci. Technol.* **2021**, *107*, 201–212. [CrossRef]

37. Kato, Y.; Kishi, Y.; Okano, Y.; Kawai, M.; Shimizu, M.; Suga, N.; Yakemoto, C.; Kato, M.; Nagata, A.; Miyoshi, N. Methylglyoxal Binds to Amines in Honey Matrix and 2′-Methoxyacetophenone Is Released in Gaseous Form into the Headspace on the Heating of Manuka Honey. *Food Chem.* **2021**, *337*, 127789. [CrossRef]

38. Johnston, M.; McBride, M.; Dahiya, D.; Owusu-Apenten, R.; Nigam, P.S. Antibacterial Activity of Manuka Honey and Its Components: An Overview. *AIMS Microbiol.* **2018**, *4*, 655–664. [CrossRef] [PubMed]

39. Louveaux, J.; Maurizio, A.; Vorwohl, G. Methods of Melissopalynology. *Bee World* **1978**, *59*, 139–157. [CrossRef]

40. Bogdanov, S.; Martin, P.; Lullmann, C.; Borneck, R.; Flamini, C.; Morlot, M.; Lheritier, J.; Vorwohl, G.; Russmann, H.; Persano, L.; et al. Harmonised Methods of the European Honey Commission. *Apidologie* **1997**, *28*, 1–59.

41. Szczęsna, T.; Waś, E.; Semkiw, P.; Skubida, P.; Jaśkiewicz, K.; Witek, M. Changes of Physicochemical Properties of Starch Syrups Recommended for Winter Feeding of Honeybees during Storage. *Agriculture* **2021**, *11*, 374. [CrossRef]

42. Moar, N.T.; Wilmshurst, J.M.; McGlone, M.S. Standardizing Names Applied to Pollen and Spores in New Zealand Quaternary Palynology. *N. Z. J. Bot.* **2011**, *49*, 201–229. [CrossRef]

43. Li, X.; Prebble, J.G.; de Lange, P.J.; Raine, J.I.; Newstrom-Lloyd, L. Discrimination of Pollen of New Zealand Mānuka (*Leptospermum Scoparium* agg.) and Kānuka (*Kunzea* spp.) (Myrtaceae). *PLoS ONE* **2022**, *17*, e0269361. [CrossRef] [PubMed]

44. Hegazi, N.M.; Elghani, G.E.A.; Farag, M.A. The Super-Food Manuka Honey, a Comprehensive Review of Its Analysis and Authenticity Approaches. *J. Food Sci. Technol.* **2022**, *59*, 2527–2534. [CrossRef] [PubMed]
45. Belay, A.; Haki, G.D.; Birringer, M.; Borck, H.; Lee, Y.-C.; Kim, K.-T.; Baye, K.; Melaku, S. Enzyme Activity, Amino Acid Profiles and Hydroxymethylfurfural Content in Ethiopian Monofloral Honey. *J. Food Sci. Technol.* **2017**, *54*, 2769–2778. [CrossRef]
46. Zappalà, M.; Fallico, B.; Arena, E.; Verzera, A. Methods for the Determination of HMF in Honey: A Comparison. *Food Control* **2005**, *16*, 273–277. [CrossRef]
47. Korkmaz, S.D.; Küplülü, Ö. Effects of Storage Temperature on HMF and Diastase Activity of Strained Honeys. *Ank. Univ. Vet. Fak. Derg.* **2017**, *64*, 281–287. [CrossRef]
48. da Silva, P.M.; Gauche, C.; Gonzaga, L.V.; Costa, A.C.O.; Fett, R. Honey: Chemical Composition, Stability and Authenticity. *Food Chem.* **2016**, *196*, 309–323. [CrossRef]
49. Kowalski, S.; Lukasiewicz, M.; Duda-Chodak, A.; Zięć, G. 5-Hydroxymethyl-2-Furfural (HMF)—Heat-Induced Formation, Occurrence in Food and Biotransformation—A Review. *Pol. J. Food Nutr. Sci.* **2013**, *63*, 207–225. [CrossRef]
50. Shapla, U.M.; Solayman, M.; Alam, N.; Khalil, M.I.; Gan, S.H. 5-Hydroxymethylfurfural (HMF) Levels in Honey and Other Food Products: Effects on Bees and Human Health. *Chem. Cent. J.* **2018**, *12*, 35. [CrossRef]
51. Ćirić, J.; Sando, D.; Spirić, D.; Janjić, J.; Bošković, M.; Glišić, M.; Baltić, M.Ž. Characterisation of Bosnia and Herzegovina Honeys According to Their Physico-Chemical Properties during 2016-2017. *Meat Technol.* **2018**, *59*, 46–53. [CrossRef]
52. Makhloufi, C.; Taïbi, K.; Ait Abderrahim, L. Characterization of Invertase and Diastase Activities, 5-Hydroxymethylfurfural Content and Hydrogen Peroxide Production of Some Algerian Honeys. *Iran. J. Sci. Technol. Trans. Sci.* **2020**, *44*, 1295–1302. [CrossRef]
53. Mavric, E.; Wittmann, S.; Barth, G.; Henle, T. Identification and Quantification of Methylglyoxal as the Dominant Antibacterial Constituent of Manuka (*Leptospermum scoparium*) Honeys from New Zealand. *Mol. Nutr. Food Res.* **2008**, *52*, 483–489. [CrossRef] [PubMed]

 *foods*

Review

The Importance of Testing the Quality and Authenticity of Food Products: The Example of Honey

Natalia Żak * and Aleksandra Wilczyńska

Department of Quality Management, Gdynia Maritime University, ul. Morska 81-87, 81-225 Gdynia, Poland; a.wilczynska@wznj.umg.edu.pl
* Correspondence: n.zak@wznj.umg.edu.pl

Abstract: The aim of this study was to review methods of honey testing in the assessment of its quality and authenticity. The quality of honey, like other food products, is multidimensional. This quality can be assessed not only on the basis of the characteristics evaluated by the consumer during purchase and consumption, but also on the basis of various physicochemical parameters. A number of research methods are used to verify the quality of honeys and to confirm their authenticity. Obligatory methods of assessing the quality of honey are usually described in legal acts. On the other hand, other, non-normative methods of honey quality assessment are used worldwide; they can be used to determine not only the elementary chemical composition of individual types of honey, but also the biological activity of honey and its components. However, so far, there has been no systematization of these methods together with a discussion of problems encountered when determining the authenticity of honeys. Therefore, the aim of our study was to collect information on the methods of assessing the quality and authenticity of honeys, and to identify the problems that occur during this assessment. As a result, a tabular summary of various research methods was created.

Keywords: quality and authenticity assessment; honey

1. Honey Quality and Authenticity

Many attempts have been made to define food quality. One of them is based on the theory of a multidimensional set of features that can be objective (measurable, taking into account research results) or subjective (immeasurable, taking into account the opinion of the consumer) [1,2]. A universal definition of quality has been proposed in an international standard on quality management: "Quality is the degree to which a set of inherent properties meets the requirements" [3]. This definition can also be applied to food, and the requirements mentioned here can be both requirements contained in legal acts and consumer requirements. Food product quality is a concept that corresponds to a set of many attributes (e.g., product specific features, product safety, acceptance by the consumer). One of the attributes of food quality is its authenticity. Food authentication and traceability are current topics in the food sector since they enable food quality and safety control [4,5]. According to researchers, the authenticity of the product is understood as confirmation of the requirements for ensuring quality, composition, safety, usability, brand, and origin along with the information/declaration provided to the consumer by the manufacturer. It determines whether the product is really what the manufacturer declared [6–10]. The global definition of food authentication is problematic. There is no clear definition of this concept in legal acts covering the US, EU countries and cooperating countries. The definition of food authenticity is also not included in the Codex Alimentarius, which introduces and promotes definitions and requirements for food that facilitate the harmonization of international food circulation. Only the definition of contamination is formulated here, but there is no definition of authenticity or the adulteration of food [11].

The definition of authenticity has evolved over time and with the development of production, research infrastructure and research. Initially, this phenomenon was associated

Citation: Żak, N.; Wilczyńska, A. The Importance of Testing the Quality and Authenticity of Food Products: The Example of Honey. *Foods* **2023**, *12*, 3210. https://doi.org/10.3390/foods12173210

Academic Editor: Evandro Bona

Received: 29 June 2023
Revised: 16 August 2023
Accepted: 24 August 2023
Published: 25 August 2023

only with food counterfeiting and misleading the consumer. The composition of the product was changed without informing consumers. For example, more valuable ingredients were replaced with less valuable ones. The counterfeit product resembled the original product, but its quality was lower [12]. Further actions violating the authenticity of the products and intentionally misleading the consumer concerned their improper labeling. For example, terms such as "bio", "eco", "protected designation of origin" and "protected geographical indication" were used unlawfully without identifying the origin of the raw materials [5,8,13,14].

In the case of food, it is much more often said that it is adulterated than authentic. According to Spink and Moyer [15], food adulteration is: "A collective term that includes knowingly and intentionally substituting, adding, tampering with, or misrepresenting food, food ingredients, or food packaging: false or misleading claims about a product for economic gain". However, according to Everestine et al. [16], "food is adulterated intentionally for financial gain".

It is also a great challenge to clearly define the quality and authenticity of products, taking into account their multi-criteria parameters. For this, fast and reliable methods must be available, which will be supported by specific and reliable markers. All this is aimed at withdrawing counterfeit products from the market, but also at preventing similar accidents [17]. Below, Table 1 presents the factors that define the quality and authenticity of food products and the methods of their assessment.

Table 1. Factors that define the quality of food products and the methods of their assessment.

Name	Parameters	Assessment Methods
Sensory and organoleptic attributes	- Color; - Taste; - Smell; - Texture; - Structure; - Appearance.	1. Objective measurements of physicochemical features responsible for shaping organoleptic features. 2. Use of sensory panels carried out by a qualified research team.
Safety	Level of presence of toxic substances; organic and inorganic food additives; and microbiological, biochemical, chemical, physical and technological contamination.	Chemical, microbiological and biochemical analysis.
Health and nutritional value	The nutrient and non-nutrient content of the product and its energy value. In addition, it indicates the presence, assimilation and impact on the body of food additives, often with health-promoting effects, such as probiotics, polyphenolic compounds or vitamins.	Chemical and biochemical analysis of the composition of products and based on biological experiments.
Functional features	They mainly concern related aspects such as the ease of use of ingredients for processing, but also the size of the portions; in addition, their range is responsible for the characteristics of resistance to damage and storage stability.	Physical, biochemical and chemical analysis of raw materials and finished products.
Psychological parameters	The use of such features as convenience and ease of use at the appropriate price level and the level of novelty and attractiveness, taking into account the individual characteristics and needs of the consumer, makes the product habituated.	1. Market and consumer behavior research. 2. The study of physiological reactions to stimuli and related behaviors

Source: [2,4,18–20].

The evaluation of the quality of food products through the performance of a series of analyses and tests is a requirement resulting from legal acts. The main purpose of these rules is to ensure food safety. In addition, food products must meet the requirements of consumers and companies [21,22]. The quality of food products is also treated as an element of marketing, competitiveness and company prestige, which translates into an increase in profits [12].

Being a natural product, honey is also considered to be one of the most frequently adulterated products. Therefore, issues related to ensuring its quality and safety have put it at the forefront of the mind of global trading concerns and food regulatory agencies [23]. The literature gives the opportunity to indicate many practices used by beekeepers and honey producers that distort the authenticity of honey, such as the following:

- Mixing honey with water and sugar or selling solutions of water, sugar and flour, and boiled flowers;
- Mixing varieties;
- The sale of imported honeys (often of lower quality, not meeting the requirements as to the composition and properties) or their mixture with domestic honeys;
- The addition of imported honeys containing residues of drugs prohibited in EU countries due to their toxic effects (e.g., chloramphenicol—an antibiotic found in honeys from China);
- Placing incorrect data on the botanical and geographic origin of the product;
- Added sugar syrups (glucose–fructose);
- The addition of potato and beetroot syrup;
- The addition of molasses;
- Adding inverts to honey in order to increase its commercial weight and achieve quick profits (an illegal practice and foreign to beekeeping ethics);
- Feeding bees with sugar during the nectar period of plants;
- The repeated heating of honey in order to decrystallize it;
- Harvesting honey before its maturity;
- The overuse of veterinary drugs and antibiotics [12,23–32].

Honey is subject to the general requirements of the EU and national legislation. Presently, obligatory quality requirements for Polish honey are specified in the Resolution of the Ministry of Agriculture and Rural Development, dated 3 October 2003, regarding the detailed requirements for the commercial quality of honey. The above resolution basically corresponds with the requirements of the Worldwide Standard for honey, developed and approved by the Commission of Food Code from 2001 (Codex Alimentarius: Draft revised standard for honey 2001) [11], and to the European Directive for honey [33]. Evaluation of the quality of honey in accordance with those standards includes determining its organoleptic characteristics, distinguishing dominant pollens and indicating its basic physicochemical parameters (moisture, electrical conductivity, 5-hydroxymethyl furfural content, apparent reducing sugars, apparent sucrose, insoluble matter and diastase activity). Table 2 presents limit values for individual physicochemical parameters and an interpretations of their excess.

Table 2. Physicochemical requirements for honeys, including the interpretation of exceeded parameters.

Parameter	Limit Value	Exceeding the Limit Values of the Parameters
Water content	Not more than 20%; however, not more than: (1) 23%—in heather honey and baker's honey; (2) 25%—in heather baker's honey.	The water content is considered as an indicator of honey stability and resistance to yeast fermentation. At a high level, it causes not only the fermentation and spoilage of honey, but also loss of taste. In addition, water activity is a parameter responsible for the growth of microorganisms [34,35]. In the case of changes in the water content, especially an increase, it can be presumed that water was added to the honey or it may indicate that the honey was removed from the hive too quickly. This parameter is also influenced by weather conditions during honey harvesting, e.g., the intensity of rainfall. An increase in its value is dangerous because it can affect the development of yeast and mold in honey [30,36]. Changes in this parameter may indicate adulteration of honey by adding, for example, invert sugar or potato (starch) syrup. It is added to honey to increase its weight. Adding water to honey has the same effect. This is very unfavorable for honey, as the increased amount of water increases the tendency to ferment [37].

Table 2. *Cont.*

Parameter	Limit Value		Exceeding the Limit Values of the Parameters
Reducing sugar content (sum of fructose and glucose)	Not less than 60 g/100 g (nectar)	Not less than 45 g/100 g (nectar and honeydew)	The quantitative ratio of glucose to fructose is the main factor used to classify monofloral honeys [38]–acacia honey contains, on average, 34.6 g of fructose and 21.6 g of glucose; rapeseed honey, 37 g of fructose and 36.7 g of glucose; and dandelion honey, 35.9 g of fructose and 37.6 g of glucose [39,40]. It was observed that in honeydew honeys, the ratio of fructose to glucose content is higher than in nectar honeys. The only exception is black locust honey, in which the predominance of fructose over glucose ensures a liquid consistency for as long as several months. The preponderance of glucose is responsible for the rapid crystallization of honey, e.g., in rapeseed or dandelion honeys [38–40].
Sucrose content	Not more than 5 g/100 g, except when not more than: 1. A total of 10 g/100 g—in locust (*Robinia pseudoacacia*), alfalfa (*Medicago sativa*), firewood banksia (*Banksia menziesii*), sweetvetch (*Hedysarum*), red rubber (*Eucalyptus camaldulensis*), leatherwood (*Eucryphia lucida, Eucryphia milligani*) and *Citrus* spp. honey; 2. A total of 15 g/100 g—in lavender (*Lavandula* spp.) and borage (*Borago officinalis*) honey.		The content of saccharides in honey is determined, inter alia, by the origin of the honey, the time of harvesting and the length of the storage period; honeydew honeys contain more oligosaccharides and dextrins, while nectar honeys are dominated by simple sugars. Unripe honey contains the highest amount of sucrose. The content of individual carbohydrates may indicate the maturity of the honey or its lack [30,40]. An increased sucrose content may indicate adulteration of honey by feeding bees with sucrose or its addition to honey, but also the mixing varieties of honey [36]. Another very important indicator of honeydew honey is honeydew sugar—melezitose; this sugar is not found in nectar honeys. It may be an indicator of mixing honey varieties or a lack of varietal purity [27,30].
Free acid content	Not more than 50 mval/kg, but not more than 80 mval/kg in baking honey (industrial).		The value of the level of free acids—the level of free acids may indicate the maturity of the honey, as well as disorders related to the microbiological contamination of honey [35,41]. They can determine the taste and aroma of honey.
Diastase number (according to the Schade scale)	Not less than 8, except for baker's (industrial) honey, but not less than 3 in honey with naturally low enzyme activity and an HMF content not more than 15 mg/kg.		The diastase number depends on the type and origin of the honey. Diastase (α-amylase) and invertase—enzymes derived from the salivary glands of bees—are among the most important biological components of honey. It is their presence that determines the nutritional and health-promoting properties of honey. The diastase number is an indicator of the enzymatic activity of honey. It is expressed by the number of Schade units per 1 g of honey. The level of diastase activity is one of the most important indicators that prove the high quality of honey. In the case of a low value of the diastase number, it can be presumed that the honey was heated to a temperature above 40 °C, which may also indicate the addition of sugar syrup [42] and the long storage of honey in unfavorable conditions [43,44].
5-hydroxymethylfurfural (HMF) content	Not more than 40 mg/kg, except for baker's honey (industrial); also not more than 80 mg/kg in honey from regions with a tropical climate, and in mixtures of such honeys.		Content of 5-HMF (5-hydroxymethylfurfural)—this natural component of honey, heterocyclic 5-hydroxymethylfurfural aldehyde, is formed in an acidic environment from fructose (2-oxohexose). Natural honey does not contain 5-HMF or it is present in very small amounts (2–7 mg/kg). Its content may increase with prolonged storage and a too-high processing temperature; hence, 5-HMF is called the honey aging parameter. The content of 5-HMF proves not only the quality, but also the authenticity of the honey—the increased content of this ingredient also indicates adulteration with invert sugar or starch syrup. In the case of a very high level of this compound above 200 mg/kg, adulteration with chemical invert can be presumed [43–46]. HMF exhibits mutagenic activity and causes damage to the structure of the DNA helix [47]. HMF derivatives in the form of 5-sulfooxymethylfurfural (SOMF), 5-chloromethylfurfural and 5-hydroxymethyl-2-furancarboxylic acid (5-HMFK) have cytotoxic, genotoxic, neurotoxic, mutagenic and carcinogenic effects, which can lead to neoplastic changes in the liver, skin and lower colon tissues [42,48,49].
Proline content, mg/100 g of honey	Not less than 25 mg/100 g of honey.		Proline is an amino acid that is predominant in honey. In the case of the adulteration of natural honey with sucrose, a decrease in its content up to 10 mg/100 g of honey is observed [50]. A high proline content indicates the maturity of the honey; this indicator is often used in research as a honey quality parameter. The largest amounts of proline are found in buckwheat honey (approx. 80.8/100 g) [40].

Table 2. *Cont.*

Parameter	Limit Value	Exceeding the Limit Values of the Parameters
Conductivity	Not more than 0.8 mS/cm, except for honeys and their mixtures listed below, but not less than 0.8 mS/cm in honeydew honey, chestnut honey and their mixtures. The conductivity of originating honey is not specified from strawberry tree (*Arbutus unedo*), heather (*Erica*), eucalyptus, linden (*Tilia* spp.), common heather (*Calluna vulgaris*), manuka leptospermum and tea tree (*Melaleuca* spp.)	The electrical conductivity of honey, as one of the physicochemical parameters, can be used to characterize its botanical origin, because it depends to a large extent on the plant from which honey was made [51,52]. The value of the electrical conductivity depends on the value of the level of mineral compounds and honey acids. In the case of this parameter, nectar honeys should have a value of up to 0.8 mS $\times$ cm^{-1}, and honeydew honeys, due to a greater presence of minerals, should have this value above the level indicated above, even up to 1.5 times—up to 0.17 mS/cm^2. Any deviations from this standard may indicate the mixing of nectar and honeydew honeys. The reduced value of electrical conductivity in honeydew honeys may indicate adulteration with nectar honey. Increased sucrose content combined with a reduced electrical conductivity, diastase number and proline content may indicate the adulteration of honey with sugar syrup [43,52].
Content of insoluble substances	No more than 0.1 g/100 g, but not more than 0.5 g/100 g in pressed honey	The content of insoluble substances indicates contamination of the hive or the product itself. The presence of these substances can lead to product contamination and consumer exposure [37,52].

Source: [11,24,26,53].

Honeys adulterated with sugar syrup or other inverts may contain starch dextrins and the wrong profile of sugars. An analysis of the literature showed that a reduced level of electrical conductivity and diastase number, while increasing the sucrose content, may indicate the deliberate adulteration of honey with sugar syrup [53]. Too-low values of such parameters as ash content, enzyme activity or proline content may mean adulteration of honey with sugars [51].

The quality and quantity determination of the basic components of honey does not pose considerable difficulties because it is accomplished using simple analytical techniques. But these parameters do not allow for a precise determination of the geographical or botanical origin of the honeys. In addition to assessing the quality of the physicochemical parameters, Kowalski and Łukasiewicz indicated the most common substances added to natural honey for the purpose of adulterating it and indicators allowing for their detection [54] (Figure 1).

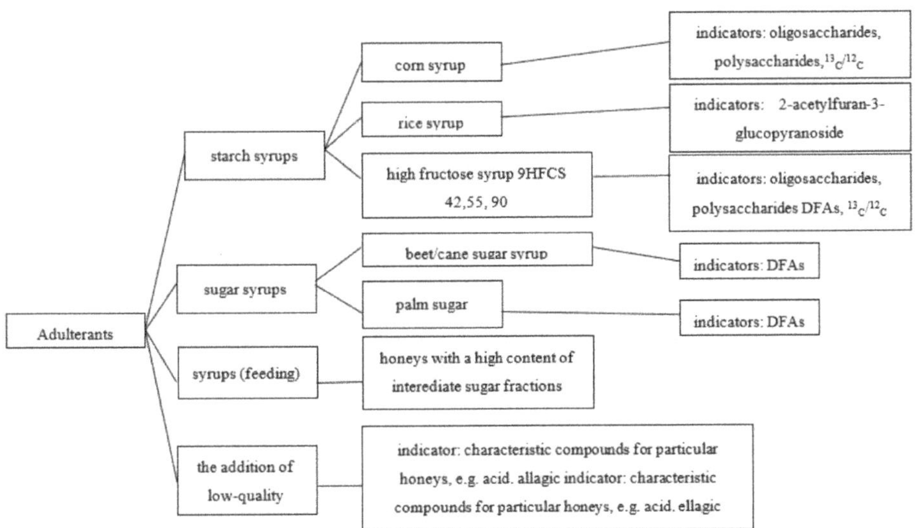

Figure 1. The most common substances added to natural honey for the purpose of adulterating it and indicators allowing for their detection. Source: [54].

The concepts of quality and the authenticity of honey are interpenetrating notions, because authenticity is one of the attributes of quality. In the case of testing the quality and authenticity of honey, the tests performed can be divided into a number of methodological groups resulting from legal acts [27–34,55] and based on modern methods used by numerous research teams. Therefore, the aim of this study was to determine the importance of the food product testing factor in assessing the quality and authenticity of honey and the systematization of this information.

2. A Review of the Methods for Assessing the Quality and Authenticity of Honeys

The selection of the appropriate research method and interpretation of the results of quality parameters in terms of confirming the authenticity of honey is not easy. It should be remembered that honey is a unique product whose quality is influenced by many parameters. In connection with the above, a review of these methods was created, from simple methods to more advanced methods. Table 6 presents problem identification factors and groups of honey authenticity problems, with the aim of facilitating the selection of research methods. In addition, methodological limitations are indicated for each method.

The research material consisted of original research articles in the field of the quality and authenticity assessment of varietal honeys, which were published up to the first half of 2023, totaling more than 160 scientific articles. The search for the articles was carried out by entering key words, such as honey, honey authenticity, geographical origin, honey quality tests, honey storage, honey color, pollen analysis, 5-HMF, diastase number, antioxidants, spectroscopy and fluorescence. A search of publications from the last 10 years was performed, but sometimes older works were included. The condition for using older publications was the well-founded knowledge contained in them. Research databases were searched, such as Google Scholar, Science Direct, MDPI, Knovel, Applied Science & Technology Source, Scopus, Taylor and Francis online and Ebsco host.

Based on the analysis of the literature, 16 methodological groups were distinguished for research in the field of honey quality and authenticity—Table 6. Below the table, there are comments on the limitations and disadvantages of the use of individual methods.

Table 3. Research methods used in assessing the quality and authenticity of varietal honeys.

Methods Used—Name	Applied Methods	Problem Identification Factors	Group of Honey Authenticity Problems
Melissopalynological analysis (honey pollen analysis)	Quantitative analysis consists of counting all plant parts (N), i.e., pollen grains, fungal spores and algae hyphae, yeast, starch grains and others in 10 g of honey. As a result, honey is assigned to one of five classes.	Identification of the leading pollen.	Identification of the botanical/geographical origin of the honey.
Sensory analysis	The bases for this form of research are the senses and feelings related to the smell, taste, color, appearance, and consistency of the product.	Identification of sensory features (color, taste, smell, appearance, consistency) characteristic of: - Varietal honeys, honey quality,—additives of prohibited substances; - Impurities - in honeys; - Signs of fermentation; - Consistency.	Identification of the botanical/geographical origin of the honey. Evaluation of the quality of bee honey. Counterfeit identification. Identification of bee honey fermentation. Identification of impurities.

Table 4. Research methods used in assessing the quality and authenticity of varietal honeys.

Methods Used—Name	Applied Methods	Problem Identification Factors	Group of Honey Authenticity Problems
Analysis of physicochemical parameters	The most useful parameters in identification are electrical conductivity, water content, total and active acidity, total ash content and sugar content, including the ratio of glucose to fructose concentration (especially important when identifying heather honey), the analysis of aromatic acids and amino acids, proline content, diastase number, proline content and pH level.	Identification of physicochemical parameters characteristic of: - Varietal honeys, - honey quality; - Additives of prohibited substances; - Impurities - in honeys; - Liquefied/heated/long-stored honeys.	Identification of the botanical/geographical origin of the honey. Evaluation of the quality of bee honey. Counterfeit identification. Identification of bee honey fermentation. Identification of impurities.
	The methods that deserve special mention are the determination of 5-HMF content and the determination of the diastase number.		Identification of heating/overheating of honey and improper storage conditions.
Measurements of color parameters in L * a * b * and X Y Z systems	Color parameters L * a * b * were determined in the international CIE system.	Identification of characteristic color parameters for varietal honeys.	Identification of too-long storage of honey.
Extraction of volatile compounds	The solid-phase microextraction (SPME) technique using gas chromatography coupled with a mass spectrometer (GC-MS).	Identification of volatile fractions of characteristic honeys for varietal honeys.	Identification of the botanical/geographical origin of the honey.
Analysis of antioxidant activity of honey and analysis of the presence of flavonoids	The botanical origin of honey significantly affects the antioxidant activity measured as the ability to scavenge DPPH• free radicals. The photochemiluminescence test (PCL).	Identification of the level of antioxidant activity and the total value of characteristic polyphenols for given honey varieties.	Identification of the botanical/geographical origin of the honey. Evaluation of the quality of bee honey. Counterfeit identification.
NMR (nuclear magnetic resonance) spectroscopic analysis	This analysis is very versatile and is used with principal component analysis (PCA). Metabolic analysis of organic extracts.	Identification and assessment of characteristic honey components for given honey varieties.	Identification of the botanical/geographical origin of the honey. Evaluation of the quality of bee honey. Counterfeit identification. Identification of impurities.

Table 5. Research methods used in assessing the quality and authenticity of varietal honeys.

Methods Used—Name	Applied Methods	Problem Identification Factors	Group of Honey Authenticity Problems
Analysis of honey microscopic image identification	This method shows the picture of honey.	Identification of additives and impurities.	Identification of honey adulteration with additives.
			Evaluation of the quality of bee honey.
			Identification of bee honey fermentation.
			Identification of impurities.
Analysis of the isotopic composition of honey using isotope ratio mass spectrometry (IRMS)	Measurement of the $^{13}C/^{12}C$ isotope ratio.	Identification of additives and impurities.	Identification of honey adulteration with additives.
Chromatographic analysis of honey composition	Chromatographic analysis using high-performance liquid chromatography (HPLC), gas chromatography (GC) and gas chromatography coupled with mass spectrometry (PTR-MS).	Identification of additives and impurities.	Identification of honey adulteration with additives.
			Evaluation of the quality of bee honey.
			Identification of bee honey fermentation.
			Identification of impurities.
Analysis of glycerin or ethanol content	Analysis of glycerin content.	Identification of characteristics for fermented honeys.	Evaluation of honey quality.
			Identification of bee honey fermentation.
Fluorescence spectroscopy research	The advantage of fluorescence spectroscopy is the high sensitivity and specificity of classification.	Identification and assessment of honey authenticity.	Identification of the botanical/geographical origin of the honey.
Infrared spectroscopic analysis	Infrared spectroscopy covers the spectrum of electromagnetic radiation in the range between the visible region and the microwave region (14,300 and 200 cm^{-1}; 700–50,000 nm).	Identification and assessment of honey authenticity.	Identification of honey adulteration with additives.
		Identification of ingredients determining the quality of natural bee honeys	Including, in particular, the adulteration of honey with sugar syrup from C4 plants. Identification of the botanical/geographical origin of the honey.
			Evaluation of the quality of bee honey.
Research on electrical properties	The electrical properties of materials (impedance, permittivity and dielectric loss factor) describe the behavior of the material in an electric field. The molecular structure of the material is responsible for the physical and chemical properties, so there is a relationship between the electrical properties of a given material and its physical and chemical parameters.	Identification and assessment of honey authenticity. Identification of characteristics for fermented honeys.	Identification of honey adulteration with additives.
			Evaluation of honey quality.
			Identification of the botanical/geographical origin of the honey.

Table 6. Research methods used in assessing the quality and authenticity of varietal honeys.

Methods Used—Name	Applied Methods	Problem Identification Factors	Group of Honey Authenticity Problems
Analysis of the microbiological purity of honey	The examination of the microbiological contamination of honey is aimed at assessing its quality; the parameters usually determined are coliform bacteria, sulfite-reducing *Clostridium*, yeasts and molds, aerobic mesophilic bacteria, *Salmonella* spp. and *Bacillus spp.*	Identification of characteristics for fermented honeys. Identification and assessment of honey authenticity.	Identification of honey adulteration with additives. Identification of the botanical/geographical origin of the honey.
Research on rheological properties of honeys	The crystal structure is a valuable source of information about honey. The rheological properties of honey indicate the characteristics of their origin and quality.	Identification and assessment of honey authenticity.	Identification of honey adulteration with additives. Evaluation of honey quality.

Source: Own research.

2.1. Melissopalynological Analysis (Honey Pollen Analysis)

Quantitative analysis consists of counting all plant parts (N), i.e., pollen grains, fungal spores and algae hyphae, yeast, starch grains and others in 10 g of honey. It allows honey to be assigned to one of five classes.

The qualitative analysis determines the varieties of honey, with particular emphasis on honey and their additives from other climatic zones. It is the basis for the determination and classification of the nectar plants involved in the production of honey. It consists of counting pollen grains in a microscope preparation, and then comparing them with the provisions regulating the content of guiding pollen in varietal honeys.

The minimum percentages of guiding pollen for honeys are as follows: rapeseed—45%, acacia—30%, linden—20%, buckwheat—45%, heather—45%, and polyfloral—none.

This method is a classic approach to confirming the botanical origin of honey. It is useful in the control and classification of honeys of individual varieties and those imported from different regions of the world. However, it is a time-consuming method and depends on the expert's experience. This method also allows nectar honey to be distinguished from honeydew [26,55–65].

It is based on the assumption that certain types of pollen are present in a given area, which makes it possible to determine the origin of honey on this basis [65].

2.2. Sensory Analysis

The bases for this form of research are the senses and feelings related to the smell, taste, color, appearance, and consistency of the product. This is a characterization analysis for honey varieties and their geographical origin, but also contributes to determination of their quality.

It is used to control the quality level and classification of honeys of particular varieties and to detect changes in physicochemical and biological parameters. The method is dependent on the experience of the assessment team and must be supported by physicochemical tests. The result of the study depends on the experience of the research team [66–72].

This method is subjective and unreliable when examining less well-known honeys, because there is no reference point [65].

Exceeding the limit values of the sensory parameters.

The color, smell and taste of honey may depend on many factors, such as the origin of pollen, climate, weather conditions and storage time and conditions. Any change in these factors may result in organoleptic characteristics different than those standard for a given variety. The composition of the colored substances depends on the botanical origin of the honey and the place where the melliferous plants grow [54]. The content of aromatic substances decreases during heating and long storage [34,35].

Honey in its fresh and mature form should be a clear, highly hygroscopic liquid with a density of 1.38–1.45 g/cm^3. The concentration of sugars (especially invert sugar and sucrose) affects the viscosity and density of honey in direct proportion [53].

Viscosity and crystal formation—crystallization is a natural process in honey. This process does not reduce the quality of the honey, but consumers prefer liquid honey. Glucose is responsible for the crystallization of honey, which is in a supersaturated state and therefore tends to reach equilibrium by crystallizing. Honeys with a predominance of fructose over glucose crystallize more slowly or not at all, e.g., acacia honeys. Fructose concentrates the solution, along with other sugars, and increases its viscosity, which makes it difficult for honey to crystallize. On the other hand, honeys with a predominance of glucose crystallize faster (rapeseed honey and dandelion honey). This parameter is not regulated by legal acts, but the very appearance of honey can indicate whether the honey is of the right variety and whether it has been heated [52].

2.3. Analysis of Physicochemical Parameters

The physicochemical parameters of honey quality are the basis for identifying the authenticity and adulteration of honeys.

The most useful parameter in identification is electrical conductivity. The use of this parameter makes it possible to distinguish nectar honeys of some varieties in relation to multifloral nectar honeys and, above all, the group of honeydew nectar honeys.

Other methods of differentiating honey varieties are as follows:

- Determination of the water content;
- Determination of the total and active acidity;
- Determination of the total ash content;
- Determination of the sugar content, including the ratio of glucose to fructose concentration (especially important when identifying heather honey);
- Analysis of aromatic acids and amino acids;
- Determination of the proline content;
- Determination of the diastase number;
- Determination of the proline content;
- Determination of the pH.

However, these methods used alone do not allow for the unambiguous differentiation of varietal honeys into particular types and varieties.

The methods that deserve special mention are as follows:

- Determination of the 5-HMF content;
- Determination of the diastase number.

These parameters make it possible to determine the level of honey aging and errors related to improper storage and thermal processing. The values of the above parameters change with the time of honey storage [61,69–84].

According to Popek [66], these methods cannot be considered fully reliable because the parameters change over time. These methods are time-consuming, cost-intensive, and their result does not provide unambiguous information about the authenticity of the honey. There are often problems with the interpretation and reproducibility of the results. Furthermore, the amount of reagent used to determine one sample has a negative impact on the environment.

*2.4. Measurements of Color Parameters in L * a * b * and X Y Z Systems*

The color of honey is one of the first features assessed by consumers.

Tristimulus colorimetry was instrumentally used to assess the color of honey. Color parameters L * a * b * were determined in the international CIE (Commission Internationale de l'Éclairage) system. The color is expressed in the CIE L * a * b * system, where L * is the lightness, and the a * and b * coordinates indicate the contribution of green (negative a * values), red (positive a *), blue (negative b * values) and yellow (b * values positive) [55,85–87].

This method requires properly prepared honey, which should be liquid with no signs of crystallization; otherwise, the test results may be different. The test is quick and easy to perform, but one should remember about the cost of purchasing the equipment and ensuring the repeatability of the test [85,86].

2.5. Extraction of Volatile Compounds

The suitability of the solid-phase microextraction (SPME) technique using gas chromatography coupled with mass spectrometry (GC-MS) is still being tested in determining the botanical authenticity of honeys. The qualitative analysis of volatile compound profiles is used to determine the botanical authenticity of nectar honeys [59,69,88–97].

2.6. Analysis of the Antioxidant Activity of Honey and Analysis of the Presence of Flavonoids

Polyphenolic compounds are among the most active antioxidants present in food. The use of spectral methods was aimed at determining the profiles of polyphenolic compounds, as well as assessing the antioxidant potential of individual varieties of bee honey. The botanical origin of honey significantly affects the antioxidant activity measured as the ability to scavenge DPPH• free radicals [75,98–102].

The use of the photochemiluminescence (PCL) test consists of the optical excitation of a UV sensitizer, which is responsible for generating free radicals, partially eliminated by the antioxidants present in the sample. Other radicals cause luminescence of the detected substance. The function of the sensitizer and detector is performed by the same compound, luminol. The measurement is fast and accurate as it only takes a few minutes to calculate based on the calibration curve, performed automatically by the software [84,103–107].

2.7. Nuclear Magnetic Resonance Spectroscopic Analysis

This analysis is very versatile, not only to assess the identification of the botanical origin of honey, but also to determine the composition and quality of the honey ingredients. For example, the diversity of honey components, including saccharides and all kinds of amino acids, is determined, which confirms their grouping according to the origin of the honey (using principal component analysis—PCA) [108–117].

Nuclear magnetic resonance (NMR) provides structural information with high reproducibility and accuracy. The time to obtain a 1H NMR spectrum is short (less than 5 min) and does not require calibration or standards. The advantage of the NMR method is the simultaneous detection of organic compounds in an unchanged state and conformation. The use of low-field 1 H NMR allows, based on increasing relaxation times, the detection of the addition of HFCS syrup in honey. The effective use of the NMR technique to identify honey adulteration is a very promising direction of research, but due to the relatively small number of reports, it is necessary to create an appropriate spectrum database, allowing for the quick interpretation of the results. Another factor limiting the use of NMR in analytics is the very high cost of the equipment [54].

2.8. Analysis of Honey Microscopic Image Identification

This is a method used when sweeteners are added to honey, e.g., sucrose by feeding bees or the adulteration of honey with the addition of sugar cane or fructose. In addition, this method shows the picture of honey impurities, e.g., nanoparticles or the presence/amount of yeast in the case of honeys with a high water content (e.g., added water). This test does not indicate unequivocal fermentation [118].

2.9. Analysis of the Isotopic Composition of Honey Using $13^C/12^C$ Isotope-Ratio Mass Spectrometry Measurement

One of the spectrometric methods used to detect the adulteration of honey with the addition of cane or corn sugar and the incorrect declaration of the origin of honey is isotope-ratio mass spectrometry (IRMS) analysis. Isotope content is related to latitude, i.e., the climate prevailing in the place where the honey is obtained. This method is based on the use of proportions of isotopes characteristic of particular plant species. Its task is to estimate the amount or ratio of isotopes of one of the three basic elements ($13^C/12^C$, $180/16O$, $2H/1H$) and compare them with standard values [119–123].

This is a method with great potential, but it requires expensive equipment and specialized staff. In addition, its universal application does not guarantee the correctness of the results [124].

2.10. Chromatographic Analysis of Honey Composition

Chromatographic analysis using high-performance liquid chromatography (HPLC), gas chromatography (GC) and gas chromatography coupled with mass spectrometry (PTR-MS) is the basis for determining the composition and quality of bee honey. The results of these analyses are interpreted using statistical tools. The determination is quick, easy and effective, but costly. This method has many advantages: speed of measurement, low cost and use of a small amount of the test sample, which will not be destroyed. Rich libraries of spectra facilitate the identification of unknown substances. In the spectrum, different peaks may overlap, which can make interpretation difficult. In addition, the cost of purchasing equipment is high [125–129].

These methods allow for the differentiation of the botanical origin of monofloral and polyfloral honeys. However, HPLC shows an advantage over PTR-MS by providing much better differentiation of all analyzed types of honey. Chromatographic fingerprints recorded at 210 nm allow for the best classification of honey. Mass spectrometry with the proton transfer reaction is useful for distinguishing buckwheat honey [125–129].

2.11. Analysis of Glycerin or Ethanol Content

One of the methods used in assessing the authenticity (distinguishing natural from artificial honey) and freshness of honeys is the analysis of glycerin content. Glycerin is the result of metabolic processes caused by microorganisms present in the liquid collected by bees [130].

2.12. Fluorescence Spectroscopy Research

The advantage of fluorescence spectroscopy is the high sensitivity and specificity of classification. Fluorescence spectroscopy requires only minimal sample preparation. The results of the above studies confirmed that single synchronous fluorescence spectra of different honeys differ significantly due to their different physicochemical properties and provide sufficient data to clearly differentiate between groups of honeys. Studies have shown that this method is a valuable and promising technique for honey authentication. Honeys are well known to contain numerous fluorophores, such as polyphenols and amino acids. Some of them have been proposed as markers for monofloral honeys—ellagic acid for heather honey; hesperetin for citrus honey; phenylalanine and tyrosine, which turned out to be characteristic of lavender honey and made it possible to distinguish it from eucalyptus honey; and tryptophan and glutamic acid, which turned out to be useful for differentiating honeydew and flower honeys. Due to the presence of such powerful fluorophores, fluorescence spectroscopy can be helpful in confirming the botanical origin of honey.

In addition, these tests can be the basis for identifying honey overheating and identifying the botanical origin of filtered honeys, in which pollen analysis is not possible [131–142].

The limitation of this method is building a database of spectra characteristic for honeys. The method is fast, cheap and without a negative impact on the environment (no reagents are used) [139].

2.13. Infrared Spectroscopic Analysis

The authenticity of different types of honey can also be confirmed by infrared spectroscopy.

Infrared spectroscopy covers the spectrum of electromagnetic radiation in the range between the visible region and the microwave region (14,300–200 cm^{-1}; 700–50,000 nm). Depending on the wavelength, it is divided into the following types of spectroscopy:

- Near infrared (NIR) spectroscopy, 14,300–4000 cm^{-1} (700–2500 nm);
- Mid (proper) infrared (MIR) spectroscopy, 700–4000 cm^{-1} (2500–14,300 nm);
- Far infrared (FIR) spectroscopy, 700–200 cm^{-1} (14,300–50,000 nm).

Quick quantitative and qualitative determination of the individual parameters that determine the quality of natural bee honeys is possible thanks to the use of spectroscopy in the NIR range. Through the basic analysis of the spectra, it is possible to distinguish between honeydew, artificial and nectar honeys, while using chemometrics, it is possible to determine the varieties of nectar honeys.

Studies conducted over the years have also shown the possibility of the presence of corn fructose in honey [78,79,127,133,143–145].

This method has many advantages: speed of measurement, low cost and use of a small amount of the test sample, which will not be destroyed. Rich libraries of spectra facilitate the identification of unknown substances. However, it should be remembered that different peaks in the spectrum may overlap, which can make interpretation difficult [143–147].

2.14. Research on Electrical Properties

The electrical properties of materials (impedance, permittivity and dielectric loss factor) describe the behavior of the material in an electric field. The molecular structure of the material is responsible for the physical and chemical properties, so there is a relationship between the electrical properties of a given material and its physical and chemical parameters. It is possible to use different dielectric quantities (electric permittivity, dielectric loss coefficient and conductivity) to differentiate honey varieties, and additives such as water and sugar are still being researched. In addition, these tests can also describe the level of honey overheating and the degree of crystallization [148–156].

2.15. Analysis of the Microbiological Purity of Honey

The examination of microbiological contamination of honey is aimed at assessing its quality. The parameters usually determined are coliform bacteria, sulfite-reducing *Clostridium*, yeasts and molds, aerobic mesophilic bacteria, *Salmonella* spp. and *Bacillus* spp. [73,113,152,153,156,157].

This method is time-consuming and not always effective. Its reproduction in the same conditions is impossible [73,156,157].

2.16. Research on Rheological Properties of Honeys

The crystal structure is a valuable source of information about honey. The rheological properties of honey indicate the characteristics of their origin and quality. The possibility of crystal formation and the observation of their behavior with the use of birefractive interferometry and computer image analysis can present the quantitative characteristics of the honey crystal structure used for its assessment. The analyses take into account standard stereological parameters, such as the number of identified objects and average values— surface area, circumference and maximum diameter of crystals. In addition, the numerical distribution of crystals with regard to the maximum diameter is analyzed. They can be used both to identify its origin and for other purposes, e.g., to determine the rheological properties of crystallized honey [157–159].

3. Conclusions

The authenticity of food products is rarely defined in the literature. However, there is some agreement in terms of nomenclature and a set of features that make it possible to recognize that a food product is what it should be, as declared by the manufacturer. Food adulteration is a frequent phenomenon, which should be considered a significant threat to every consumer—it is a violation of consumer rights, but also often leads to an increase in risk associated with the consumption of food products. Therefore, it is necessary to develop tools that will protect the consumer against abuse from dishonest producers.

Honey is often mentioned as an example of a product that can be counterfeited in various ways—its composition is changed (e.g., by adding sweeteners), and in recent times, consumers are often misled by giving false information about the geographical or botanical origin of honey. There are many research methods that are used to assess the authenticity of honeys, but above all, they ensure the confirmation of their quality. Regardless of the method adopted, the goal is always to determine whether the tested product is manufactured fairly and meets all legal criteria, or whether there has been a violation of the law. In the field of food authentication, there are targeted and non-targeted analytic methods that are proven by many studies [160]. For each of the products, a number of research methods can be indicated, but as in the case of assessing the quality and authenticity of honey, there is no combination of methods that would be 100% effective [54,161,162]. Currently, honey authenticity tests are difficult because not only local honeys are sold on domestic markets, but also foreign honeys from other continents, which are characterized by a different chemical composition and properties. Therefore, researchers are still looking for an ideal method that will not leave even 1% uncertainty in confirming the quality and authenticity of honeys [16]. The above list shows how active various researchers are in the search for a universal and relatively simple method of confirming the authenticity of honeys.

The present work provides a review not only of research methods, but also of their practical use. The methodological limitations and strengths of each method have been indicated. The work can be used as a resource and a quick path to finding an appropriate research method to determine the quality and authenticity of honeys.

In the example of honey quality assessment, it can be seen that single-component or multi-component analyses of honey quality parameters do not lead to obtaining unambiguous information on the botanical origin of honeys. However, they are helpful in more accurately indicating the place of origin and quality of honey. However, there is still no interdisciplinary, fast, effective and cheap method that can confirm the authenticity of honey, taking into account its quality characteristics.

The above analysis of the literature illustrates the pace at which the research and the methods used in the assessment of the quality of honey have developed. It demonstrates their differentiation, taking into account the aspects of varietal authenticity and the composition of honeys.

According to Kowalski and Łukasiewicz, the introduction of new, hitherto unknown techniques and the improvement of many measurement techniques (increasing their sensitivity and precision) are creating more and more tools and opportunities to identify falsifications, even at a very low level. However, it still requires an appropriate approach and experience from the analytical side, because honey is a product with a complicated analytical matrix [54]. This paper presents a review of the possible applications of methods with a high level of efficiency. Methods such as fluorescence, NIR and Raman spectroscopy seem to be multifaceted. However, they have limitations because they require a complex mathematical apparatus to interpret the results [54].

Summarizing this review of research methods, it was not possible to identify a method that is unequivocally the most effective. However, the analysis of research methods allowed us to identify 16 methodological groups in the field of honey quality and authenticity. The review of methods also allowed for the extraction of parameters indicating changes not only in the quality, but also in the authenticity of the honeys, along with the interpretation of exceeding the limit values of the parameters. The intention of the authors was to create a tool to help in selecting the most effective research methods, but also to combine several methods in order to obtain a reliable result.

Author Contributions: A.W. developed the conception of this work and developed the final version of this article. N.Ż. conducted the literature review and analyzed the results. All authors have read and agreed to the published version of the manuscript.

Funding: This research received no external funding.

Data Availability Statement: No new data were created.

Acknowledgments: The authors acknowledge the financial support of Gdynia Maritime University (project: WZNJ/2023/PZ/01, WZNJ/2023/PI/01).

Conflicts of Interest: The authors declare that there are no conflict of interest regarding the publication of this paper.

References

1. Haska, A.; Martyniuk, E. Wybrane metody wyróżniania produktów spożywczych na rynku. *Żywn. Nauka Technol. Jakość* **2019**, *26*, 18–31. [CrossRef]
2. Downey, G. *Advances in Food Authenticity Testing*; Elsevier: Amsterdam, The Netherlands, 2016.
3. *EN ISO 9001:2015*; Quality Management Systems—Requirements. International Organization for Standardization: Genève, Switzerland, 2015.
4. Galanakis, C.M. (Ed.) *Innovative Food Analysis*; Academic Press: Cambridge, MA, USA; Elsevier: Amsterdam, The Netherlands, 2020.
5. Medina, S.; Pereira, J.A.; Silva, P.; Perestrelo, R.; Câmara, J.S. Food fingerprints–A valuable tool to monitor food authenticity and safety. *Food Chem.* **2019**, *278*, 144–162. [CrossRef] [PubMed]
6. Cubero-Leon, E.; De Rudder, O.; Maquet, A. Metabolomics for organic food authentication: Results from a long-term field study in carrots. *Food Chem.* **2018**, *239*, 760–770. [CrossRef] [PubMed]
7. Le Gall, G.; Puaud, M.; Colquhoun, I.J. Discrimination between orange juice and pulp wash by 1H nuclear magnetic resonance spectroscopy: Identification of marker compounds. *J. Agric. Food Chem.* **2001**, *49*, 580–588. [CrossRef]
8. Rubert, J.; Lacina, O.; Zachariasova, M.; Hajslova, J. Saffron authentication based on liquid chromatography high resolution tandem mass spectrometry and multivariate data analysis. *Food Chem.* **2016**, *204*, 201–209. [CrossRef]
9. Tena, N.; Aparicio-Ruiz, R.; Koidis, A.; García-González, D.L. Analytical tools in authenticity and traceability of olive oil. In *Food Traceability and Authenticity*; CRC Press: Boca Raton, FL, USA, 2017; pp. 232–260.
10. Spink, J.W. *Food Fraud Prevention: Introduction, Implementation and Management*; Springer: New York, NY, USA, 2019; p. 25.
11. Codex Standard for Honey, European Regional Standard CXS-12-1981, Codex Alimentarius, International Food Standards, Rev. 1. 1987, Rev. 2 2001. FAO, WHO, 2019, p. 52. Available online: www.fao.org/input/download/standards/310/cxs_012e.pdf (accessed on 16 February 2021).
12. Śmiechowska, M. Wybrane problemy autentyczności i identyfikowalności żywności ekologicznej. *J. Res. Applic. Agric. Eng.* **2007**, *52*, 80–88.
13. Rozporządzenie Ministra Rolnictwa i Rozwoju Wsi z dnia 23 Grudnia 2014 r. *W Sprawie Znakowania Poszczególnych Rodzajów Środków Spożywczych*; Polish Minister: Warsaw, Poland, 2015; p. 29.
14. Rozporządzenie Ministra Rolnictwa i Rozwoju Wsi z dnia 29 maja 2015 r. *Zmieniające Rozporządzenie w Sprawie Szczegółowych Wymagań w Zakresie Jakości Handlowej Miodu*; Polish Minister: Warsaw, Poland, 2015; p. 850.
15. Spink, J.; Moyer, D.C. Defining the public health threat of food fraud. *J. Food Sci.* **2011**, *76*, R157–R163. [CrossRef]
16. Everstine, K.; Spink, J.; Kennedy, S. Economically motivated adulteration (EMA) of food: Common characteristics of EMA incidents. *J. Food Prot.* **2013**, *76*, 723–735. [CrossRef]
17. Llano, S.M.; Muñoz-Jiménez, A.M.; Jiménez-Cartagena, C.; Londoño-Londoño, J.; Medina, S. Untargeted metabolomics reveals specific withanolides and fatty acyl glycoside as tentative metabolites to differentiate organic and conventional Physalis peruviana fruits. *Food Chem.* **2018**, *244*, 120–127. [CrossRef] [PubMed]
18. Zawirska-Wojtasiak, R. Methods for sensory analysis. In *Food Flavours: Chemical, Sensory and Technological Properties*; Jeleń, H., Ed.; CRC Press: Boca Raton, FL, USA; Taylor & Francis Group: New York, NY, USA, 2012; pp. 439–456.
19. Wijayaa, C.H.; Wijaya, W.; Mehta, B.M. General Properties of Major Food Components. In *Handbook of Food Chemistry*; Springer: Berlin/Heidelberg, Germany, 2015; pp. 1–32.
20. Piotrowska-Puchała, A. Preferencje konsumentów, jakość i bezpieczeństwo nabywanej przez nich żywności. In *Jakość i Zarządzanie w Agrobiznesie Wybrane Aspekty*; Czernyszewicz, E., Kołodziej, E., Eds.; Uniwersytet Rolniczy im Hugona Kołłątaja w Krakowie: Kraków, Poland, 2018; pp. 84–93.
21. Niemczas-Dobrowolska, M. *Jakość i Bezpieczeństwo Żywności. Systemy, Postawy, Konsumenci*; Wydawnictwo Naukowe PTTŻ: Kraków, Poland, 2021.

22. Grzybowska-Brzezińska, M. Preferencje konsumentów wobec atrybutów produktów żywnościowych. *Handel. Wewn.* **2018**, *3*, 184–196.
23. Tiwari, K.; Tudu, B.; Bandyopadhyay, R.; Chatterjee, A.; Pramanik, P. Voltammetric sensor for electrochemical determination of the floral origin of honey based on a zinc oxide nanoparticle modified carbon paste electrode. *J. Sens. Sens. Syst.* **2018**, *7*, 319–329. [CrossRef]
24. *PN-88 A-77626*; Miód Pszczeli. Polski Komitet Normalizacyjny: Warsaw, Poland, 1988.
25. Bogdanov, S.; Gallmann, P. *Authenticity of Honey and Other Bee Products: State of the Art*; Agroscope Liebefeld-Posieux Swiss Federal Research Station for Animal Production and Dairy Products (ALP): Bern, Switzerland, 2008.
26. *Council Directive 2014/63/EC of the European Parliament and of the Council of 15 May 2014 Amending Council Directive 2001/110/EC Relating to Honey*; European Parliament: Luxembourg, 2014.
27. Guler, A.; Bakan, A.; Nisbet, C.; Yavuz, O. Determination of important biochemical properties of honey to discriminate pure and adulterated honey with sucrose (*Saccharum officinarum* L.) syrup. *Food Chem.* **2007**, *105*, 1119–1125. [CrossRef]
28. Szczęsna, T. Problemy z jakością miodu na rynku krajowym. *Pasieka* **2003**, *3*, 5–7.
29. Piotraszewska-Pająk, A.; Gliszczyńska-Świgło, A. Directions of colour changes of nestar honeys depending on honey type and storage conditions. *J. Apic. Sci.* **2015**, *59*, 51–61.
30. Soares, S.; Amaral, J.S.; Oliveira, M.B.P.P.; Mafra, I. A Comprehensive review on the main honey. Authentication issues: Production and Origin. *Compr. Rev. Food Sci. Food Saf.* **2017**, *16*, 1072–1100. [CrossRef]
31. Dżugan, M.; Tomczyk, M.; Sowa, P.; Grabek-Lejko, D. Antioxidant activity as biomarker of honey variety. *Molecules* **2018**, *23*, 2069. [CrossRef]
32. Kemsley, E.K.; Defernez, M.; Marini, F. Multivariate statistics: Considerations and confidences in food authenticity problems. *Food Control* **2019**, *105*, 102–112. [CrossRef]
33. *Council Directive 2001/110/EC of 20 December 2001 Relating to Honey*; European Parliament: Luxembourg, 2001.
34. Tornuk, F.; Karaman, S.; Ozturk, I.; Toker, O.S.; Tastemur, B.; Sagdic, O.; Kayacier, A. Quality characterization of artisanal and retail Turkish blossom honeys: Determination of physicochemical, microbiological, bioactive properties and aroma profile. *Ind. Crops Prod.* **2013**, *46*, 124–131. [CrossRef]
35. Pentoś, K.; Łuczycka, D.; Wysoczański, T. Dielectric properties of selected wood species in Poland. *Wood Res.* **2017**, *62*, 727–736.
36. Laaroussi, H.; Bouddine, T.; Bakour, M.; Ousaaid, D.; Lyoussi, B. Physicochemical Properties, Mineral Content, Antioxidant Activities, and Microbiological Quality of Bupleurum spinosum Gouan Honey from the Middle Atlas in Morocco. *J. Food Qual.* **2020**, *2022*, 7609454. [CrossRef]
37. Wilczyńska, A. *Jakość Miodów w Aspekcie Czynników Wpływających na ich Właściwości Przeciwutleniające*; Wydawnictwo Akademii Morskiej w Gdyni: Gdynia, Poland, 2012.
38. Cianciosi, D.; Forbes-Hernández, T.Y.; Afrin, S.; Gasparrini, M.; Reboredo-Rodriguez, P.; Manna, P.P.; Battino, M. Phenolic compounds in honey and their associated health benefits: A review. *Molecules* **2018**, *23*, 2322. [CrossRef] [PubMed]
39. Ruoff, K.; Luginbühl, W.; Kilchenmann, V.; Bosset, J.O.; von Der Ohe, K.; von Der Ohe, W.; Amadò, R. Authentication of the botanical origin of honey using profiles of classical measurands and discriminant analysis. *Apidologie* **2007**, *38*, 438–452. [CrossRef]
40. Jędrusek-Golińska, A.; Szymandera-Buszka, K.; Hęś, M. Gospodarcze i prozdrowotne znaczenie miodu. *Zag. Doradz. Rol.* **2023**, *112*, 42–53.
41. Terrab, A.; González, A.G.; Díez, M.J.; Heredia, F.J. Characterisation of Moroccan unifloral honey using multivariate analysis. *Eur. Food Res. Technol.* **2003**, *218*, 88–95. [CrossRef]
42. Pasias, I.N.; Kiriakou, I.K.; Proestos, C. HMF and diastase activity honeys: A fully validated approach and a chemometric analysis for identification of honey freshness and Adulteration. *Food Chem.* **2017**, *229*, 425–431. [CrossRef]
43. Kędzierska-Matysek, M.; Wolanciuk, A.; Florek, M.; Skałecki, P.; Litwińczuk, A. Hydroxymethylfurfural content, diastase activity and colour of multifloral honeys in relation to origin and storage time. *J. Cent. Europ. Agri.* **2017**, *18*, 657–668. [CrossRef]
44. Wesołowska, M.; Dżugan, M. Aktywność i stabilność termiczna diastazy występującej w podkarpackich miodach odmianowych. *Żywn. Nauka Technol. Jakość* **2017**, *24*, 103–112. [CrossRef]
45. Al- Diab, D.; Jarkas, B. Effect of storage and termal treatment on the quality of some locals brands of honey from Latakia markets. *J. Entomol. Zool. Stud.* **2015**, *3*, 328–334.
46. Kursa, K.; Popek, S. Próba zastosowania oceny poziomu 5-HMF jako wskaźnika jakości miodu pszczelego typu spadziowego. *Zesz. Nauk. Uniw. Ekon. Poz.* **2011**, *196*, 84–90.
47. Teixido, E.; Nunez, O.; Santos, F.J.; Galceran, M.T. 5-Hydroxymethylfurfural content in foodstuffs determined by micellar electrokinetic chromatography. *Food Chem.* **2011**, *126*, 1902–1908. [CrossRef]
48. Nikolov, P.Y.; Yaylayan, V.A. Reversible and covalent binding of 5-(hydroxymethyl)-2-furaldehyde (HMF) with lysine and selected amino acids. *J. Food Agric. Food Chem.* **2011**, *59*, 6099–6107. [CrossRef] [PubMed]
49. Śliwińska, A.; Przybylska, A.; Bazylak, G. Wpływ zmian temperatury przechowywania na zawartość 5-hydroksymetylofurfuralu w odmianowych i wielokwiatowych miodach pszczelich. *Bromat. Chem. Toksykol.—XIV* **2012**, *3*, 271–279.
50. Kursa, K. Zawartość proliny jako wskaźnik autentyczności miodów. *Zesz. Nauk. Akad. Morskiej Gdyni* **2015**, *88*, 172–176.
51. Majewska, E.; Drużyńska, B.; Kowalska, J.; Wołosiak, R.; Ciecierska, M.; Derewiaka, D. Zastosowanie metod fizykochemicznych i chemometrycznych do oceny jakości i autentyczności botanicznej miodów gryczanych. *Zesz. Prob. Postępów Nauk Rol.* **2017**, *589*, 59–68. [CrossRef]

52. Tichonow, A.I.; Bondarenko, L.A.; Jarnych, T.G.; Szpyczak, O.S.; Kowal, W.M.; Skrypnik–Tichonow, R.I. *Miód Naturalny w Medycynie i Farmacji (Pochodzenie, Właściwości, Zastosowanie, Preparaty Lecznicze)*; Gospodarstwo Pasieczne Sądecki Bartnik: Stróże, Poland, 2017.

53. Zhu, X.; Li, S.; Zhang, Z.; Li, G.; Su, D.; Liu, F. Detection of adulterants such as sweeteners materials in honey using near-infrared spectroscopy and chemometrics. *J. Food Eng.* **2010**, *101*, 92–97. [CrossRef]

54. Kowalski, S.; Łukasiewicz, M. Zafałszowania i autentyczność miodu–metody identyfikacji. In Proceedings of the Pszczoły Ludziom, Ludzie Pszczołom, Kraków, Poland, 13 October 2018; Volume 25.

55. Kuś, P.M.; Jerković, I.; Marijanović, Z.; Kranjac, M.; Tuberosod, C.I.G. Unlocking Phacelia tanacetifolia Benth. honey characterization through melissopalynological analysis, color determination and volatiles chemical profiling. *Food Res. Int.* **2018**, *106*, 243–253. [CrossRef] [PubMed]

56. Flaczyk, E.; Górecka, D.; Korczak, J. *Towaroznawstwo Produktów Spożywczych*; Akademia Rolnicza: Poznań, Poland, 2006.

57. Louveaux, J.; Maurizio, A.; Vorwohl, G. Methods of melissopalynology. *Bee World* **1978**, *59*, 139–157. [CrossRef]

58. Thakodee, T.; Deowanish, S.; Duangmal, K. Melissopalynological analysis of stingless bee (*Tetragonula pagdeni*) honey in Eastern Thailand. *J. Asia-Pac. Entom* **2018**, *21*, 620–630. [CrossRef]

59. Karabagias, I.K.; Badeka, A.V.; Kontakos, S.; Karabournioti, S.; Kontominas, M.G. Botanical discrimination of Greek unifloral honeys with physico-chemical and chemometric analyses. *Food Chem.* **2014**, *165*, 181–190. [CrossRef]

60. Puścion-Jakubik, A.; Brawska, M.H. Odmianowe miody pszczele—pyłki główne i towarzyszące jako podstawa ich zaklasyfikowania. *Probl. Hig. Epidemiol.* **2016**, *97*, 275–278.

61. Bodó, A.; Radványi, L.; Kőszegi, T.; Csepregi, R.; Nagya, D.U.; Ágnes, F.; Kocsis, M. Melissopalynology, antioxidant activity and multielement analysis of two types of early spring honeys from Hungary. *Food Biosci.* **2020**, *35*, 100587. [CrossRef]

62. Yang, Y.; Battesti, M.-J.; Gjabou, N.; Muselli, A.; Paolini, J.; Tomi, P.; Costa, J. Melissopalynological origin determination and volatile composition analysis of Corsican "chestnut grove" honeys. *Food Chem.* **2012**, *132*, 2144–2154. [CrossRef]

63. Mureșan, C.I.; Cornea-Cipcigan, M.; Suharoschi, R.; Erler, S.; Mărgăoan, R. Honey botanical origin and honey-specific protein pattern: Characterization of some European honeys. *LWT* **2022**, *154*, 112883. [CrossRef]

64. Serra Bonvehi, J.; Gomez Pajuelo, A. Evaluation of honey by organoleptical analysis. *Apiacta* **1988**, *23*, 103.

65. Popek, S. *Studium Identyfikacji Miodów Odmianowych i Metodologii Oceny Właściwości Fizykochemicznych Determinujących ich Jakość*; Zeszyty Naukowe/Akademia Ekonomiczna w Krakowie: Kraków, Poland, 2001.

66. Kortesniemi, M.; Rosenvald, S.; Laaksonena, O.; Vanaga, A.; Ollikka, T.; Vene, K.; Yanga, B. Sensory and chemical profiles of Finnish honeys of different botanical origins and consumer preferences. *Food Chem.* **2018**, *246*, 351–359. [CrossRef]

67. Vieira da Costa, A.C.; Batista Sousa, J.M.; Pereirada Silva, M.A.A.; dos Santos Garrut, D.; Madruga, M.S. Sensory and volatile profiles of monofloral honeys produced by native stingless bees of the brazilian semiarid region. *Food Res. Int.* **2018**, *105*, 110–120. [CrossRef] [PubMed]

68. Rosiak, E.; Jaworska, D. Właściwości probiotyczne i prebiotyczne miodów Pszczelich w aspekcie ich jakości i bezpieczeństwa zdrowotnego. *Żywn. Nauka Technol. Jakość* **2019**, *26*, 36–48. [CrossRef]

69. Tischer Seraglio, S.K.; Bergamo, G.; Molognoni, L.; Daguer, H.; Silva, B.; Gonzaga, L.V.; Fett, R.; Oliviera Costa, A.C. Quality changes during long-term storage of a peculiar Brazilian honeydew honey: "Bracatinga". *J. Food Comp. Anal.* **2021**, *97*, 103769. [CrossRef]

70. Srinual, K.; Intipunya, P. Effects of crystallization and processing on sensory and physicochemical qualities of Thai sunflower honey. *Asian J. Food Agro-Ind.* **2009**, *2*, 749–754.

71. Popek, S. Electrical conductivity as an indicator of the quality of nectar honeys. *Forum Ware* **1998**, *1*, 75–79.

72. Popek, S.; Halagarda, M.; Kursa, K. A new model to identify botanical origin of Polish honeys based on the physicochemical parameters and chemometric analysis. *LWT* **2017**, *77*, 482–487. [CrossRef]

73. Mohamat, R.N.; Noor, N.R.A.M.; Yusof, Y.A.; Sabri, S.; Zawawi, N. Differentiation of High-Fructose Corn Syrup Adulterated Kelulut Honey Using Physicochemical, Rheological, and Antibacterial Parameters. *Foods* **2023**, *12*, 1670. [CrossRef] [PubMed]

74. El Sohaimy, S.A.; Masry, S.H.D.; Shehata, M.G. Physicochemical characteristics of honey from different origins. *Ann. Agric. Sci.* **2015**, *60*, 279–287. [CrossRef]

75. Chan, B.K.; Haron, H.; Talib, R.A.; Subramaniam, P. Physical properties, antioxidant content and anti-oxidative activities of Malaysian stingless Kelulut (*Trigona* spp.). *Honey J. Agric. Sci.* **2017**, *9*, 32–40.

76. Bakar, M.A.; Sanusi, S.B.; Bakar, F.A.; Cong, O.J.; Mian, Z. Physicochemical and antioxidant potential of raw unprocessed honey from Malaysian stingless bees. *Pak. J. Nutr.* **2017**, *16*, 888–894. [CrossRef]

77. Bogdanov, S.; Ruoff, K.; Oddo, L.P. Physico-chemical methods for the characterisation of unifloral honeys: A review. *Apidologie* **2004**, *35* (Suppl. 1), 4–17. [CrossRef]

78. Acquarone, C.; Buera, P.; Elizalde, B. Pattern of pH and electrical conductivity upon honey dilution as a complementary tool for discriminating geographical origin of honeys. *Food Chem.* **2007**, *101*, 695–703. [CrossRef]

79. Ruoff, K.; Luginbuhl, W.; Bogdanov, S.; Bosset, J.O.; Estermann, B.; Ziolko, T.; Amadò, R. Authentication of the botanical origin of honey by near-infrared spectroscopy. *J. Agric. Food Chem.* **2006**, *54*, 6867–6872. [CrossRef]

80. Majewska, E.; Kowalska, J. Badanie korelacji pomiędzy przewodnością elektryczną i zawartością popiołu w wybranych miodach pszczelich. *Acta Agrophysica* **2011**, *17*, 369–376.

81. Sykut, B.; Kowalik, K.; Hus, W. Badanie jakości i zafałszowań miodów naturalnych. *Postępy Tech. Przetw. Spoż.* **2018**, *1*, 60–64.

82. Mădaş, N.M.; Mărghitaş, L.A.; Dezmirean, D.S.; Bonta, V.; Bobiş, O.; Fauconnier, M.-L.; Francis, F.; Haubruge, E.; Nguyen, K.B. Volatile Profile and Physico-Chemical Analysis of Acacia Honey for Geographical Origin and Nutritional Value Determination. *Foods* **2019**, *8*, 445. [CrossRef]

83. Karabagias, I.K.; Vavoura, M.V.; Nikolaou, C.; Badeka, A.V.; Kontakos, S.; Kontominas, M.G. Floral authentication of Greek unifloral honeys based on the combination of phenolic compounds, physicochemical parameters and chemometrics. *Food Res. Int.* **2014**, *62*, 753–760. [CrossRef]

84. Giemza, M.A. *Znaczenie Barwy w Ocenie Jakości Produktów na Przykładzie Miodów Odmianowych*; Akademia Ekonomiczna: Kraków, Poland, 1999.

85. Wilczyńska, A. Wpływ procesów technologicznych na jakość miodów pszczelich—zmiany parametrów barwy oraz zawartości HMF pod wpływem przechowywania i ogrzewania. *Zesz. Nauk. Uniw. Ekon. Pozn.* **2011**, *196*, 91–98.

86. Szabó, R.T.; Mézes, M.; Szalai, T.; Zajácz, E.; Weber, M. Colour identification of honey and methodical development of its instrumental measuring. Columella. *J. Agric. Environ. Sci.* **2016**, *3*, 29–36.

87. Radovic, B.S.; Careri, M.; Mangia, A.; Musci, M.; Gerboles, M.; Anklam, E. Contribution of dynamic headspace GC-MS analysis of aroma compounds to authenticity testing of honey. *Food Chem.* **2001**, *72*, 511–520. [CrossRef]

88. Verzera, A.; Campisi, S.; Zappala, M.; Bonaccorsi, I. SPME-GC/MS Analysis of honey volatile components for the characterization of different floral origin. *Am. Lab.* **2001**, *33*, 18–21.

89. Majewska, E.; Delmanowicz, A. Profile związków lotnych wybranych miodów pszczelich. *Żywn. Nauka Technol. Jakość* **2007**, *14*, 247–259.

90. Glory-Cuevas, L.F. A review of volatile analytical methods for determining the botanical origin of honey. *Food Chem.* **2007**, *103*, 1032–1043. [CrossRef]

91. Jasicka-Misiak, I.; Kafarski, P. Chemiczne markery miodów odmianowych. *Wiad Chem.* **2011**, *65*, 823–837.

92. Majewska, E. *Studia nad Wykorzystaniem Wybranych Parametrów Fizyko-Chemicznych i Związków Lotnych do Określania Autentyczności Polskich Miodów Odmianowych*; Wydawnictwo SGGW: Warszawa, Poland, 2013.

93. Escriche, I.; Sobrino-Gregorio, L.; Conchado, A.; Juan-Borrás, M. Volatile profile in the accurate labelling of monofloral honey. The case of lavender and thyme honey. *Food Chem.* **2017**, *226*, 61–68. [CrossRef]

94. Morais da Silva, P.L.; de Lima, L.S.; Kaminski Caetano, Í.; Reyes Torres, Y. Comparative analysis of the volatile composition of honeys from Brazilian stingless bees by static headspace GC–MS. *Food Res. Int.* **2017**, *102*, 536–543. [CrossRef]

95. Karabagias, I.K.; Badeka, A.V.; Kontominas, M.G. A decisive strategy for monofloral honey authentication using analysis of volatile compounds and pattern recognition techniques. *Microchem. J.* **2020**, *152*, 104263. [CrossRef]

96. Rodríguez-Flores, M.S.; Falcão, S.I.; Escuredo, O.; Seijo, M.C.; Vilas-Boas, M. Description of the volatile fraction of Erica honey from the northwest of the Iberian Peninsula. *Food Chem.* **2021**, *336*, 127758. [CrossRef] [PubMed]

97. Meda, A.; Euloga Lamiec, C.; Romito, M.; Millogo, J.; Nacoulma, O. Determination of the total phenolic, flavonoid and proline contents in Burkina Fasan honey, as well as their radical scavenging activity. *Food Chem.* **2005**, *91*, 571–577. [CrossRef]

98. Wilczyńska, A. Phenolic content and antioxidant activity of different types of polish honey—A short report. *Pol. J. Food Nutr. Sci.* **2010**, *60*, 309–313.

99. Borawska, M.H.; Piekut, J. Wartość liczby diastazowej, potencjał antyoksydacyjnego i zawartość polifenoli w miodach pszczelich z regionu Podlasia. *Brom. Chem. Toksyk.* **2006**, *39*, 373–376.

100. Braghini, F.; Biluca, F.C.; Ottequir, F.; Gonzaga, V.L.; da Silva, M.; Vitali, L.; Micke, G.A.; Costa, A.C.O.; Fetta, R. Effect of different storage conditions on physicochemical and bioactive characteristics of thermally processed stingless bee honeys. *LWT* **2020**, *131*, 10972. [CrossRef]

101. Da Silva, P.M.; Gonzaga, L.V.; Biluca, F.C.; Schulz, M.; Vitali, L.; Micke, G.A.; Costa, A.C.O.; Fetta, R. Stability of Brazilian Apis mellifera L. honey during prolonged storage: Physicochemical parameters and bioactive compounds. *LWT* **2020**, *129*, 109521. [CrossRef]

102. Halagarda, M.; Groth, S.; Popek, S.; Rohn, S.; Pedan, V. Antioxidant Activity and Phenolic Profile of Selected Organic and Conventional Honeys from Poland. *Antioxidants* **2020**, *9*, 44. [CrossRef]

103. Popov, I.; Lewin, G. Antioxidative homeostasis: Characterization by means of chemiluminescent technique. *Methods Enzym.* **1999**, *300*, 437–456.

104. Dżugan, M.; Kisała, J. Application of the Photochem system in agricultural investigations. In *Modern Methods in Analysis of Agricultural Raw Materials*; Puchalski, C., Bartosz, G., Eds.; University of Rzeszow Publishing Office: Rzeszów, Poland, 2011; pp. 193–203.

105. Sarmento Silva, T.M.; Santos, F.P.; Evangelista-Rodrigues, A.; Sarmentoda Silva, E.M.; Sarmento da Silva, G.; Santosde Novais, J.; de Assis, F.; dos Santos, R.; Camaraa, C.A. Phenolic compounds, melissopalynological, physicochemical analysis and antioxidant activity of jandaíra (*Melipona subnitida*) honey. *J. Food Comp. Anal.* **2013**, *29*, 10–18. [CrossRef]

106. Sidor, A.; Gramza-Michałowska, A.; Drgas, M.; Korczak, J.; Skręty, J. Evaluation of chokeberry preparations antioxidant activity with use of the photochemilumiescence (PCL) assay. *Probl. Hig. Epidemiol.* **2013**, *94*, 835–838.

107. Wesołowska, M.; Dżugan, M. The use of the photochem device in evaluation of antioxidant activity of polish honey. *Food Anal. Methods* **2017**, *10*, 1568–1574. [CrossRef]

108. Cazor, A.; Deborde, C.; Moing, A.; Rolin, D.; This, H. Sucrose, glucose, and fructose extraction in aqueous carrot root extracts prepared at different temperatures by means of direct NMR measurements. *J. Agric. Food Chem.* **2006**, *54*, 4681–4686. [CrossRef] [PubMed]
109. Donarski, J.A.; Roberts, D.P.T.; Charlton, A.J. Quantitative NMR spectroscopy for the rapid measurement of methylglyoxal in manuka honey. *Anal. Methods* **2010**, *2*, 1479–1483. [CrossRef]
110. Boffo, E.F.; Tavares, L.A.; Tobias, A.C.T.; Ferreira, M.M.C.; Ferreira, A.G. Identification of components of Brazilian honey by 1H NMR and classification of its botanical origin by chemometric methods. *LWT Food Sci. Technol.* **2012**, *49*, 55–63. [CrossRef]
111. Consonni, R.; Cagliani, L.R.; Cogliati, C. NMR characterization of saccharides in Italian honeys of different floral sources. *J. Agric. Food Chem.* **2012**, *60*, 4526–4534. [CrossRef]
112. Zheng, X.; Zhao, Y.; Wu, H.; Dong, J.; Feng, J. Origin Identification and Quantitative Analysis of Honeys by Nuclear Magnetic Resonance and Chemometric Techniques. *Food Anal. Methods* **2016**, *9*, 1470–1479. [CrossRef]
113. Schievano, E.; Piana, L.; Tessari, M. Automatic NMR-based protocol for assessment of honey authenticity. *Food Chem.* **2023**, *420*, 136094. [CrossRef]
114. Chaji, S.; Olmo-García, L.; Serrano-García, I.; Carrasco-Pancorbo, A.; Bajoub, A. Metabolomic approaches applied to food authentication: From data acquisition to biomarkers discovery. In *Food Authentication and Traceability*; Academic Press: Cambridge, MA, USA, 2021; pp. 331–378.
115. Zuccato, V.; Finotello, C.; Menegazzo, I.; Peccolo, G.; Schievano, E. Entomological authentication of stingless bee honey by1H NMR-based metabolomics approach. *Food Control* **2017**, *82*, 145–153. [CrossRef]
116. Schievano, S.; Stocchero, M.; Zuccato, V.; Conti, I.; Piana, L. NMR assessment of European acacia honey origin and composition of EU-blend based on geographical floral markers. *Food Chem.* **2019**, *288*, 96–101. [CrossRef] [PubMed]
117. Gerginova, D.; Simova, S.; Popova, M.; Stefova, M.; Stanoeva, J.P.; Bankova, V. NMR profiling of North Macedonian and Bulgarian honeys for detection of botanical and geographical origin. *Molecules* **2020**, *25*, 4687. [CrossRef] [PubMed]
118. Kerkvliet, J.D.; Meijer, H.A.J. Adulteration of honey: Relation between microscopic analysis and delta C-13 easurements. *Apidologie* **2000**, *31*, 717–726. [CrossRef]
119. Przetaczek-Rożnowska, I.; Rosiak, M. Wykrywanie zafałszowań żywności. *Przem. Spoż* **2011**, *65*, 20–24.
120. Lipiński, Z.; Czajkowska, K. Metody nowoczesnego wykrywania zafałszowań miodu syropami cukrowymi ze skrobi, Higiena żywności i pasz. *Życie Wet.* **2012**, *87*, 417–418.
121. She, S.; Chen, L.; Song, H.; Lin, G.; Li, Y.; Zhou, J.; Liu, C. Discrimination of geographical origins of Chinese acacia honey using complex 13C/12C, oligosaccharides and polyphenols. *Food Chem.* **2019**, *272*, 580–585. [CrossRef]
122. Bong, J.; Middleditch, M.; Stephens, J.M.; Loomes, K.M. Analiza proteomiczna miodu: Profilowanie peptydów jako nowatorskie podejście do uwierzytelniania miodu Mānuka (*Leptospermum scoparium*) w Nowej Zelandii. *Żywność* **2023**, *12*, 1968.
123. Wang, X.; Yan, S.; Zhao, W.; Wu, L.; Tian, W.; Xue, X. Comprehensive study of volatile compounds of rare Leucosceptrum canum Smith honey: Aroma profiling and characteristic compound screening via GC–MS and GC–MS/MS. *Food Res. Int.* **2023**, *169*, 112799. [CrossRef] [PubMed]
124. Yu, W.; Zhang, G.; Wu, D.; Guo, L.; Huang, X.; Ning, F.; Luo, L. Identification of the botanical origins of honey based on nanoliter electrospray ionization mass spectrometry. *Food Chem.* **2023**, *418*, 135976. [CrossRef] [PubMed]
125. Cabanero, A.I.; Recio, J.L.; Ruperez, M. Liquid chromatography coupled to isotope ratio mass spectrometry: A new perspective on honey adulteration detection. *J. Agric. Food Chem.* **2006**, *54*, 9719–9727. [CrossRef] [PubMed]
126. Kuś, P.M.; van Ruth, S. Discrimination of Polish unifloral honeys using overall PTR-MS and HPLC fingerprints combined with chemometrics. *LWT–Food Sci. Technol.* **2015**, *62*, 69–75. [CrossRef]
127. Nasaba, S.G.; Yazd, M.; Marini, F.; Nescatelli, R.; Biancolillo, A. Classification of honey applying high performance liquid chromatography, near-infrared spectroscopy and chemometrics. *Chemom. Intell. Lab. Syst.* **2020**, *202*, 104037. [CrossRef]
128. Wang, Y.; Xing, L.; Zhang, J.; Ma, X.; Weng, R. Determination of endogenous phenolic compounds in honey by HPLC-MS/MS. *LWT* **2023**, *183*, 114951. [CrossRef]
129. Beckh, G.; Wessel, P.; Lullmann, C. Contribution to yeasts and their metabolisms products as natural components of honey—Part 3: Contents of ethanol and glycerol as quality parameters. *Dtsch. Leb. Rundsch* **2005**, *101*, 1–6.
130. Zucchi, P.; Marcazzan, G.L.; Dal Pozzo, M.; Sabatini, A.G.; Desalvo, F.; Floris, I. Il contenuto di etanolo nel miele per la valutazione di processi fermentativi. *APOidea* **2006**, *3*, 18–26.
131. Gębala, S. Measurements of solution fluorescence—A new concept. *Opt. Appl.* **2009**, *39*, 391–399.
132. Karoui, R.; Dufour, E.; Bosset, J.-O.; De Baerdemaeker, J. The use of front face fluorescence spectroscopy to classify the botanical origin of honey samples produced in Switzerland. *Food Chem.* **2007**, *101*, 314–323. [CrossRef]
133. Lenhardt, L.; Bro, R.; Zekovic, I.; Dramicanin, T.; Dramicanin, M.D. Fluorescence spectroscopy coupled with PARAFAC and PLS DA for characterization and classification of honey. *Food Chem.* **2015**, *175*, 284–291. [CrossRef]
134. Lenhardt, L.; Zekovic, I.; Dramicanin, T.; Dramicanin, M.D.; Bro, R. Determination of the botanical origin of honey by front-face synchronous fluorescence spectroscopy. *Appl. Spectrosc.* **2014**, *68*, 557–563. [CrossRef] [PubMed]
135. Dramićanin, T.; Lenhardt, L.; Ackovic, L.; Zekovic, I.; Dramicanin, M.D. Detection of Adulterated Honey by Fluorescence Excitation-Emission Matrices. *J. Spectrosc.* **2018**, *2018*, 8395212. [CrossRef]
136. Wilczyńska, A.; Żak, N. The Use of Fluorescence Spectrometry to Determine the Botanical Origin of Filtered Honeys. *Molecules* **2020**, *25*, 1350. [CrossRef] [PubMed]

137. Parri, E.; Santinami, G.; Domenici, V. Front-Face Fluorescence of Honey of Different Botanic Origin: A Case Study from Tuscany (Italy). *Appl. Sci.* **2020**, *10*, 1776. [CrossRef]

138. Xagoraris, M.; Revelou, P.-K.; Alissandrakis, E.; Tarantilis, P.A.; Pappas, C.S. The Use of Right Angle Fluorescence Spectroscopy to Distinguish the Botanical Origin of Greek Common Honey Varieties. *Appl. Sci.* **2021**, *11*, 4047. [CrossRef]

139. Żak, N.; Wilczyńska, A.; Przybyłowski, P. Zastosowanie spektroskopii fluoroscencyjnej do oceny stopnia podgrzania miodu. *Folia Pomeranae Univ. Technol. Stetin.* **2018**, *340*, 131–142.

140. Truong, H.T.D.; Reddy, P.; Reis, M.M.; Archer, R. Internal reflectance cell fluorescence measurement combined with multi-way analysis to detect fluorescence signatures of undiluted honeys and a fusion of fluorescence and NIR to enhance predictability. *Spectrochim. Acta Part A Mol. Biomol. Spectrosc.* **2023**, *290*, 122274. [CrossRef] [PubMed]

141. Ruoff, K.; Iglesias, M.T.; Luginbuehl, W.; Jacques-Olivier, B.; Stefan, B.; Amado, R. Quantitative analysis of physical and cemical measurands in honey by mid-infrared spectrometry. *Eur. Food Res. Technol.* **2005**, *223*, 22–29. [CrossRef]

142. Ruoff, K.; Karoui, R.; Dufour, E.; Luginbuhl, W.; Bosset, J.O.; Bogdanov, S.; Amadò, R. Authentication of the botanical origin of honey by front-face fluorescence spectroscopy, a preliminary study. *J. Agric. Food Chem.* **2005**, *53*, 1343–1347. [CrossRef]

143. Ruoff, K.; Luginbühl, W.; Künzli, R.; Bogdanov, S.; Bosset, J.O.; von der Ohe, K.; von der Ohe, W.; Amadò, R. Authentication of the botanical and geographical origin of honey by front-face fluorescence spectroscopy. *J. Agric. Food Chem.* **2006**, *54*, 6858–6866. [CrossRef]

144. Kelly, J.D.; Petisco, C.; Downey, G. Potential of near infrared transflectance spectroscopy to detect adulteration of Irish honey by beet invert syrup and high fructose corn syrup. *J. Near Infrared Spectr.* **2006**, *14*, 139–146. [CrossRef]

145. Woodcock, T.; Downey, G.; O'Donnell, C.O. Near infrared spectral fingerprinting for confirmation of claimes PDO provenance of honey. *Food Chem.* **2009**, *114*, 742–746. [CrossRef]

146. Szterk, A.; Lewicki, P. Spektroskopia NIR on-line w kontroli procesów produkcji żywności. *Przem. Spoż* **2010**, *64*, 26–30.

147. Piekut, J. *Zastosowanie Spektroskopii w Bliskiej Podczerwieni (NIR) Do Analizy Wybranych Parametrów Jakościowych Naturalnych Miodów Pszczelich*; Politechnika Białostocka, Zakład Chemii: Białystok, Poland, 2011.

148. Łuczycka, D. Właściwości dielektryczne wybranych odmian miodu. *Inżynieria Rol.* **2010**, *5*, 137–142.

149. Pentoś, K.; Łuczycka, D.; Kapłon, T. The identification of relationships between selected honey parameters by extracting the contribution of independent variables in a neural network model. *Eur. Food Res. Technol.* **2015**, *241*, 793–801. [CrossRef]

150. Pentoś, K.; Łuczycka, D.; Wróbel, R. The identification of the relationship between chemical and electrical parameters of honeys using artificial neural networks. *Comp. Biol. Med.* **2014**, *53*, 244–249. [CrossRef]

151. Łuczycka, D.; Pentoś, K. The use of dielectric honey features for overheating diagnostics. *Acta Aliment.* **2019**, *48*, 28–36. [CrossRef]

152. Łuczycka, D.; Pentoś, K.; Wysoczański, T. The influence of crystallization and temperature on electrical parameters of honey. *Zesz. Probl. Postępów Nauk Rol.* **2016**, *586*, 59–68.

153. Vozáry, E.; Ignáczk, K.; Gillay, B. Dielectrical properties of Hungarian acacia honeys. *Prog. Agric. Eng. Sci.* **2020**, *16* (Suppl. S1), 131–139. [CrossRef]

154. Pentoś, K. The methods of extracting the contribution of variables in artificial neural network models—Comparison of inherent instability. *Comp. Electr. Agric.* **2016**, *127*, 141–146. [CrossRef]

155. Pentoś, K.; Łuczycka, D. Dielectric properties of honey—The potential usability for quality assessment. *Eur. Food Res. Technol.* **2018**, *244*, 873–880. [CrossRef]

156. Bakier, S. *Badania Właściwości Reologicznych Miodu w Postaci Skrystalizowanej*; Wydawnictwo SGGW: Warszawa, Poland, 2008.

157. Addo, M.G.; Mutala, A.H.; Badu, K. A Comparative Study on the Antimicrobial Activity of Natural and Artificial (Adulterated) Honey Produced in Some Localities in Ghana. *Int. J. Curr. Microbiol. App. Sci.* **2020**, *9*, 962–972. [CrossRef]

158. Yilmaz, M.T.; Tatlisu, N.B.; Toker, O.S.; Karaman, S.; Dertli, E.; Sagdic, O.; Arici, M. Steady, dynamic and creep rheological analysis as a novel approach to detect honey adulteration by fructose and saccharose syrups: Correlations with HPLC-RID results. *Food Res. Int.* **2014**, *64*, 634–646. [CrossRef] [PubMed]

159. Kamboj, U.; Mishra, S. Prediction of adulteration in honey using rheological parameters. *Int. J. Food Prop.* **2015**, *18*, 2056–2063. [CrossRef]

160. Karoui, R. Food authenticity and fraud. In *Chemical Analysis of Food*; Academic Press: Cambridge, MA, USA, 2020; pp. 579–608.

161. Brar, D.S.; Pant, K.; Krishnan, R.; Kaur, S.; Rasane, P.; Nanda, V.; Gautam, S. A comprehensive review on unethical honey: Validation by emerging techniques. *Food Control* **2023**, *145*, 109482. [CrossRef]

162. Valverde, S.; Ares, A.M.; Elmore, J.S.; Bernal, J. Recent trends in the analysis of honey constituents. *Food Chem.* **2022**, *387*, 132920. [CrossRef] [PubMed]

 *foods*

Editorial

Reinforce Bee Product Quality Evaluation to Protect Human Health

Qiangqiang Li and Liming Wu *

State Key Laboratory of Resource Insects, Institute of Apicultural Research, Chinese Academy of Agricultural Sciences (CAAS), Beijing 100093, China; liqiangqiang@caas.cn
* Correspondence: apiswu@126.com

1. Introduction

The quality of bee products is directly related to the health of consumers. Ensuring that bee products meet the required standards is essential to improve their quality. Osaili et al. conducted a study examining the physicochemical characteristics of honey imported into the United Arab Emirates (UAE) through Dubai ports from 2017 to 2021, analyzing 1330 samples for sugar components, moisture, hydroxymethylfurfural (HMF) content, free acidity, and diastase number [1]. The results showed that while 79.2% of honey samples met Emirates honey standard requirements, 20.8% did not comply due to possible adulteration or improper storage or heat treatment. Raweh et al.'s investigation into local and imported honey samples revealed significant differences in their physicochemical properties such as moisture level, color, electrical conductivity (EC), free acidity (FA), pH value, diastase activity, HMF content and individual sugar content [2]. These studies emphasize the importance of comprehensive analysis of physicochemical parameters for standardizing bee products and increasing awareness regarding quality investigations.

2. Honey Quality Evaluation

The identification of honey adulteration with various sugar syrups poses a complex and challenging issue, which is crucial for ensuring equitable trade and protecting the interests of beekeepers. Yan et al. developed an efficient approach to detect adulterants in honey by employing fluorescence spectroscopy to analyze the emission spectra and frequency doubled peak (FDP) intensity at 740 nm, which exhibit distinguishable characteristics between honey and sugar syrups [3]. Furthermore, Hao et al. utilized fluorescence spectroscopy combined with chemometrics to investigate a rapid and non-destructive method for identifying corn sugar syrup as an adulterant in wolfberry honey [4].

Additionally, the authentication of honey is crucial for ensuring quality control. Żak et al. collected data on various techniques used to evaluate the quality and authenticity of honey, while also highlighting the challenges encountered during this evaluation process [5]. Furthermore, Dżugan et al. utilized SDS-PAGE and HPTLC techniques to identify distinctive protein and polyphenolic profiles as a means of authenticating goldenrod honey [6]. Sęk et al. conducted a study focusing on two important factors in honey quality control: the diastase number (DN) and HMF content, alongside melissopalynological analysis of manuka honey. The investigation revealed significant fluctuations in the percentage of *Leptospermum scoparium* pollen found in manuka honey. Moreover, a considerable proportion of manuka honey exhibited low DN levels [7].

The storage conditions also influence the quality of honey. Živkov Baloš et al. conducted a study to evaluate the quality and stability of sunflower honey during storage by analyzing its physicochemical parameters such as water content, HMF content, diastase activity, pH value, and free acidity [8]. The findings revealed that after being stored at room temperature for 18 months, there was a significant decrease in diastase activity but a

Citation: Li, Q.; Wu, L. Reinforce Bee Product Quality Evaluation to Protect Human Health. *Foods* **2023**, *12*, 4143. https://doi.org/10.3390/foods12224143

Received: 1 November 2023
Accepted: 6 November 2023
Published: 16 November 2023

notable increase in HMF content. Moreover, the pH value of honey decreased from 3.66 to 3.56, despite relatively stable levels of water content and free acidity.

3. Bee Pollen Quality Evaluation

The nutritional significance of bee pollen is paramount, as it encompasses a diverse range of bioactivities that contribute to disease prevention. Varied species of bee pollen possess distinct nutrient profiles and bioactive compounds, thereby exhibiting differential bioactivity. Oroian et al. conducted a comprehensive analysis of various parameters including organic acids, total phenols, total flavonoids, individual phenolic compounds, fatty acids and amino acids in Romanian bee pollen [9]. Additionally, Rojo et al. assessed the differences in levels of total phenols and flavonoids as well as antioxidant activity among bee pollen samples sourced from different botanical origins within Galicia, northwest Spain [10]. Qi et al. identified L-theanine (864.83–2204.26 mg/kg) and epicatechin gallate (94.08–401.82 mg/kg) as exclusive markers for the quality evaluation of *Camellia* bee pollen [11]. The results demonstrated the extensive botanical diversity and nutrient richness of bee pollen, thereby enhancing its potential as a valuable nutritional supplement.

However, there is a concern regarding the safety of bee pollen due to its potential to trigger food-related allergies. Tao et al. conducted a study employing cellulase, pectinase, and papain for the hydrolysis of allergens in bee pollen and investigated the impact of enzyme treatment on allergic reactions in BALB/c mice [12]. The results demonstrated that enzyme-treated bee pollen effectively reduced scratching frequency in mice, mitigated tissue damage caused by allergies, decreased serum IgE levels, and regulated bioamine production. Furthermore, it was observed that enzyme-treated bee pollen influenced metabolic pathways and modulated gut microbiota composition in mice. These findings indicate that enzyme-treated bee pollen has the potential to be utilized as a hypoallergenic product for consumer consumption.

4. Royal Jelly Quality Evaluation

Royal jelly is highly valued by consumers for its health benefits attributed to its abundant protein, peptides, lipids, and other bioactive nutrients. Consequently, the quality assessment of royal jelly plays a pivotal role in ensuring the market's sustainable and healthy development. Considering that the lipid content in royal jelly originates from bee-collected pollen, and different plant sources of pollen can influence the lipid composition of royal jelly, Zhou et al. conducted an investigation to explore how feeding honeybees with various types of pollen affects the lipid composition of royal jelly [13]. The findings unveiled substantial variations in the phospholipid and fatty acid content in royal jelly produced by honeybees fed with pollen containing diverse lipid components, thereby exerting an impact on its overall quality. This research offers valuable guidance for producing premium-grade royal jelly.

5. Pesticide Detection in Bee Products

Honeybees are exposed to a significant amount of environmental pollutants during their collection of nectar and pollen from plants. Consequently, upon entering the beehive, a substantial portion of these pollutants is inevitably transferd to bee products. Kasiotis et al. employed two effective multi-residue analysis methods, namely HPLC-ESI-MS/MS and GC-MS/MS, to ascertain the presence of more than 130 pesticides and their metabolites in 109 honey and bee pollen samples collected between 2015 and 2020 [14]. The findings revealed that the pesticides detected in honey and pollen with higher residue levels primarily comprised coumaphos, imidacloprid, acetamiprid, amitraz metabolites (DMF and DMPF), as well as tau-fluvalinate. Therefore, it is imperative to conduct pesticide residue detection and risk assessment in bee products for the preservation of human health.

Finally, we express our sincere gratitude to the authors who have made significant contributions to this volume through their unwavering dedication and exceptional expertise.

We cordially invite readers of this article to provide valuable feedback and constructive suggestions on the content under consideration, including any potential typographical errors.

Author Contributions: Writing—review and editing, Q.L.; supervision, L.W. All authors have read and agreed to the published version of the manuscript.

Conflicts of Interest: The authors declare no conflict of interest.

References

1. Osaili, T.M.; Odeh, W.A.M.B.; Al Sallagi, M.S.; Al Ali, A.A.S.A.; Obaid, R.S.; Garimella, V.; Bakhit, F.S.B.; Hasan, H.; Holley, R.; El Darra, N. Quality of Honey Imported into the United Arab Emirates. *Foods* **2023**, *12*, 729. [CrossRef] [PubMed]
2. Raweh, H.S.A.; Badjah-Hadj-Ahmed, A.Y.; Iqbal, J.; Alqarni, A.S. Physicochemical Composition of Local and Imported Honeys Associated with Quality Standards. *Foods* **2023**, *12*, 2181. [CrossRef] [PubMed]
3. Yan, S.; Sun, M.; Wang, X.; Shan, J.; Xue, X. A Novel, Rapid Screening Technique for Sugar Syrup Adulteration in Honey Using Fluorescence Spectroscopy. *Foods* **2022**, *11*, 2316. [CrossRef] [PubMed]
4. Hao, S.; Yuan, J.; Wu, Q.; Liu, X.; Cui, J.; Xuan, H. Rapid Identification of Corn Sugar Syrup Adulteration in Wolfberry Honey Based on Fluorescence Spectroscopy Coupled with Chemometrics. *Foods* **2023**, *12*, 2309. [CrossRef] [PubMed]
5. Żak, N.; Wilczyńska, A. The Importance of Testing the Quality and Authenticity of Food Products: The Example of Honey. *Foods* **2023**, *12*, 3210. [CrossRef] [PubMed]
6. Dżugan, M.; Miłek, M.; Kielar, P.; Stępień, K.; Sidor, E.; Bocian, A. SDS-PAGE Protein and HPTLC Polyphenols Profiling as a Promising Tool for Authentication of Goldenrod Honey. *Foods* **2022**, *11*, 2390. [CrossRef] [PubMed]
7. Sęk, A.; Porębska, A.; Szczęsna, T. Quality of Commercially Available Manuka Honey Expressed by Pollen Composition, Diastase Activity, and Hydroxymethylfurfural Content. *Foods* **2023**, *12*, 2930. [CrossRef] [PubMed]
8. Živkov Baloš, M.; Popov, N.; Jakšić, S.; Mihaljev, Ž.; Pelić, M.; Ratajac, R.; Ljubojević Pelić, D. Sunflower Honey—Evaluation of Quality and Stability during Storage. *Foods* **2023**, *12*, 2585. [CrossRef] [PubMed]
9. Oroian, M.; Dranca, F.; Ursachi, F. Characterization of Romanian Bee Pollen—An Important Nutritional Source. *Foods* **2022**, *11*, 2633. [CrossRef] [PubMed]
10. Rojo, S.; Escuredo, O.; Rodríguez-Flores, M.S.; Seijo, M.C. Botanical Origin of Galician Bee Pollen (Northwest Spain) for the Characterization of Phenolic Content and Antioxidant Activity. *Foods* **2023**, *12*, 294. [CrossRef] [PubMed]
11. Qi, D.; Lu, M.; Li, J.; Ma, C. Metabolomics Reveals Distinctive Metabolic Profiles and Marker Compounds of Camellia (*Camellia sinensis* L.) Bee Pollen. *Foods* **2023**, *12*, 2661. [CrossRef] [PubMed]
12. Tao, Y.; Zhou, E.; Li, F.; Meng, L.; Li, Q.; Wu, L. Allergenicity Alleviation of Bee Pollen by Enzymatic Hydrolysis: Regulation in Mice Allergic Mediators, Metabolism, and Gut Microbiota. *Foods* **2022**, *11*, 3454. [CrossRef] [PubMed]
13. Zhou, E.; Wang, Q.; Li, X.; Zhu, D.; Niu, Q.; Li, Q.; Wu, L. Effects of Bee Pollen Derived from *Acer mono* Maxim. or *Phellodendron amurense* Rupr. on the Lipid Composition of Royal Jelly Secreted by Honeybees. *Foods* **2023**, *12*, 625. [CrossRef] [PubMed]
14. Kasiotis, K.M.; Zafeiraki, E.; Manea-Karga, E.; Anastasiadou, P.; Machera, K. Pesticide Residues and Metabolites in Greek Honey and Pollen: Bees and Human Health Risk Assessment. *Foods* **2023**, *12*, 706. [CrossRef] [PubMed]

MDPI

St. Alban-Anlage 66

4052 Basel

Switzerland

www.mdpi.com

Foods Editorial Office

E-mail: foods@mdpi.com

www.mdpi.com/journal/foods